Insulation
and Interiors

Dearborn™

Home Inspection

This publication is designed to provide accurate and authoritative information in regard to the subject matter covered. It is sold with the understanding that the publisher is not engaged in rendering legal, accounting, or other professional service. If legal advice or other expert assistance is required, the services of a competent professional person should be sought.

President: Roy Lipner
Publisher and Director of Distance Learning: Evan M. Butterfield
Senior Development Editor: Laurie McGuire
Content Consultant: Alan Carson
Acting Editorial Production Manager: Daniel Frey
Creative Director: Lucy Jenkins
Graphic Design: Neglia Design Inc.

INTRODUCTION

Welcome! This home inspection training program has two primary goals:

- To provide you with a sound introduction to the components, materials and mechanics of house systems that you will encounter and evaluate as a home inspector;

- To provide you with a solid understanding of inspection processes, strategies and standards of practice that will help define the scope of your inspections.

We hope you enjoy this training program and develop a good understanding of the various house systems as you proceed. Good luck!

FEATURES OF THIS PROGRAM

This program is structured to help you learn and retain the key concepts of home inspection. It also will help you form a set of best practices for conducting inspections. A number of features are included to help you master the information and put your knowledge into practice:

- Topics are organized into evenly paced Study Sessions. Each Session begins with learning objectives and key words to set up the important concepts you should master by the end of the Session. Each Session concludes with Quick Quizzes to help you test your understanding. Answers to Quick Quizzes are provided so you can check your results.

- Scope and Introduction sections present the ASHI® (American Society of Home Inspectors) Standards of Practice for each major topic. Standards help you define a professional, consistent depth and breadth for your inspections.

- An Inspection Checklist at the end of each section summarizes the important components you will be inspecting and their typical problems. You can use this as a set of field notes during your own inspections.

- The Inspection Tools list will help you build your toolkit of "must have" and optional tools for the job.

- An Inspection Procedures section provides some general guidelines to conducting your inspection of each major house system. This feature will help you develop a methodology to complement your technical knowledge.

- Field Exercises give you an opportunity to turn your knowledge into real world experience.

SUMMARY

The road we have paved for you is designed to be easy and enjoyable to follow. We trust it will lead you quickly to your destination of success in the home inspection profession.

ACKNOWLEDGMENTS

Thanks to Kevin O'Malley for his inspiration, advice and guidance. Thanks also to James Dobney for his invaluable input and encouragement. Special thanks are extended to Dan Friedman for his numerous and significant contributions.

We are grateful for the contributions of: Duncan Hannay, Richard Weldon, Peter Yeates, Tony Wong, Graham Clarke, Ian Cunliffe, Joe Seymour, Charles Gravely, Graham Lobban, Dave Frost, Gerard Gransaull, Jim Stroud, Diana DeSantis, David Ballantyne, Shawn Carr and Steve Liew.

Special thanks are also extended to Susan Bonham, Dearbhla Lynch, Lucia Cardoso-Tavares, Jill Brownlee, Ida Cristello and Rita Minicucci-Colavecchia who have brought everything together. Thanks also to Jim Lingerfelt for his invaluable editing assistance.

► INSULATION AND VENTILATION

► INTERIORS

1

INSULATION AND VENTILATION

Insulation & Interiors

MODULE

► TABLE OF CONTENTS

FIELD EXERCISE 1

► 1.0 OBJECTIVES

In this Module, you will learn how insulation, air/vapor barriers and ventilation systems work to keep houses comfortable and structurally sound. While the goal for most homeowners is reduced heating and cooling costs, the home inspector's goal on insulation, air/vapor barriers, and ventilation systems is to identify conditions that may ruin the house.

Insulation And Moisture Control

We'll discuss the basics of how insulation works and describe some of the common materials and their characteristics. We'll talk about moisture control, a very important and poorly understood component. We'll talk about ventilation in two different senses and put the whole package together to look at the house as a group of interrelated systems.

Insulating House Components

We'll discuss how roofs, walls, floors, basements and crawlspaces are typically insulated and ventilated, and we'll discuss where problems typically crop up.

Ventilation And Air Quality

We'll conclude by discussing some of the ventilation approaches that help maintain good air quality in buildings.

Inspection To The Standards

Our goal is to enable you to inspect insulation and ventilation systems according to our Standards.

Not The Final Word

This section is not technically exhaustive and won't qualify you to install or design these components. As always, there are many places to go to expand your knowledge, and we encourage you to do this throughout your home inspection career.

Problem Identification

By the time you have finished this section, you'll be able to spot commonperformance-related problems and understand their implications. You'll be able to make appropriate recommendations. The inspection of insulation and ventilation systems is challenging because so little of these systems can be seen. In many cases, you will rely on indirect or incomplete evidence.

Insulation & Interiors
MODULE

STUDY SESSION 1

1. This Session covers the ASHI® Standards for inspecting insulation systems. It also covers a brief introduction to the Insulation inspection.

2. At the end of this Study Session you should be able to –
  • identify three components that have to be observed during the Insulation inspection.
  • identify two components that have to be reported as a result of your Insulation inspection.
  • define in one sentence the terms **insulation** and **vapor retarder**.
  • describe in one sentence the implications of inadequate insulation.
  • describe in one sentence the implications of inadequate air/vapor barriers.
  • define **degree-day** in two sentences.
  • name two house systems that affect moisture control in houses.
  • name two kinds of house ventilation.

3. Before starting this section, read the brief introduction to the Insulation chapter of **The Home Reference Book**.

4. This Session may take roughly one half hour.

5. Quick Quiz 1 is at the end of this Session.

Key Words:
  • *Insulation*
  • *Vapor retarder*
  • *Air/vapor barrier*
  • *Moisture control*
  • *Degree-days*
  • *Ventilation – unconditioned spaces*
  • *Ventilation – fresh air for occupants*

► 2.0 SCOPE AND INTRODUCTION

2.1 SCOPE

THE ASHI® STANDARDS OF PRACTICE

The following are excerpted from the ASHI® Standards of Practice, effective January 1, 2000.

2. PURPOSE AND SCOPE

2.1 The purpose of these Standards of Practice is to establish a minimum and uniform standard for private, fee-paid home *inspectors* who are members of the American Society of Home Inspectors. *Home Inspections* performed to these Standards of Practice are intended to provide the client with information regarding the condition of the systems and *components* of the home as inspected at the time of the *Home Inspection*.

2.2 The inspector shall:

A. *inspect*:

1. *readily accessible systems* and components of homes listed in these Standards of Practice.
2. *installed systems* and *components* of homes listed in these Standards of Practice.

B. report:

1. on those *systems* and *components* inspected which, in the professional opinion of the inspector, are *significantly deficient* or are near the end of their service lives.
2. a reason why, if not self-evident, the *system* or *component* is *significantly deficient* or near the end of its service life.
3. the inspector's recommendations to correct or monitor the reported deficiency.
4. on any *systems* and *components* designated for inspection in these Standards of Practice which were present at the time of the *Home Inspection* but were not inspected and a reason they were not inspected.

2.3 These Standards of Practice are not intended to limit inspectors from:

A. including other inspection services, systems or *components* in addition to those required by these Standards of Practice.
B. specifying repairs, provided the *inspector* is appropriately qualified and willing to do so.
C. excluding systems and *components* from the inspection if requested by the client.

11. INSULATION AND VENTILATION

11.1 The inspector shall:

A. *inspect*:

    1. the insulation and vapor retarders in unfinished spaces.
    2. the ventilation of attics and foundation areas.
    3. the mechanical ventilation *systems*.

B. describe:

    1. the insulation and vapor retarders in unfinished spaces.
    2. the absence of insulation in unfinished spaces at conditioned surfaces.

11.2 The *inspector* is NOT required to:

A. disturb insulation or vapor retarders.

B. determine indoor air quality.

13. GENERAL LIMITATIONS AND EXCLUSIONS

13.1 General limitations:

A. Inspections performed in accordance with these Standards of Practice

    1. are not *technically exhaustive*.
    2. will not identify concealed conditions or latent defects.

B. These Standards of Practice are applicable to buildings with four or fewer dwelling units and their garages or carports.

13.2 General exclusions:

A. The *inspector* is not required to perform any action or make any determination unless specifically stated in these Standards of Practice, except as may be required by lawful authority.

B. *Inspectors* are NOT required to determine:

    1. the condition of *systems* or *components* which are not *readily accessible*.
    2. the remaining life of any *system* or *component*.
    3. the strength, adequacy, effectiveness, or efficiency of any *system* or *component*.
    4. the causes of any condition or deficiency.
    5. the methods, materials, or costs of corrections.
    6. future conditions including, but not limited to, failure of *systems* and *components*.
    7. the suitability of the property for any specialized use.
    8. compliance with regulatory requirements (codes, regulations, laws, ordinances, etc.).
    9. the market value of the property or its marketability.
    10. the advisability of the purchase of the property.

11. the presence of potentially hazardous plants or animals including, but not limited to wood destroying organisms or diseases harmful to humans.
12. the presence of any environmental hazards including, but not limited to toxins, carcinogens, noise, and contaminants in soil, water and air.
13. the effectiveness of any *system installed* or methods utilized to control or remove suspected hazardous substances.
14. the operating costs of *systems* or *components*.
15. the acoustical properties of any system or component.

C. Inspectors are NOT required to offer:

1. or perform any act or service contrary to law.
2. or perform *engineering* services.
3. or perform work in any trade or any professional service other than home *inspection*.
4. warranties or guarantees of any kind.

D. *Inspectors* are NOT required to operate:

1. any *system* or *component* which is *shut down* or otherwise inoperable.
2. any *system* or *component* which does not respond to *normal operating controls*.
3. shut-off valves.

E. *Inspectors* are NOT required to enter:

1. any area which will, in the opinion of the inspector, likely be dangerous to the *inspector* or other persons or damage the property or its *systems* or *components*.
2. The *under-floor crawl* spaces or attics which are not *readily accessible*.

F. *Inspectors* are NOT required to *inspect*:

1. underground items including, but not limited to underground storage tanks or other underground indications of their presence, whether abandoned or active.
2. *systems* or *components* which are not *installed*.
3. *decorative* items
4. *systems* or *components* located in areas that are not entered in accordance with these Standards of Practice.
5. detached structures other than garages and carports.
6. common elements or common areas in multi-unit housing, such as condominium properties or cooperative housing.

G. *Inspectors* are NOT required to:

1. perform any procedure or operation which will, in the opinion of the *inspector*, likely be dangerous to the *inspector* or other persons or damage the property or its *systems* or *components*.
2. move suspended ceiling tiles, personal property, furniture, equipment, plants, soil, snow, ice, or debris.
3. *dismantle* any *system* or *component*, except as explicitly required by these Standards of Practice.

GLOSSARY OF ITALICIZED TERMS

Alarm Systems
Warning devices, installed or free-standing, including but not limited to; carbon monoxide detectors, flue gas and other spillage detectors, security equipment, ejector pumps and smoke alarms

Architectural Service
Any practice involving the art and science of building design for construction of any structure or grouping of structures and the use of space within and surrounding the structures or the design for construction, including but not specifically limited to, schematic design, design development, preparation of construction contract documents, and administration of the construction contract

Automatic Safety Controls
Devices designed and installed to protect *systems* and *components* from unsafe conditions

Component
A part of a *system*

Decorative
Ornamental; not required for the operation of the essential *systems* and *components* of a home

Describe
To *report* a *system* or *component* by its type or other observed, significant characteristics to distinguish it from other *systems* or components

Dismantle
To take apart or remove any component, device or piece of equipment that would not be taken apart or removed by a homeowner in the course of normal and routine homeowner maintenance

Engineering Service
Any professional service or creative work requiring engineering education, training, and experience and the application of special knowledge of the mathematical, physical and engineering sciences to such professional service or creative work as consultation, investigation, evaluation, planning, design and supervision of construction for the purpose of assuring compliance with the specifications and design, in conjunction with structures, buildings, machines, equipment, works or processes

Further Evaluation
Examination and analysis by a qualified professional, tradesman or service technician beyond that provided by the home inspection

Home Inspection
The process by which an *inspector* visually examines the *readily accessible systems* and *components* of a home and which *describes* those *systems* and *components* in accordance with these Standards of Practice

Household Appliances
Kitchen, laundry, and similar appliances, whether *installed* or free-standing

Inspect
To examine *readily accessible systems* and *components* of a building in accordance with these Standards of Practice, using *normal operating controls* and opening *readily openable access panels*

Inspector
A person hired to examine any *system or component* of a building in accordance with these Standards of Practice

Installed
Attached such that removal requires tools

Normal Operating Controls
Devices such as thermostats, switches or valves intended to be operated by the homeowner

Readily Accessible
Available for visual inspection without requiring moving of personal property, dismantling, destructive measures, or any action which will likely involve risk to persons or property

Readily Openable Access Panel
A panel provided for homeowner inspection and maintenance that is within normal reach, can be removed by one person, and is not sealed in place

Recreational Facilities
Spas, saunas, steam baths, swimming pools, exercise, entertainment, athletic, playground or other similar equipment and associated accessories

Report
To communicate in writing

Representative Number
One *component* per room for multiple similar interior *components* such as windows and electric outlets; one *component* on each side of the building for multiple similar exterior *components*

Roof Drainage Systems
Components used to carry water off a roof and away from a building

Significantly Deficient
Unsafe or not functioning

Shut Down
A state in which a *system* or *component* cannot be operated by *normal operating controls*

Solid Fuel Burning Appliances
A hearth and fire chamber or similar prepared place in which a fire may be built and which is built in conjunction with a chimney; or a listed assembly of a fire chamber, its chimney and related factory-made parts designed for unit assembly without requiring field construction

Structural Component
A *component* that supports non-variable forces or weights (dead loads) and variable forces or weights (live loads)

System
A combination of interacting or interdependent *components*, assembled to carry out one or more functions

Technically Exhaustive
An investigation that involves dismantling, the extensive use of advanced techniques, measurements, instruments, testing, calculations, or other means

Under-floor Crawl Space
The area within the confines of the foundation and between the ground and the underside of the floor

Unsafe
A condition in a *readily accessible*, *installed system* or component which is judged to be a significant risk of personal injury during normal, day-to-day use. The risk may be due to damage, deterioration, improper installation or a change in accepted residential construction standards

Wiring Methods
Identification of electrical conductors or wires by their general type, such as "non-metallic sheathed cable" ("Romex"), "armored cable" ("bx") or "knob and tube", etc.

► NOTES ON THE STANDARDS

Inspect The Standards are clear on the meaning of **inspect**. When we inspect we have to look at and test the components listed in the Standards. We look at them if they are **readily accessible** or if we can get at them through **readily openable access panels**. These are panels designed for the homeowner to remove. They are within normal reach, can be removed by one person, and are not sealed in place.

Testing We test components and systems by using their **normal operating controls**, but not the safety controls. We turn thermostats up or down, open and close doors and windows, turn light switches and water faucets on and off, flush toilets, etc. We do not test heating systems on high limit switches, test pressure relief valves on water heaters and boilers, overload electrical circuits to trip breakers, etc.

Systems Shut Down We do not start up systems that are shut down. If the furnace pilot is off, we don't light it. If the electricity, water or gas is shut off in the home, we don't turn it on. If the disconnect for the air conditioner is off, we don't turn it on.

Accessible We have to inspect house components that are **readily accessible**. That means we don't have to move furniture, lift carpets or ceiling tiles, dismantle components, damage things or do something dangerous. The exception is covers that would normally be **removed by homeowners during routine maintenance**. The furnace fan cover is a good example because homeowners remove this to change the furnace filter. Many inspectors use tools as the threshold. If tools are required to open or dismantle the component, it is not considered **readily accessible**.

Installed We only have to inspect things that are **installed** in homes. This means we don't have to inspect window air conditioners or portable heaters, for example.

Deficiencies We have to report on systems that are **significantly deficient**. This means they are unsafe or not performing their intended function. Although the Standards are not explicit, we are not required to identify every minor defect in a home. Failing to report a sticking door latch or cracked pane of glass would not be a meaningful breach of the Standards. Some common sense is needed here, determining the effect the issue will have on the safety, usability and durability of the home.

End Of Life We are required to report on any system or component that in our professional opinion is **near the end of its service life**. This is tricky since we don't know whether inspectors will be held accountable for failed components on the basis that they should have known the component was near the end of its life. With the wisdom of hindsight, it may be hard to argue that the component could not have been expected to fail, when in fact, it did. Time will tell. The situation is also tricky because it includes not only **systems** but individual **components** as well. For many systems there are broadly accepted life expectancy ranges, but these aren't available for some individual components. A reasonable criteria may also be the apparent condition of the component.

Remaining Life We are not required to determine the **remaining life** of systems or components. This is related to, but different than, the **end of service life** issue. If the item is new or in the middle part of its life, we don't have to predict service life, even though the same broadly accepted life expectancy ranges would apply. It's only when the item is near the end, in your opinion, that you have to report it.

Reporting Implications We have to tell people in writing the **implications** of conditions or problems unless they are self-evident. A cracked heat exchanger on a furnace has a very different implication for a homeowner than a cracked windowpane, for example. It's not enough to tell a client that they have aluminum wiring. We have to tell them of the potential fire risk.

Tell Client What To Do We have to tell the client in the report what to do about any conditions we found. We might recommend they repair, replace, service or clean the component. We might advise them to have a specialist further investigate the condition. It's all right to tell the client to monitor a situation, but we can't tell them that their roof shingles are curled and leave it at that. We have to tell them what to do about the aluminum wire to reduce the fire risk.

What We Left Out We have to report anything that we would usually inspect but didn't. We also have to include in our report why we didn't inspect it. The reasons may be that the component was inaccessible, unsafe to inspect or was shut down. It may also be that the occupant or the client asked us not to inspect it.

Insulation And Vapor Retarders We are expected to inspect the **insulation** and **vapor retarders** in unfinished spaces. Any exposed insulation in finished spaces should also be inspected. Our approach is to inspect any of the insulation and vapor retarder system we can. Most people consider vapor retarders the same as **air/vapor barriers** or **vapor barriers**. All three terms mean essentially the same thing to most homeowners.

What Is Insulation? The primary function of insulation is to control heat loss. Brick, for example, has some ability to control heat loss, but that is not its primary function. Fiberglass batts are primarily designed to control heat loss, so they are insulation.

Vapor Retarder A vapor retarder is designed to restrict vapor diffusion. An air/vapor barrier restricts vapor diffusion and air leakage.

A vapor retarder, or air/vapor barrier as we will call it throughout this section, can be several things but, traditionally, is thought of as polyethylene film. In older construction, kraft paper was used as an air/vapor barrier. Again, there are several building materials that act as air/vapor barriers, but that is not their primary function. A painted sheet of drywall is a fairly good air/vapor barrier; so is a sheet of Styrofoam™ insulation, plywood or waferboard, for example.

Ventilation Of Attics And Foundations We have to look at the ventilation of attics and foundation areas. We expand this to include spaces where there may be no attic, flat roof spaces, knee wall areas, crawlspace areas and wall cavities.

Flush Out Moisture The purpose of these ventilation systems is to flush out moisture, preventing condensation that can result in rot damage to the building.

S1

Heat Control

The other function of this type of ventilation is to control heat in these unconditioned spaces. During the summer months, especially in the southern parts of North America, we want to flush air through these compartments to remove built-up heat. This helps reduce cooling costs and improves comfort.

The Standards require us to inspect mechanical ventilation systems including kitchen and bathroom exhaust fans and laundry venting systems. These systems have two functions:

Kitchen, Bathroom And Laundry Venting Systems

- To eliminate moisture, which may otherwise condense in wall systems, causing the damage we talked about earlier.
- To remove indoor air pollutants and draw an adequate supply of fresh air into the house.

We'll get into this in more detail a little later.

What We Have To Report

The Standards ask us to describe the insulation and vapor retarder materials and report where they are missing.

What We Don't Have To Do

The Standards don't ask us to disturb insulation or vapor retarders. We do not have to evaluate indoor air quality.

We are required to inspect kitchen, bathroom and laundry venting systems, but we do not have to report on their absence. You can decide how far to go in your inspection of venting systems.

2.2 INTRODUCTION

Perspective

Let's start by putting insulation and ventilation systems in proper perspective. Houses in various parts of North America have different emphases on insulation. In very cold climates, good insulation may be critical to ensure the building is comfortable. In very hot climates, good insulation helps keep the building cool and comfortable, although it's far less critical than preventing someone from freezing to death.

Know Your Levels

Recommended levels of insulation in roofs, walls and floors vary depending on your area. You should learn what is recommended in new construction in your area. Many older homes have insulation levels far below what is recommended in new construction.

What Do You Tell Your Clients?

How important is insulation and what should home inspectors report to clients? Attic insulation, while not a luxury, can be improved to at least modest levels, relatively inexpensively in most cases. A lack of insulation rarely makes a house impossible to live in. For these reasons, we hate to see clients walk away from a house because there isn't enough insulation. We put this in perspective for our clients by advising that insulation improvements are discretionary, not an immediate priority, and are an improvement rather than a repair.

Moisture
Control

We feel differently about the moisture control systems, such as air/vapor barriers and ventilation. While poor levels of insulation have no immediate health and safety effects, missing or ineffective air/vapor barriers and ventilation can result in damage to the building. We are concerned with the building's ability to get rid of moisture and heat without damaging the structure.

Some Clients
Want To Know

The energy efficiency of a home is a matter of great interest to some clients. A detailed evaluation of the energy-related performance of a building is beyond the scope of a home inspection. Even the most sophisticated analyses are only best guesses. The energy performance of a building depends on the lifestyle and number of the occupants, among other things.

Check
Heating Bills

Clients who are very interested in energy performance can refer to the heating and cooling costs for the building over the last few years. These numbers may be available from the seller and are usually available from the utility. Some clients compare the costs for this house to others to establish some kind of ranking. This is a guess, too, since the energy consumption will vary depending on the number and lifestyle of the occupants and the heating and cooling loads, which vary from year to year.

Family Needs
Vary

A family with infants and teenagers will want to keep the house warm and will use a lot of hot water. A working couple out of the house all day often keeps the house cooler and uses considerably less hot water.

Degree-Days

Many utilities report on the **degree-days** for heating and cooling. These can be averaged for a given location, but vary from year to year, often dramatically. The number of degree-days indicated the demand over a heating season. The number is calculated as follows:

- The average temperature for a given day is subtracted from 65°F. If the average temperature on a given day is 45°F, there would be 20 degree-days attributed to that day.
- This calculation is done for each day of the heating season, and the number of degree-days are added. The total provides the number of degree-days for that season.

The Basics

Many people don't understand how heat moves into, through and out of a home. Similarly, the movement of moisture through a home can be a difficult concept. We have to understand how heat and moisture cause problems before we discuss how to control them. In some ways, the Insulation and Ventilation Section might be better titled **Heat and Moisture Control for Homes**. In the first part of this section, we'll cover the basics of heat and moisture movement, within the context of single family homes.

Insulation

Insulation is not as simple as it first seems. We're going to talk about the goals of insulation. We'll discuss how insulation is able to meet its goals and we'll look at a variety of insulation materials.

Moisture Control

We'll look at why moisture control is important and discuss the two major house systems that affect moisture control:

• Air/vapor barriers, and
• Ventilation.

Two Kinds Of Ventilation

We will address two types of ventilation. One type flushes moisture from unconditioned spaces in the winter, and flushes heat from these spaces in the summer. The second type provides fresh air for the occupants of the home. The second type of ventilation ensures good air quality, rather than protecting against building deterioration. Not surprisingly, the two different types of ventilation sometimes overlap.

The Building Envelope

A house is a complex set of interrelated systems and occupants. Some people use the concept of **the house as a system** or **the building envelope**. These phrases help remind us that we can't look at insulation without looking at ventilation, climate, occupant lifestyles and so on. A change in one system can affect the performance of the entire home.

Insulation And Ventilation Applications

We'll discuss how insulation is added to buildings. We'll also look at how air/vapor barriers and ventilation systems are installed. We'll address roofs, walls, floors, basements and crawlspaces.

Ventilation For People

We'll also look at the ventilation systems that help provide fresh air for homes. Now that we've set the stage, let's start with the basics.

Insulation & Interiors
MODULE

QUICK QUIZ 1

☑ INSTRUCTIONS

- You should finish Study Session 1 before doing this Quiz.
- Write your answers in the spaces provided.
- Check your answers against ours at the end of this Section.
- If you have trouble with the Quiz, re-read the Study Session and try the Quiz again.
- If you did well, it's time for Study Session 2.

1. List three areas that have to be observed during the Insulation part of an inspection to the Standards.

    un finised Areas   Insulation and Vapor retarders in the ventilazation of attics and foundation areas the mechanical ventelation systems

2. List two items that have to be described in your report.

    the insulation and vapor retarders in unfinished spaces the absences of insulation in unfinished spaces at conditioned surfaces

3. List two items that you are not required to report on as part of your Insulation inspection.

    Significantly Deficient - Unsafe or not functioning System or Component that cannot be operated by Normal operating controls

4. The primary function of insulation is to

Control Heat Loss

5. A vapor retarder is designed to

restrict vapor diffusion

6. Give four examples of unconditioned spaces:

attics and foundation areas
Flat roof spaces
Knee wall areas
Crawl space areas and wall cavities

7. Describe two functions of kitchen and bathroom venting systems:

To eliminate moisture, To remove indoor pollutants
and draw adequate supply of fresh air into the house

8. What are the implications of inadequate insulation?

Increased Heating and cooling costs

9. What are the implications of an inadequate air/vapor barrier?

Mold, pealing wallpaper, damage to the
building, Health risks

10. Define **degree-day**.

during the Heating season the average is 65°F you
take the difference for each day. the number of degree days are add
the total provides the number of degree-days for that seaso

11. Name two house systems that affect moisture control in houses.

Air/Vapor barriers
Ventilation

12. Name two purposes of house ventilation.

One flushes moisture from unconditioned spaces in the winter and flushes heat from these spaces in the summer the other provides fresh air for the occupants of the home. and ensures good air quality.

If you didn't have any difficulty with the Quiz, then you are ready for Study Session 2.

Key Words:

- *Insulation*
- *Vapor retarder*
- *Air/vapor barrier*
- *Moisture control*
- *Degree-days*
- *Ventilation – unconditioned spaces*
- *Ventilation – fresh air for occupants*

Insulation & Interiors
M O D U L E

STUDY SESSION 2

1. You should have finished Study Session 1 and Quick Quiz 1 before starting this Session.

2. This Session covers heat and temperature, heat transfer, the control of heat flows, how insulation works, and the control of air leakage.

3. By the end of this Session, you should be able to –
- define **heat** and **temperature**, giving the units for each.
- define **sensible heat**.
- define **latent heat of vaporization**.
- define **latent heat of fusion**.
- define thermal **conductivity (k)**.
- define **thermal conductance (C)**.
- list six **thermal conductors**.
- list three **thermal insulators**.
- describe in one sentence the role of air in insulation .
- define **thermal resistance (R)**.
- name one good thing about air leakage.
- name one bad thing about air leakage.
- define **wind washing**.

4. This Session may take roughly 45 minutes.

5. Quick Quiz 2 is at the end of this Session.

Key Words:

- *Heat*
- *Temperature*
- *British Thermal Unit (BTU)*
- *Sensible heat*
- *Latent heat of vaporization*
- *Latent heat of fusion*
- *Conduction*
- *Radiation*
- *Convection*
- *Evaporation*
- *Thermal conductivity*
- *Thermal conductance*
- *Thermal conductors*
- *Thermal insulators*
- *Thermal resistance*
- *Air leakage*
- *Wind washing*

► 3.0 THE BASICS

3.1 HEAT – WHAT IT IS AND HOW IT MOVES

Heat Defined

Heat is the thermal energy of a body. Solids, liquids and gases all have thermal energy or heat. Heat is the kinetic energy of the molecules in the substance. Even solids are moving at the molecular level.

Heat Versus Temperature

Temperature is not the same as heat. Heat is the amount of thermal energy in a body. The temperature is the level of the thermal energy in the body. **Temperature is measured in degrees Fahrenheit or Celsius**. Something can have a very high temperature in degrees, but not a lot of heat. Something can also have a lot of heat, but be at a fairly low temperature. A match has a very high temperature, but doesn't have a lot of heat. You couldn't heat a house with a match. The water in a hot water heating system may only be 160°F, but can heat an entire house.

BTU

Heat is measured in BTUs or British Thermal Units. One BTU is the amount of heat required to raise the temperature of one pound of water by one degree Fahrenheit. A match has roughly one BTU of thermal energy.

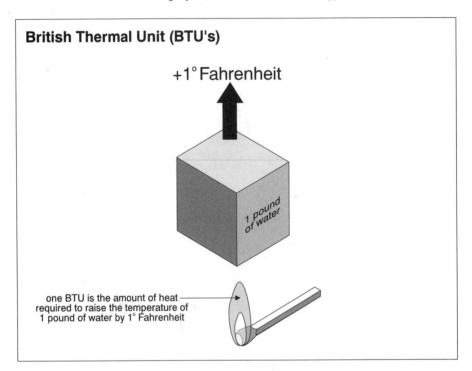

British Thermal Unit (BTU's)

+1° Fahrenheit

1 pound of water

one BTU is the amount of heat required to raise the temperature of 1 pound of water by 1° Fahrenheit

Sensible And Latent Heat

Sensible heat causes the temperature to change. For example, if we turn up the thermostat on a home heating system and change the room temperature from 70° to 80°F, we've changed the sensible heat. Sensible heat does not change the state of an object. For example, we can heat one pound of water from 32°F to 212°F and it will still be a liquid. We have to add 180 BTUs to the water to do this.

Latent Heat Of Vaporization

Water at 212°F can be converted to steam at 212°F, but only if we add another 970 BTUs. That's why "a watched pot never boils"! This is the **latent heat of vaporization.** Latent heat is the energy used to change the state of a substance without changing the temperature. One pound of ice at 32°F can be changed to one pound of water at 32°F by adding 144 BTUs. This is known as the **latent heat of fusion**. Latent heat works in both directions. Steam at 212°F **releases** 970 BTUs as it is converted to water at 212°F. One pound of water at 32°F **releases** 144 BTUs as it freezes to become ice at 32°F.

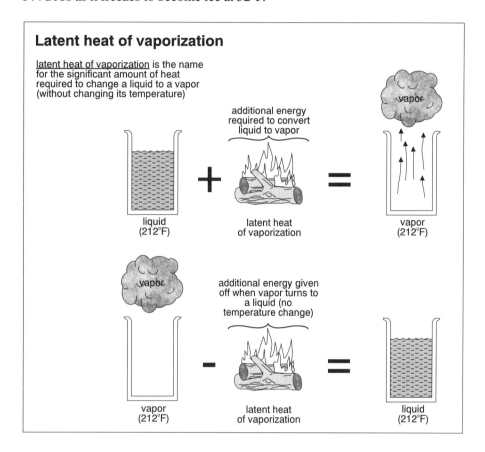

Latent heat of vaporization

latent heat of vaporization is the name for the significant amount of heat required to change a liquid to a vapor (without changing its temperature)

liquid
(212°F)

additional energy required to convert liquid to vapor

latent heat of vaporization

vapor

vapor
(212°F)

vapor

additional energy given off when vapor turns to a liquid (no temperature change)

vapor
(212°F)

latent heat of vaporization

liquid
(212°F)

Moving Heat

Heat moves through solids, liquids, gases and even through a vacuum. Thermal energy or heat **always** flows from an area of high energy to an area of low energy. Warm materials always transfer their heat to cooler materials. Cool does not flow. Heat flows.

Mechanisms Of Heat Transfer

Heat can flow by **conduction, radiation**, or **convection**. When we're talking about people, heat can also be transferred by **evaporation**. This mechanism uses the latent heat of vaporization, incidentally. People sweat to cool their bodies. Sweat glands create moisture on the surface of the skin, so the moisture will evaporate into the air around it. As the water is evaporated, there is a loss of heat from the body. This makes us feel cooler. The more humid it is outside, the less moisture will be evaporated from your skin. With less evaporation, there is less cooling and you feel warmer. This helps explain why dry heat is more comfortable than hot, humid weather.

Let's take a closer look at the three conventional types of heat transfer:

• Conduction
• Radiation
• Convection

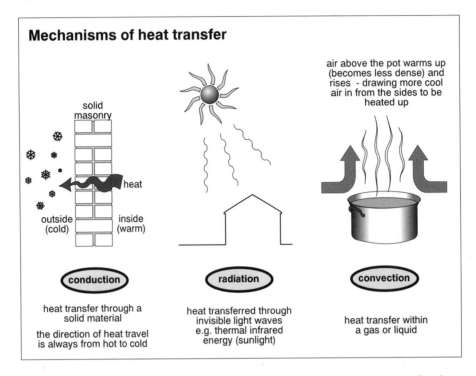

Mechanisms of heat transfer

air above the pot warms up (becomes less dense) and rises - drawing more cool air in from the sides to be heated up

solid masonry

heat

outside (cold)   inside (warm)

conduction

heat transfer through a solid material

the direction of heat travel is always from hot to cold

radiation

heat transferred through invisible light waves e.g. thermal infrared energy (sunlight)

convection

heat transfer within a gas or liquid

Conduction

Conduction is the transfer of heat between two bodies by direct contact. Putting a metal rod in the flame of a fireplace will heat up the rod by conduction. Putting your cold hands on someone's cheeks will cool their cheeks by conduction. Heat flows from the cheeks into your hands.

Sand Heats Feet

On a hot summer day, the sand on the beach can feel like it's burning your feet. It is conduction that transfers heat from the sand to your feet.

Radiation

Radiation is the transfer of heat through invisible light waves from one body to another. Radiant heat can move through air or through a vacuum. Heat travels in a straight line from the source to a solid body that receives the heat. The sun and the earth are good examples. The sun radiates its heat through about 93 million miles to warm the earth. Infrared lamps in restaurants radiate heat onto the food to keep it warm. Overhead radiant heaters in arenas, factories and warehouses at the ceiling level radiate heat to the people and objects below. These heaters don't warm the air directly.

Sun Heats Sand

Let's go back to the beach. How did the sand get hot in the first place? The sun heated the sand by radiation. The sand heats your feet by conduction.

SECTION ONE: INSULATION AND VENTILATION

S2

Convection

Convection is the movement of heat through liquids or gases that will flow, such as air or water. Warmer air or water is lighter than cool air or cool water. If a bundle of air is heated by conduction, for example, it will rise. As it rises, cool, dense bundles of air will move in to take its place. As the warmed air that has risen cools because it is no longer close to a heat source, it will fall. This sets up what are known as convective loops. It is this concept of convection that has led to the phrase "heat rises."

Convection
At The Beach

Let's go back to the hot summer day at the beach. The sun heats up the sand (radiation) much more quickly than it does the water (convection), because of differences in specific heat. As the sand gets hot, it warms the air above it. The air above the beach rises (convection) and the cooler air over the water moves in to take its place. This results in an onshore breeze during a hot summer day. This breeze is caused by convection. An **onshore** breeze often feels cool because the air over the water is not heated as much as the air over the sand.

Cool Breezes

Air moving across your skin cools you. On a calm day, as droplets of moisture are evaporated off your skin, they saturate the air immediately adjacent to your skin. This slows the rate of evaporation because the air can't hold any more moisture. If that air is quickly replaced by relatively dry air, evaporation of moisture from your skin is faster and you feel cooler. When it's hot, you feel cooler if there is a breeze or a fan, because of the increased rate of evaporation of perspiration.

Wind Chill

On a cold winter day, a similar process takes place. It doesn't necessarily involve evaporation (a latent heat), but it does involve the carrying away of heat from the body (sensible heat). Heat moves from warm bodies to cold bodies faster if there is a big temperature difference between the two bodies. Your skin temperature may be considerably warmer than the outdoor air. Heat will flow from your skin to the air. The colder it is, the faster the heat will be lost. On a calm day, your skin transfers heat into the air next to your skin. As this air is warmed, the rate of heat transfer slows. When it's windy, the air that is warmed by your skin is quickly removed and replaced by cold air. This accelerates the rate of heat loss from your body. That's why you feel colder on a windy winter day.

People
Releasing Heat

We've talked about heat moving from our skin into the air around us. Someone sleeping might release heat at a rate of 300 BTUs per hour. Someone sitting might release 400 BTUs per hour. Someone walking three miles per hour might release about 1,000 BTUs per hour. Walking upstairs quickly can release about 4,000 BTUs per hour.

By comparison, an average house might lose 60,000 BTUs per hour when the outdoor temperature is 30°F.

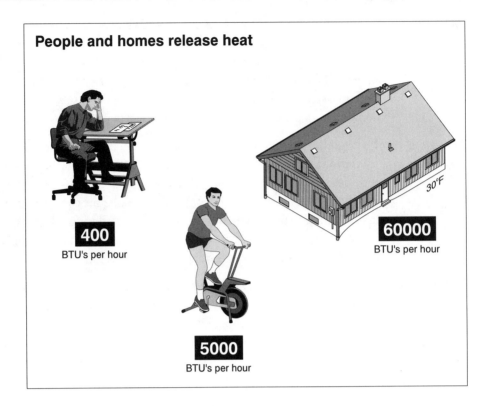

People and homes release heat

400
BTU's per hour

5000
BTU's per hour

30°F

60000
BTU's per hour

Measuring The Movement Of Heat

The rate of heat transfer can be measured in BTUs per hour. For individual materials, we can identify specific thermal conductivities. The **thermal conductivity (k)** of a homogeneous substance is defined as the number of BTUs per hour that will move through one square foot of a block of this material that is one inch thick, when the temperature difference from one side of the material to the other is one degree F. Concrete, for example, has a thermal conductivity of 12.0. This means a one-foot by one-foot by one-inch-thick concrete slab would allow 12 BTUs to flow through it in one hour if the temperature was 70°F on one side and 69°F on the other. To give you a comparison, a urethane foam block of the same size under the same conditions would only allow 0.16 BTUs to pass in one hour. The conductivity (k) of urethane foam is 0.16.

Thermal conductivity

the thermal conductivity (k) of a <u>homogeneous</u> material is equal to the number of BTUs that will pass through one square foot of the material (that is 1" thick) over the course of 1 hour (with a 1 °F temperature difference across the material)

in this case, the thermal conductivity (k) of the concrete slab is 12

concrete slab

T °F

$T+1$ °F

1 square foot

1" thick

12 BTUs

Conductance   Some materials are measured by **conductance (C)**. This applies to materials that are not homogeneous or have large air voids. A concrete block is a good example. We can't measure the conductivity of the concrete block because some slices of the block would be mostly concrete and some would be mostly air. They will have different thermal conductivities. What we can do, however, is define the material and find out how many BTUs will pass through one square foot of this material over the course of an hour with a one degree F temperature difference from one side to the other. The conductance (C) of an eight-inch hollow concrete block wall, for example, is 0.90.

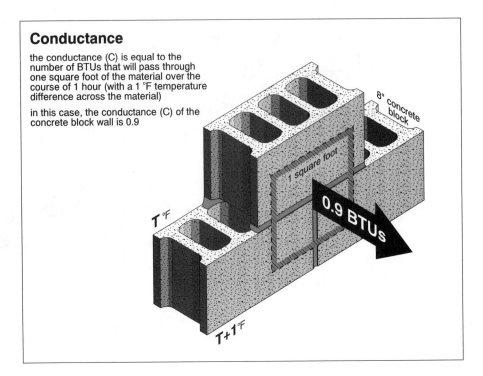

Conductance

the conductance (C) is equal to the number of BTUs that will pass through one square foot of the material over the course of 1 hour (with a 1 °F temperature difference across the material)

in this case, the conductance (C) of the concrete block wall is 0.9

8" concrete block

1 square foot

T °F

0.9 BTUs

T+1°F

Summary

In this discussion, we have defined heat and talked about how it's measured. We've talked about temperature and how it's measured. We've talked about the way heat moves (conduction, radiation, convection and evaporation), we've touched on human comfort issues, and we've discussed how we measure the movement of heat. Now, let's look at controlling the heat flow.

3.2 CONTROLLING HEAT FLOW

People in houses want to control the flow of heat. During the winter, we want to keep the heat in one area (the house) from moving to another area (outdoors). During the summer, we also want to control the flow of heat, but in a different way. We want to keep the heat in one area (outdoors) from moving to another area (the house). Remember, heat always flows from warm to cold, not the other way around.

Two Comfort Strategies   Let's look at two ways we can make a house comfortable. Consider two identical houses side by side. The first house has very little insulation and the second has a lot. Both houses have good heating systems that keep their houses at 70°F, no matter how cold it gets. How can this be?

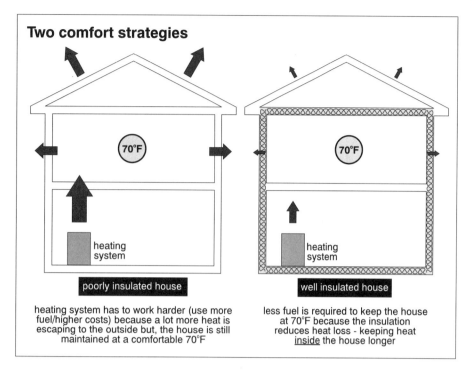

Two comfort strategies

poorly insulated house

heating system has to work harder (use more
fuel/higher costs) because a lot more heat is
escaping to the outside but, the house is still
maintained at a comfortable 70°F

well insulated house

less fuel is required to keep the house
at 70°F because the insulation
reduces heat loss - keeping heat
inside the house longer

Poorly Insulated House

The uninsulated house has much more heat flowing to the cold outdoors. However, the heating system is able to replace the lost heat and maintain a comfortable temperature.

Well-Insulated House

The second house is well insulated and loses heat to the outdoors at a lower rate than the house next door. The heating system has to add less heat to keep its occupants comfortable. The uninsulated house uses more fuel to generate heat. The owner will spend more on gas, oil or electricity.

Heat Flow

Both houses are comfortable to live in, although one is more expensive to heat. Heat flow is not directly related to comfort.

What Stops Heat?

We've talked about how heat moves. Now, let's talk about what keeps heat from moving.

Stopping Heat Flow

We used the word **insulation** in the previous discussion. Insulation is something that slows the rate of heat flow. Nothing can stop heat flow, but some materials slow it down more than others do. Materials like steel, lead, concrete, copper, plaster, stone, glass and clay have relatively high thermal conductivity. They aren't good at slowing heat flow. Materials such as these are **thermal conductors**. Materials that have low thermal conductivity are good at slowing heat flow. These are considered **thermal insulators.** These include cork, sawdust, and some plastics.

*Insulators
Contain Air*

Good thermal conductors are dense. Good thermal insulators have low density. Surprisingly, air is a really good insulator. Low-density materials have a lot of air in them. Good thermal insulators depend on trapped air to stop heat flow. We said glass was a thermal conductor, but fiberglass is an insulator! It's the same material, but its thermal properties are very different because of the air in the fiberglass. As a simple rule, the more air it has in it, the better insulator it's likely to be.

*Keep The
Air Still*

Life is never as simple as we'd like. A good insulator needs more than just a lot of air. Every room is filled with air; so is the great outdoors. Are these good insulators? Not necessarily. Air is a very good insulator, but unfortunately air moves around a lot. Air can slow heat transfer very well. However, if a warm bundle of air moves from inside the house to outside, it takes the heat with it. A good thermal insulator needs to have two characteristics:

• It has a lot of air
• It can hold the air still in small, individual bundles

*Convective
Loops*

A large volume of air is difficult to hold still. Convective loops are set up, even in a sealed glass bottle, for example. These loops move heat from warm to cool areas. This is not what we want insulation to do. Good insulators stop convective loops by holding air in little compartments. If the air is held relatively still and can't communicate with adjacent bundles of air, we've got a good insulator.

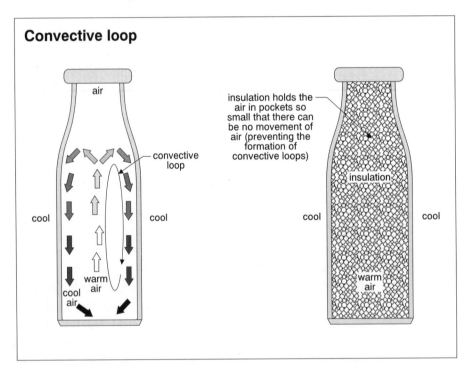

*Nothing Is
Perfect*

Insulators **slow** the rate of heat transfer. Even the best insulators **don't stop** the rate of heat flow.

Insulation Controls Radiation

We've been talking about using insulation to slow heat flow by conduction. The small pockets of air work well here. Insulation should also reduce heat flow due to radiation. We said earlier that heat moves easily through air or a vacuum by radiation. Radiation is slowed by solid surfaces. Insulation must be solid enough to restrict radiation. You should not be able to look through insulation.

Measuring Insulators

We've already talked about measuring heat flow by conductivity or conductance. The higher the number for a given material, the better thermal conductor it is. The lower the thermal conductivity, the better insulation a material is. When we look for good insulators, we look for materials with very low conductivity.

Flipping Things Around

People don't like to compare small numbers. If the numbers are less than one, they involve decimal points and are confusing. As a result, we don't use thermal conductivity to describe the quality of insulators. We use the concept of **thermal resistance (R)**. The resistance of a material is the inverse or the reciprocal of its conductivity. If the conductivity of a material is 0.5, the resistance of the material is 1/0.5 or 2.0. The units of resistance are 1.0 divided by BTUs per hour per square foot per inch per degree F. The units are so complicated, nobody pays any attention. We just refer to the R-value.

R-Value Per Inch

The thermal resistance of materials is usually measured per inch of thickness. For example, the R-value of a fiberglass insulation might be 3.0 per inch. The value of a 4-inch fiberglass batt is 3.0 x 4 = 12. All materials have R-values. Good insulation materials have R-values as high as R-7 or R-8 per inch. Thermal conductors have very low R-values because they are very poor thermal insulators. For example, the R-value of a one-inch-thick concrete slab is 0.08. One inch of stucco has an R-value of 0.2. A one-inch pine board has an R-value of 1.25.

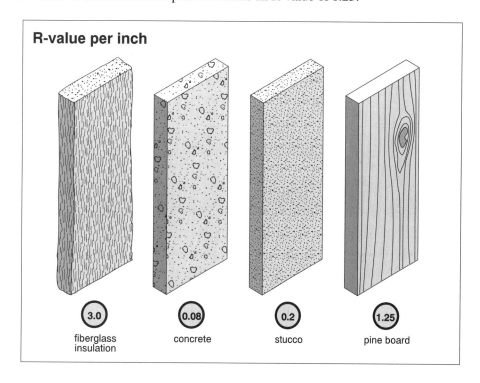

R-value per inch

3.0	0.08	0.2	1.25
fiberglass insulation	concrete	stucco	pine board

How Much Insulation Is Enough?

Depending on climate, cost of heating fuels and a number of other factors, the recommended levels of insulation vary across North America. In the northern half of North America, attic insulation levels of R-25 to R-40, wall insulation levels of R-12 to R-20 and floor insulation levels (over unheated spaces) of R-20 to R-30 are commonly recommended. Find the recommended levels in your area by looking at codes and talking to authorities. Levels may depend on the heating fuel. In some area, more insulation is recommended with electric heat because it's a more expensive fuel than others.

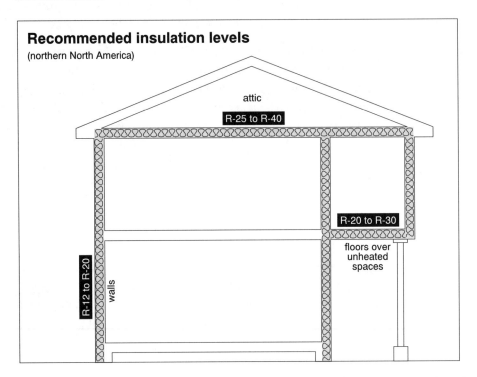

Recommended insulation levels
(northern North America)

attic

R-25 to R-40

R-20 to R-30

floors over
unheated
spaces

R-12 to R-20

walls

Consider The Entire Assembly

The insulation levels of wall, roof and floor assemblies are higher than the value of the insulation. On a wall, for example, the drywall, air/vapor barrier, sheathing, sheathing paper and siding all contribute to restricting heat flow. The air film on either side of the assembly also helps increase the R-value. The total R-value of the wall assembly will be slightly greater than the value of the insulation. If a wall has an R-12 batt insulation, the total R-value for the wall might be R-14.

Air Leakage

Now let's consider the loss of heat that is carried by the air.

Heat Flow Through Air

Movement

Air is a good insulator if we keep it in small bundles. However, air can also be the enemy when we're talking about heat flow. Air carries heat as it moves, so any air that leaks from the house is lost heat. If we didn't replace the air that leaked out of the house, the house would eventually collapse. As warm air leaks out of the house, it has to be replaced by cold, outside air. Air moving into houses can create cold drafts, reducing comfort.

Need Some Air Leakage

If we lose too much warm air from the house, we lose a lot of heat (and can create moisture problems, which we'll talk about shortly). On the other hand, if we don't change the air in the house often enough, we end up with stale, polluted air and an unhealthy environment inside the house.

Balanced Air Changes

Striking a balance between too many and too few air changes has received a lot of attention over the last 30 years, and remains a tricky issue. We've seen many solutions that are not completely satisfactory for a number of reasons. Houses are complex, and this issue has several variables. We'll get into this more, later on.

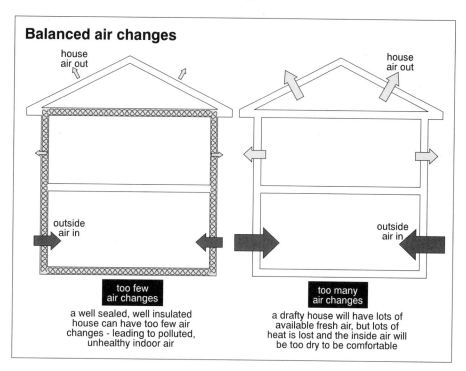

Balanced air changes

house air out — house air out

outside air in — outside air in

too few air changes
a well sealed, well insulated house can have too few air changes - leading to polluted, unhealthy indoor air

too many air changes
a drafty house will have lots of available fresh air, but lots of heat is lost and the inside air will be too dry to be comfortable

The Effect Of Wind

Speaking of things complex, let's examine another area where air movement can be troublesome. Many siding systems (aluminum, vinyl, wood shingle) are loose and wind can move through these readily. Wind moving through a roof or wall assembly can reduce the R-value of the insulation dramatically because, as we've said, insulation works by keeping air pockets still. This depends on the type of insulation and the ease with which air can be blown through the wall assembly. Fiberglass insulation is vulnerable to this problem, sometimes referred to as **wind washing**.

Controlling Air Leakage

Wind washing is not as important as air leakage. We have recognized air leakage is a significant issue with respect to controlling heat loss and gain. The traditional techniques used to restrict air leakage include polyethylene **air barriers** near the inside of wall assemblies, typically just behind the drywall.

Polyethylene
Air/Vapor
Barrier

These barriers can reduce heat loss by stopping the air from escaping through the walls and ceilings of the building. They also help prevent moisture damage to the building. In practice, air barriers are less than perfect. They are not usually continuous from basement to the roof. They are not continuous across doors and windows. Pipes, wires, chimneys and other penetrations in the wall pierce them. Joints and edges are often not well sealed.

There are some different approaches to creating air barriers, but since these have implications with respect to moisture control, we'll hold off on our discussion of airflow until we get the moisture part of the picture filled in.

Summary

We've talked about controlling heat flow with –

• insulation to reduce heat loss due to conduction and radiation, and
• air barriers to reduce heat loss due to air leakage out of the building.

Insulation
& Interiors
M O D U L E

QUICK QUIZ 2

☑ INSTRUCTIONS

- You should finish Study Session 2 before doing this Quiz.
- Write your answers in the spaces provided.
- Check your answers against ours at the end of this Section.
- If you have trouble with the Quiz, re-read the Study Session and try the Quiz again.
- If you did well, it's time for Study Session 3.

1. Define **heat** and give its imperial units.

   _____

   _____

2. Define **temperature** and give its imperial units.

   _____

   _____

3. Define **sensible heat**.

   _____

   _____

4. Define **latent heat of vaporization**.

   _____

5. Define **latent heat of fusion**.

_____

6. List three types of heat transfer.

_____

_____

_____

7. Define **thermal conductivity**.

_____

_____

8. Define **conductance.**

_____

_____

9. List six thermal conductors.

_____

_____

_____

_____

_____

_____

10. List three thermal insulators.

_____

_____

_____

11. What role does air play in insulation materials?

_____

_____

12. Define **thermal resistance**.

_____

_____

13. Name one good thing about air leakage.

_____

14. Name one bad thing about air leakage.

_____

15. Define **wind washing**.

_____

_____

If you had no trouble with this Quiz, you are ready for Study Session 3.

Key Words:

- **Heat**
- **Temperature**
- **British Thermal Unit (BTU)**
- **Sensible heat**
- **Latent heat of vaporization**
- **Latent heat of fusion**
- **Conduction**
- **Radiation**
- **Convection**
- **Evaporation**
- **Thermal conductivity**
- **Thermal conductance**
- **Thermal conductors**
- **Thermal insulators**
- **Thermal resistance**
- **Air leakage**
- **Wind washing**

Insulation & Interiors
MODULE

STUDY SESSION 3

1. You should have finished Study Session 2 and Quick Quiz 2 before starting this Session.

2. This Session covers moisture flow in houses, including moisture sources, how moisture moves, condensation and its effect in buildings, how air leakage affects moisture flow and moisture flow in homes and hot climates.

3. At the end of this Session, you should be able to –
- list eight sources of moisture in house air.
- define **absolute** and **relative humidity**.
- define **saturated air**.
- list four ways that moisture moves through homes.
- describe in three sentences **stack effect**.
- describe in two sentences **neutral pressure plane**.
- define **dew point temperature**.
- rank the importance of vapor diffusion and air leakage in condensation problems in homes.
- describe **drying potential** in two sentences.
- explain in three sentences why newer homes are more likely to have condensation problems.
- list five possible ways to control moisture and condensation problems.
- describe in three sentences how condensation problems are different in hot, humid climates.

4. This Session may take you roughly one hour to complete.

5. Quick Quiz 3 is at the end of this Session.

Key Words:

- *Condensation*
- *Absolute humidity*
- *Relative humidity*
- *Stack effect*
- *Neutral pressure plane*
- *Dew point temperature*
- *Vapor diffusion*
- *Drying potential*
- *Flushing*

3.3 MOISTURE FLOW

While heat loss makes a building more expensive to operate and may make it less comfortable, it doesn't damage the building. The same can be said of unwanted heat gain in the summer, up to a point. However, the flow of moisture in a building can cause considerable damage that is often concealed for some time. Let's take a look at how moisture flows in buildings and how damage can result.

Moisture As A Vapor

Let's consider moisture as vapor in the house air. We're not talking about water that comes into the building through leaks in roofs, walls or subgrade areas, nor from plumbing leaks, sewer backups or leaks from water heating systems. In short, we're talking about water you can't see.

Condensation Causes Damage

Moisture in vapor form in the air causes no harm. When this moisture condenses to a liquid, damage can occur. The tricky part is this often happens in areas difficult or impossible to see.

Climate Problems

Our discussion will focus on problems in heating climates. At the end of the discussion, we'll examine cooling climate problems that involve the same principles but in a different sense.

Sources Of Moisture

Where does the moisture in a house come from? Let's assume we have no leaks from roofs, walls, windows, foundation, plumbing or heating systems. Moisture in the air comes from –

- people washing their face and hands and brushing their teeth
- taking showers and baths
- cooking and washing dishes
- washing clothes
- washing floors, walls and furniture
- watering plants
- people breathing and perspiring
- damp soil in subgrade spaces (crawlspaces)
- firewood (one cord can release one gallon of water per day as it dries)
- pets

A family of four can generate 10 to 12 gallons of water a week through their normal household activities. Unprotected soil in a subgrade area under a house can contribute 10 gallons a **day**.

Measuring Moisture

While special equipment is needed to determine the actual amount of moisture in the air, we can discuss the common terms used to describe moisture in air.

Humidity

The generic term for moisture in the air is **humidity**. There are two types of humidity of interest: absolute **humidity** and **relative humidity**.

Absolute Humidity

Absolute humidity is a measure of the actual amount of moisture in the air. This can be measured in grains (0.002285 ounces) of moisture per pound of dry air. This is less important and is less commonly referred to than relative humidity. For example, 70°F air that is **saturated** contains 108 grains of moisture per pound of dry air. The term saturated refers to the amount of moisture air can hold without forming condensation. If more moisture is added to the air, you'll get rain or condensation.

Saturation Levels Change

Air can hold more moisture at higher temperatures. For example, we said that 70°F air can hold a little more than 100 grains of moisture per pound of dry air. Saturated 40°F air can only hold about 36 grains. This is an important concept. **As air warms up, it can hold more moisture. As air cools down, it's able to hold less moisture**. Condensation usually forms as air is cooled.

Relative Humidity

Relative humidity (RH) is given as a percentage. It is the amount of moisture in the air relative to the amount of moisture the air could hold if saturated. For example, since air at 70°F can hold 108 grains of moisture per pound of dry air, a sample with 54 grains of moisture has a relative humidity of (54/108) 50%. Air at 70°F can hold twice as much moisture as air at 50°F. If we cool a sample of air at 70°F with a 50% relative humidity down to 50°F, the relative humidity will rise to 100%. If the temperature falls to 49°F, we'll get condensation.

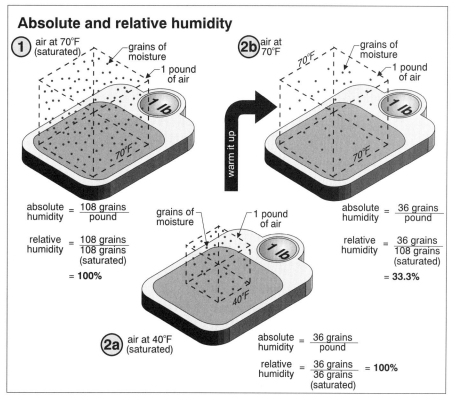

Here's another example: Let's heat a bundle of air with 50% RH from 50°F to 70°F. The relative humidity will drop to 25%. The absolute amount of moisture in the air is the same, but the amount of moisture *relative to how much total moisture the air could hold* drops.

Summary

To review, cold air can hold very little moisture. Warm air can hold a lot of moisture. If the air is nearly saturated at any given temperature, cooling the air causes condensation.

Real World Examples

This explains why a glass of lemonade will *sweat*. On a hot summer day, the air temperature may be 80°F with a 70% RH. The lemonade in the refrigerator is at about 40°F. As we pour the lemonade into a glass, the lemonade cools the glass (heat moves from the glass into the lemonade). The outer surface of the glass cools the air immediately around the glass, as heat flows from the air into the glass. Dropping the temperature of 80°F air at 70% RH will result in condensation as the air cools to about 68°F. Water droplets will form on the outside of the glass.

A Cold Winter Day

Weather reports will often indicate the relative humidity is 80% when it's about 30°F outside. The air is close to being saturated, yet people complain about how dry winter air is. How can both of these statements be correct? The air is fairly dry. Air at 30°F is not capable of holding very much moisture. The absolute humidity is low even though the relative humidity is high.

What happens to that cold air if it leaks into the house through an open door or window or a crack in the wall? The air is typically warmed to about 70°F. Warmer air can hold a lot more moisture. As a result, the relative humidity drops. The relative humidity drops from 80% to about 20% as the air heats from 30°F to 70°F. The 70°F air at 20% RH feels dry. This helps explain why winter air is dry, despite the high relative humidity in outdoor air.

How Much Humidity Is Enough?

People and buildings have different preferences for humidity. People like humidity in the range of 25% to 50%. However, many homes cannot tolerate moisture contents of 40% in the winter without suffering condensation problems. When it's freezing outside, buildings would fare better with 5% to 15% RH.

Optimum Humidity For People

An optimum humidity range of 38 to 58% RH minimizes the effect of bacteria, fungi, mites, respiratory infections, asthma, chemical reactions and ozone production. This level is unfortunately too high for most buildings.

The 40% Problem

The air in your house at 70°F and 40% RH will be saturated (100% relative humidity) and will produce condensation when the temperature drops to 45°F. Can you imagine any house air getting that cool? What if it leaks out of the house? It's easy to see how that air leaking into a wall or attic space will cool quickly. The outside of your walls and the underside of the roof in your attic space will be close to the outdoor temperature. As this bundle of air leaks out through the wall or roof, it will cool and, quite possibly, condense.

Moisture moves through the house in four ways:

How Moisture Moves

- Bulk moisture – these are leaks.
- Capillary action – this is wicking or rising damp.
- Air-transported moisture – the water vapor in the air moves with the air.
- Vapor diffusion – water vapor moves from high to low pressure areas without air movement.

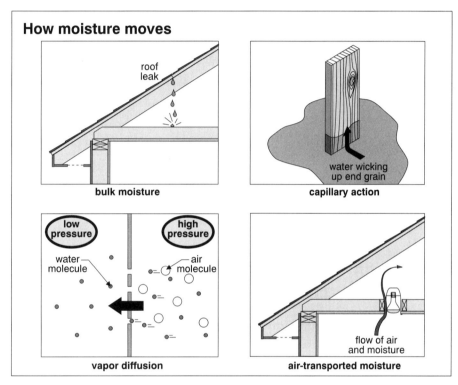

How moisture moves

roof leak

bulk moisture

water wicking up end grain

capillary action

low pressure

high pressure

water molecule

air molecule

vapor diffusion

flow of air and moisture

air-transported moisture

Air-Transported Moisture

Bulk moisture or leaks are usually identified quickly and corrected. Of the other three mechanisms, moisture that gets carried with moving air is the most important. Let's look at some ways air leaks into and out of houses:

- Air can move freely between indoors and outdoors through doors and windows.
- Air leaves houses through chimneys and vents for wood stoves, fireplaces, furnaces, boilers and water heaters.
- Air leaves the house through clothes dryer vents, bathroom exhaust fans and kitchen exhaust fans.
- Air leaks out of and into homes as a result of **stack effect**.
- Wind can cause air to be blown into a house on the windward side and drawn from the house on the leeward side.

Some of these are obvious, but some warrant a little discussion.

Stack Effect

Stack effect in a building is fairly simple. Warm air is lighter than cool air because it is less dense. The warm air rises and expands, creating a higher pressure near the top of the house. This relatively high pressure air tries to get out of the house through any cracks or holes it can find. The cooler, lower pressure air near the bottom of the house allows outdoor air in through cracks and holes.

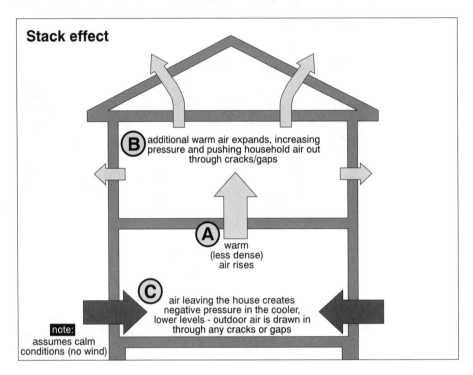

Stack Effect Great In Tall Buildings

In tall buildings, the stack effect can be a tremendous force. That's why most high rise offices have revolving doors, or vestibules. When a door is opened directly to the outdoors in a tall building, there is a tremendous amount of air drawn into the building. Revolving doors use an air lock to greatly reduce this influx of cold air.

Neutral Pressure Plane

In the top part of a building there is positive pressure relative to the outdoor air pressure. In the lower part of the house, there is negative pressure relative to the outdoor air pressure. At some point in the house, there is a level where the pressure is neither positive nor negative. This is referred to as the **neutral pressure plane**. The neutral pressure plane on a calm day in an average house might be a little more than halfway up the house. However, there are many things that can change the location of the neutral pressure plane. This is important because warm, moist air leaking out through the walls is more likely to damage a home than cold, dry air leaking in through the walls.

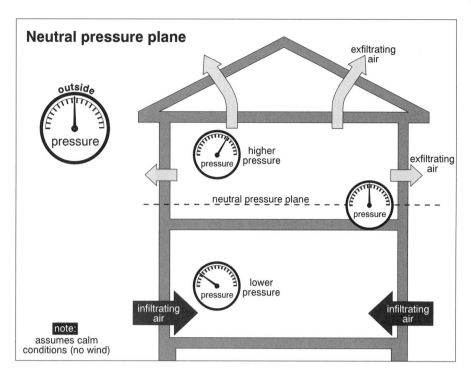

Neutral pressure plane

outside pressure

exfiltrating air

higher pressure

exfiltrating air

neutral pressure plane

pressure

lower pressure

infiltrating air

infiltrating air

note: assumes calm conditions (no wind)

The Big Points   In summary –

• Moisture in the air can condense in buildings.
• Moisture in the air moves primarily by being carried with the air that moves.

We're going to talk about the implications of moisture movement and ways to control it. Keep in mind that controlling air movement controls moisture movement, as well as heat movement.

Dew Point   We talked about air being saturated with moisture when the relative humidity is 100%. For any bundle of air, the **dew point temperature** is the temperature at which condensation will just start to occur. It's just another way of saying the humidity level is 100%.

Many people think of the dew point as a cold temperature. This isn't necessarily so. Air at 80°F and 50% RH has a dew point of 60°F. If the outdoor temperature is 80°F and the humidity is 90%, the dew point is 78°F. The dew point can be at any temperature. Any bundle of air that is almost saturated has to cool only slightly to produce condensation.

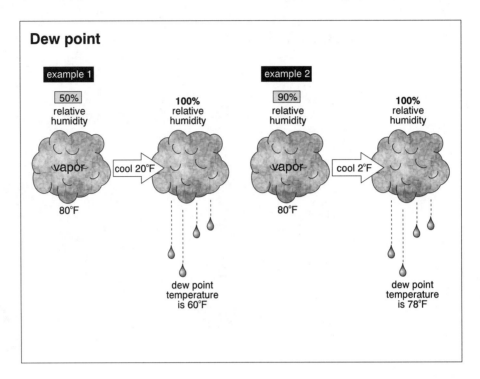

Dew point

example 1		example 2	
50% relative humidity	100% relative humidity	90% relative humidity	100% relative humidity
vapor	cool 20°F	vapor	cool 2°F
80°F	dew point temperature is 60°F	80°F	dew point temperature is 78°F

One Little Twist

We've talked about a temperature at which condensation would occur. Condensation doesn't always behave as it should. The indoor surface of a wall might be roughly 70°F and the outdoor surface could be 0°F on a winter day. The temperature in the wall drops as you move through the wall from the 70°F side to the 0°F side. The closer you are to the outside, the colder the wall is. This makes perfect sense. Partway through the wall, the temperature of the wall will be at the dew point. For example, if the indoor air is 70°F with 40% RH, the dew point is at 45°F. This will be somewhere in the middle of the wall. Strangely, this isn't necessarily where the condensation occurs.

Preferential Nucleation Sites

Condensation forms on a surface. Some surfaces are more conducive to having water droplets form on them. These are called **preferential nucleation sites**. As it turns out, fiberglass batts do not offer very favorable spots for condensation to occur. Tests have shown the condensation occurs on the wall sheathing, not in the insulation partway through the wall. This helps explain why damage does not always occur where we expect it.

Vapor Diffusion

Vapor diffusion is the movement of vapor without air movement. Many materials allow moisture through without allowing air through. Human skin is one good example. A good quality windbreaker is another. These materials are designed to stop the flow of air, but still breathe. It's important for clothing because we want to be able to get rid of the moisture that evaporates off our skin, but we don't want to be exposed to the direct effect of the wind. The people who make Gortex® have been very successful at developing products that do exactly this. Housewraps do not allow air through, but do allow moisture through.

Vapor Pressure

Water vapor will move from an area of high vapor pressure to an area of low vapor pressure. A house at 70°F and 35% relative humidity has roughly ten times the vapor pressure in the air as outdoor air at 0°F and 75% relative humidity. (There's more moisture in the 70°F air even though the **relative** humidity is lower.) Moisture will move as a result of the vapor pressure difference from indoors toward outdoors.

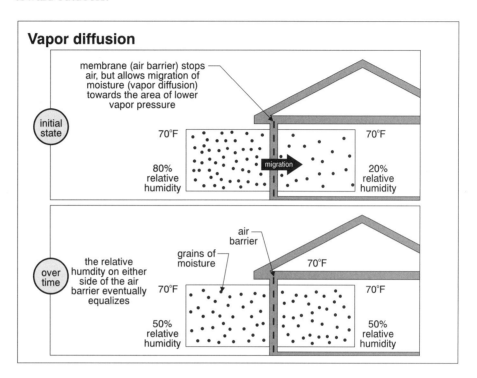

Vapor diffusion

Putting It In Perspective

The movement of air in and out of buildings is about a hundred times more important than vapor diffusion. For practical purposes, we can ignore vapor diffusion. **Vapor retarders aren't very important. Air barriers are important.**

The Problem with Moisture

Interstitial Condensation

We've been talking about moisture condensing in building components as warm, moist air cools on its way out of the building. This is sometimes called **interstitial condensation.** This causes –

• rotting wood
• expanding and shrinking wood due to changes in moisture content
• rusting steel
• spalling concrete and masonry (with freezing)
• reduced thermal resistance of insulation materials (may be temporary)

It's easy to see why we don't want condensation accumulating in roofs and walls.

*Moisture
Gets Stored*

Some building materials like wood are **hygroscopic**. This means they have fairly large surface pores and are able to absorb considerable amounts of water. Organic insulations, such as cellulose fiber and wood shavings or sawdust, are also hygroscopic. So are wood framing materials, sheathing and siding, for example. Materials like steel, glass, fiberglass and mineral wool are not hygroscopic. Hygroscopic building materials are able to absorb moisture and store it during wet periods. As the air dries out (as temperatures increase), the building materials will dry out. There is a limit to how much moisture can be stored. Wood, for example, is able to safely store up to twenty percent moisture by weight. Moisture levels above this can cause rot.

*Drying
Potential*

Some buildings are able to dry faster than others. Where air can move freely through attics or between siding and sheathing, moisture can be flushed into the outdoor air during periods of low relative humidity. Wall assemblies with wood shingles and shakes, conventional wood siding, and aluminum and vinyl siding have fairly good **drying potential**. Materials like stucco and EIFS (Exterior Insulation and Finishing Systems) do not have good drying potential. This can have an effect on whether (and how much) damage is done to structures due to trapped moisture.

*Can't See
Everything*

While the amount of hygroscopic material and the drying potential of building components has some effect on whether or not damage will be extensive, home inspectors are often not in a position to evaluate these things and certainly not in a position to predict performance. Let's look at some performance history.

*Moisture
Was Not
Always A
Problem*

North American houses have not always suffered moisture damage. Homes built into the 1950s did not suffer serious condensation damage. Was it because they produced less moisture? No. Was it because houses had sophisticated ventilation systems? No. Was it because houses were very tight and prevented warm, moist air from getting into cold wall cavities? No. Let's look at why older houses did not have problems.

*Poorly
Insulated*

Older houses were usually poorly insulated. This meant wall and roof assemblies were warm. This isn't very energy efficient, but the longer we keep the moist air warm, the less condensation we'll have.

Loosely Built

What is more important, older homes were not airtight. Moisture was flushed from the houses regularly. Houses often had several air changes per hour. The air in the house didn't stay long enough to pick up a lot of moisture. The air passed through wall cavities and roof spaces so quickly, very little condensation occurred. The condensation that did appear was stored in the building materials until the weather warmed up and dried out. At this point, the building materials would release their moisture quickly as the air flushed the moisture through the loosely built walls and ceilings.

Drafty Is Good

Those old, drafty houses often had several air changes per hour. They were neither as energy efficient nor as comfortable, but the buildings were healthy. Warm, moist air did not build up inside the house. Although moisture was produced just as it is today, the moisture levels in the air did not reach the levels in modern homes because the air was changed so frequently.

Cold air leaking into the house is very dry. It can absorb a considerable amount of moisture. If we move that air out of the house before it is nearly saturated, we're less likely to suffer condensation.

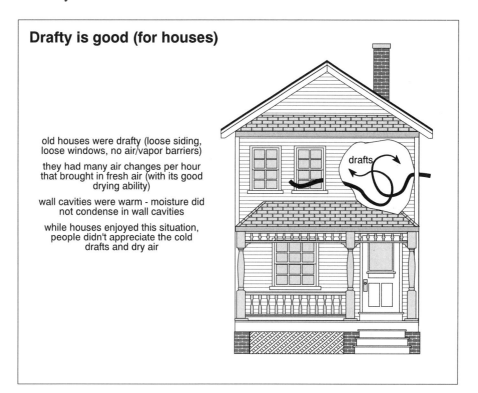

Drafty is good (for houses)

old houses were drafty (loose siding, loose windows, no air/vapor barriers)

they had many air changes per hour that brought in fresh air (with its good drying ability)

wall cavities were warm - moisture did not condense in wall cavities

while houses enjoyed this situation, people didn't appreciate the cold drafts and dry air

drafts

Summary

The flushing action, combined with the warm wall and roof cavities, meant we didn't have problems.

Energy Efficiency Is A Problem

North Americans started to worry about the conservation of gas and oil in the early 1970s. Heating costs for homes had never been an issue, but with the oil embargo and the crisis of the Middle East, we realized our supplies of gas and oil were limited. As the supply dwindled, costs rose. We became more sensitive to energy efficiency and more demanding about comfort issues. We objected to those cold drafts our parents had taken for granted.

Insulation And Air Sealing

We started adding insulation to existing homes and putting more insulation in new homes. Upgrading the insulation meant slower heat flow through roofs and walls and reduced heat loss. We found that adding insulation sometimes didn't help reduce fuel bills as much as we had hoped. We eventually figured out that warm air was still leaking through loose roofs and walls, taking heat with it.

*Reduced
Air Leakage*

We started to reduce the air leakage through buildings to slow heat loss and control drafts. Windows were made more weather-tight. Buildings were caulked, and the rate of air movement was reduced.

*Heating Costs
And Comfort*

The improved insulation and reduced air leakage meant heating costs were reduced and comfort was increased. However, we reduced the number of air changes in the house.

*More
Moisture In Air*

Air stayed in the house longer and had more time to absorb the moisture generated in the house.

Less Flushing

Slower leakage of air out of the house meant slower infiltration of fresh dry air. It also meant less flushing of floor and wall cavities.

*Colder
Cavities*

The improved insulation and reduced air leakage meant colder wall and roof cavities.

*Here's The
Catch*

Let's summarize the results:
• More insulation meant colder walls and roofs.
• Less air leakage meant the air in the house absorbed more moisture.
• Less air leakage also meant less flushing of roof and wall cavities.

These factors aren't problems by themselves, but let's add one more issue. No matter how much we controlled air leakage, there was always some air movement out of and into homes. As the air with increased moisture escaped into the colder wall and roof cavities, it moved more slowly through them. We had accidentally created a recipe for condensation.

Summary

More insulation and reduced air leakage led to –

• more moisture in the house air,
• colder walls and roofs, and
• less flushing of moist air.

This led to concealed condensation in houses. And it's getting worse. Let's look at why.

*Chimneys Are
Disappearing*

Older houses had open chimneys with natural-draft appliances. Air escaped up the chimney 24 hours a day. This helped flush moisture out of houses, although it wasn't very energy efficient. Flushing warm, moist air up the chimney meant cool, dry air was drawn in through cracks and holes. This helped keep walls and roofs dry and prevented moisture problems caused by condensation.

*Modern
Heating
Systems*

Modern heating systems have eliminated the chimney or put a sock in it. Electrically heated homes, mid- and high-efficiency furnaces, automatic vent dampers and induced-draft fans are all designed to prevent air escaping from the building. Even fireplaces have glass doors to stop air escaping up the chimney. As we reduce the amount of air escaping from the home, we also reduce the amount of moisture that escapes. This raises indoor humidity levels (tight homes don't need humidifiers) and adds to the problem. The moist air leaking into wall and roof cavities condenses. The trend to using outdoor combustion air for furnaces, water heaters and fireplaces, further reduces air changes and makes the situation worse.

*New Isn't
Always Better*

As we can see, modern construction techniques that emphasize better insulation, reduced air leakage and greater energy efficiency can lead to problems. Older houses with upgraded insulation, caulking, weather stripping and modern combustion appliances may suffer as well.

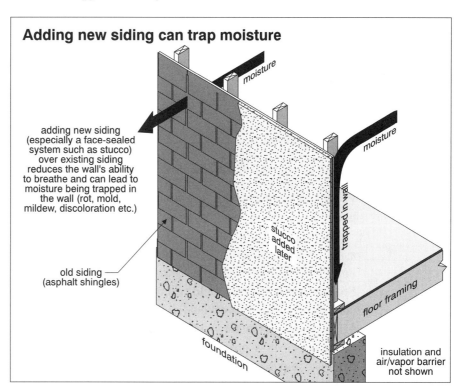

Adding new siding can trap moisture

moisture

adding new siding
(especially a face-sealed
system such as stucco)
over existing siding
reduces the wall's ability
to breathe and can lead to
moisture being trapped in
the wall (rot, mold,
mildew, discoloration etc.)

moisture

trapped in wall

stucco
added
later

old siding
(asphalt shingles)

floor framing

foundation

insulation and
air/vapor barrier
not shown

*Double Siding
And Face-
sealed Systems*

Older houses can be doubly troublesome because old siding materials are often covered with new siding materials. This creates more places for moisture to be trapped in the wall assembly. The popularity of face-sealed exterior claddings also contributes to problems. Stucco over existing siding can significantly reduce the drying potential of a wall.

Do All Homes Suffer?

All homes don't suffer, for a number of reasons:

- Air sealing is often not as successful as people expect.
- Dry climates are forgiving.
- Some lifestyles generate less moisture than others.
- Efficient fans and venting systems reduce moisture buildup.

Every home is different. And every family is different. It's difficult to predict which homes will have problems.

Evidence Of Humidity Problems In House

The following clues may indicate excess moisture levels in homes:

1. Condensation on windows
2. Staining or mold, often in bathrooms or on cold walls (such as closets) and window frames
3. Stuffy air
4. Odors that linger in the house
5. Spillage or backdraft from furnaces, boilers or water heaters
6. Backdraft at fireplaces or wood stoves

Solving the Problem

We need an approach to reducing moisture problems. Let's look at some possibilities:

- **Produce less moisture**. A house with two people will produce less moisture than a house with 8 people.
- **Keep the condensing surfaces warmer**. Insulation on the outside of the wall assembly is one approach. Double- and triple-glazed windows provide warmer inner surfaces than single-glazed windows.
- **Stop air leakage**. If we could ensure that no warm, moist air got into the cold parts of the structure, we wouldn't have condensation.
- **Flush the air out faster**. Well-ventilated loose wall and roof assemblies could let the warm, moist air out quickly. Improved attic ventilation, for example, helps accomplish this, although there are some downsides to this approach.
- **Exhaust warm, moist air directly outside**. Kitchen and bathroom exhaust fans, clothes dryer exhaust vents and central vacuum systems dump moist air outside rather than allowing it to leak into wall and roof assemblies.

All of these approaches have some validity and some limitations. There are several interesting strategies being considered for controlling the movement of moisture and air in homes. So far, no one solution has solved all the problems.

Controlling House Moisture

By now you should have a good sense of how moisture moves into and out of homes and how it can be a problem. We'll emphasize again that, for practical purposes, controlling the air movement also controls the moisture movement.

Moisture Control in Hot Climates

We've been talking about moisture control problems in cold climates during the heating season. We also run into moisture problems in warm climates during the cooling season. The mechanism is exactly the same, but the moisture source and direction of movement are different.

Moisture Source

When we talked about cold climates, we talked about getting rid of the moisture people generate inside houses. The outdoor air is dry. In hot climates, the outdoor air may be very moist. The southeastern United States, for example, is prone to high humidity with high temperatures. Hot, dry climates like Arizona are less of a problem. You know already whether you're in a relatively humid climate.

Air Conditioning Is The Problem

If you live in a hot, humid climate, many homes have central air conditioning systems – something homes have only had to deal with over the last 50 years. Air conditioning creates an indoor environment that is different from the outdoor environment. It's cooler and drier. Air conditioning systems cool and dehumidify the air to improve comfort.

Drying The House

It seems air conditioning systems should help the home because they dry the air as they cool it, but let's look at little further. Air-conditioned homes need some air changes too. Whether it's accidental air changes from leakage induced by stack effect or wind, or intentional air changes from kitchen or bathroom exhaust fans, air moves through the walls and roof.

Air Leaking Into The Building

Let's look at outdoor air leaking into the building through the walls. Outdoor air might be 85°F and have a relative humidity of 80%. If the house were not air conditioned, the indoor temperature and humidity would be similar. An air-conditioned house may have an indoor temperature of 75°F and 50% relative humidity.

An 80°F Dew Point

That warm outdoor air is almost saturated. We talked earlier about how dropping the air temperature raises the relative humidity to the saturation point (dew point). *Air at 85°F and 80% RH will hit the dew point if it's cooled to 80°F!* Outdoor air leaking into an air-conditioned home will cool as it moves through the wall. The inside surface of the wall is roughly 75°F. As the warm, moist air from outdoors moves through the wall, it will be cooled to 80°F somewhere in the wall. Since this is the dew point, we can get condensation in the walls.

All Homes Should Rot

This seems like a problem that should destroy all homes during the air conditioning season. There are many variables. We mentioned earlier that building materials can absorb and store moisture up to a point, without damage. Weather conditions change. Temperatures fluctuate and so do humidity levels. Buildings may store considerable moisture during warm, wet seasons and dry during other parts of the year. There are daily cycles, too. Hot days may be followed by cooler nights.

Amount Of Air Leakage Is A Factor

People with central air conditioning try to keep the cool air indoors and the warm air outdoors. These houses are typically sealed to some extent, to prevent air leakage. If we could stop all the warm, moist air from leaking into the building through the roof and walls, we wouldn't have much trouble.

Leakage Depends On Pressure Difference

If the air pressure in the house is higher than the outdoor air pressure, air will leak out through the building, rather than in. Cool air moving out through the walls will warm up and be able to pick up moisture, rather than drop moisture. This won't cause any damage.

If you are in a hot, humid climate, you'll want to speak to local designers (architects), builders and authorities (inspectors) to find out whether this is an issue in your area, and if so, what steps are taken to control it.

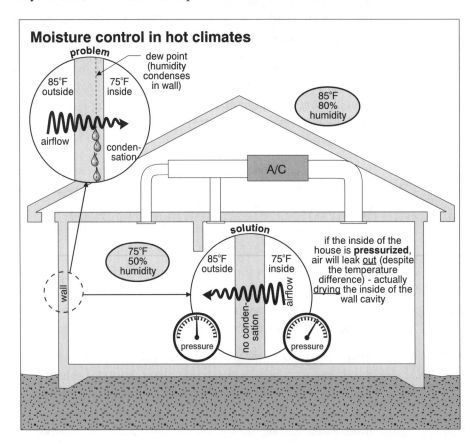

Insulation
& Interiors
M O D U L E

QUICK QUIZ 3

☑ INSTRUCTIONS

- You should finish Study Session 3 before doing this Quiz.
- Write your answers in the spaces provided.
- Check your answers against ours at the end of this Section.
- If you have trouble with the Quiz, re-read the Study Session and try the Quiz again.
- If you did well, it's time for Study Session 4.

1. List at least eight sources of moisture in house air (you may be able to come up with ten).

    _____   _____

    _____   _____

    _____   _____

    _____   _____

    _____   _____

2. Define **absolute humidity**.

    _____

    _____

3. Define **relative humidity**.

    _____

    _____

4. What is meant by **saturated air**?

     _____

5. List four ways moisture moves through homes.

     _____

     _____

     _____

     _____

6. Define **stack effect**.

     _____

     _____

7. What is meant by **neutral pressure plane**?

     _____

     _____

8. Define **dew point temperature**.

     _____

     _____

9. Rank the importance of vapor diffusion and air leakage in condensation problems in homes.

     1. _____

     2. _____

10. What is meant by **drying potential**?

     _____

     _____

11. Why are new homes more likely to have condensation problems than old homes?

_____

_____

_____

_____

12. List five possible ways to control moisture and condensation problems in homes.

_____

_____

_____

_____

_____

13. How are condensation problems different in hot, humid climates?

_____

_____

_____

_____

_____

If you had no trouble with this Quiz, you are ready for Study Session 4.

Key Words:
- *Condensation*
- *Absolute humidity*
- *Relative humidity*
- *Stack effect*
- *Neutral pressure plane*
- *Dew point temperature*
- *Vapor diffusion*
- *Drying potential*
- *Flushing*

Insulation & Interiors

M O D U L E

STUDY SESSION 4

1. You should have completed Study Session 3 and Quick Quiz 3 before starting this Session.

2. This Session covers the control of moisture flow in homes in both heating and cooling climates.

3. At the end of this Session, you should be able to –
- Rank air leakage and vapor diffusion in importance with respect to moisture problems in homes.
- List two common air/vapor barrier materials.
- Explain in two sentences how sheathing and roofing materials have affected air leakage in homes.
- Name one important difference between housewrap and building paper.
- Explain in two sentences how windows affect moisture levels in homes.
- List two advantages and two disadvantages of exhaust fans.
- Explain in two sentences how direct-vent furnaces affect moisture levels in homes.
- Describe in three sentences how heat recovery ventilators (HRVS) work.
- List six components of HRVS.
- List two different types of ventilation systems in homes.
- Describe in one sentence the source of moisture that can cause condensation damage in homes in hot, humid climates.
- Indicate where the air/vapor barrier should be in a wall system in a hot, humid climate.

4. This section may take you roughly 45 minutes to complete.

5. Quick Quiz 4 is at the end of this section.

Key Words:

* *Air/vapor barriers*

* *Sheathing*

* *Housewrap*

* *Building paper*

* *Windows*

* *Exhaust fans*

* *Direct-vent appliances*

* *Heat recovery ventilators (HRVs)*

* *Venting roofs*

* *Venting living spaces*

* *Indoor air quality (IAQ)*

* *Depressurization*

3.4 CONTROLLING MOISTURE FLOW IN THE BUILDING ENVELOPE

The concept of a house as an environmental envelope is a helpful one. A home is an artificial environment surrounded by a huge natural environment. This is a problem because –

- we can't isolate ourselves from the natural environment or we'd have no fresh air to breathe.
- we can't completely control the interaction between the indoor and outdoor environments.

This leakage of indoor air outside, and outdoor air into our homes is necessary and yet troublesome. A house is not a tightly sealed chamber isolated from the world outdoors, but a rather loose envelope between the indoor and outdoor environments.

Problem Recognized

As insulation strategies changed in the latter half of the 1900s, we recognized the need to control moisture (largely by controlling air movement). How did we approach the problem?

Vapor Retarders Or Air/Vapor Barriers

We tried to reduce the air leakage through the roofs and walls. We created **vapor retarders**, now often referred to as **air/vapor barriers**, to reduce the movement of air through the building envelope. As the name suggests, we used to think vapor diffusion was the big problem. We now know air leakage is a far bigger problem.

Poly Replaced Kraft Paper

Older air/vapor barriers were of kraft paper, often attached to fiberglass or mineral wool batts. Since these were typically laid between ceiling joists or wall studs, they were not continuous. While imperfect vapor barriers can perform their duties reasonably well, an air barrier that is not continuous is not very useful at all. Modern air/vapor barriers are typically polyethylene films. Considerable attention is now given to creating a continuous air/vapor barrier to minimize air leakage. We'll look more closely at these, later.

Ventilation Of Roofs And Walls

Despite the use of continuous air/vapor barriers, we were not 100 percent successful at stopping leakage of warm, moist air into cool wall and roof spaces. As a result, increased emphasis was placed on ventilation of roof and wall spaces. Much more attention has been paid to roofs than walls. Soffit vents, ridge vents, gable vents, and vents in the field of the roof are all designed to help flush moist air out of roof spaces, replacing it with cold, dry air. There are some problems with this approach, which we'll discuss later.

Tighter Roof Sheathing

Ventilation has also become more important as loose-laid planks and wood shingle roofs have been replaced with plywood and waferboard sheathing and asphalt shingle roofs. The old plank and shingle or shake roofs were fairly leaky and allowed moisture to be flushed out quickly. The new panel-type roof sheathings and asphalt shingles allow less air movement.

Wall Sheathing Has Changed | Wall sheathing is like roof sheathing in that the materials and methods have changed over the years. Loose plank sheathing has been replaced with plywood, gypsum board and OSB. These modern panel-type sheathings allow much less air into and out of the walls. By now you'll have a sense that this can be a good thing or a bad thing, depending on differences between the indoor and outdoor environments, and the mechanisms inducing movement of air and moisture between these environments.

Siding Treatments | In some cases, we have attempted to allow siding materials to breathe. Aluminum and vinyl siding, for example, are loose-fitting cladding materials with drainage holes in the bottom of each section. However, other siding materials do not allow much air movement and flushing of moisture. Stucco treatments and panel-type sidings (plywood, fiber-cement and OSB, for example) provide very little opportunity for flushing out moisture.

Housewrap Versus Building Paper | In many parts of North America, loosely fit building paper between the sheathing and the siding has been replaced by tightly fit housewraps. These restrict air movement while allowing vapor diffusion through. It's like putting a big windbreaker on your house. Housewrap is designed to reduce heating costs. An air barrier on the outside of the house stops wind washing in wall assemblies. We mentioned earlier that wind moving through fiberglass batt insulation can dramatically reduce its insulation value. Restricting this air movement also restricts the flushing effect as air moves through wall spaces. Three-foot-wide rolls of building paper were installed in a loosely overlapping system on the outside of sheathings. Housewrap typically comes in wide rolls capable of covering an entire story of a building without any seams. Seams of the housewrap are often taped to further restrict air leakage.

A Good Thing Or A Bad Thing? | Housewraps, like building paper, are also a second line of defense against moisture such as wind-driven rain. Water that leaks past siding is stopped by the housewrap. There is considerable discussion at the time of writing (early 1998) about whether housewrap is a big step forward or not. While we understand the issues, we are not in a position to quantify the importance of each and predict the long-term positive and negative effects of housewrap versus building paper. We suspect the results will vary case to case because there are so many factors involved in building envelope performance.

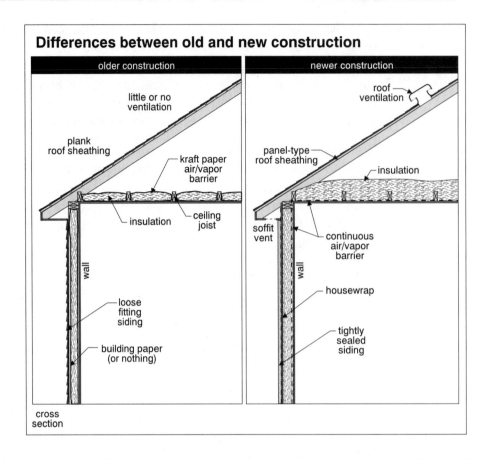

Differences between old and new construction

older construction

- little or no ventilation
- plank roof sheathing
- kraft paper air/vapor barrier
- insulation
- ceiling joist
- wall
- loose fitting siding
- building paper (or nothing)

newer construction

- roof ventilation
- panel-type roof sheathing
- insulation
- soffit vent
- continuous air/vapor barrier
- wall
- housewrap
- tightly sealed siding

cross section

Reduced Air Leaks And Improved Ventilation

So far we've talked about modern homes having reduced air leakage and improved ventilation in roofs and, in some cases, reduced ventilation in walls.

Humidifiers

Some homes try to control moisture directly with humidifiers. When homes in cold climates were very leaky, the large number of air changes meant we were always warming up cold air. This cold air had very little moisture in it and, as a result, our houses always felt dry during the winter. Humidifiers were added to furnaces to raise the humidity levels. As construction got tighter and the number of air changes per hour reduced, the need for humidifiers has, in many cases, disappeared.

New Windows May Hurt

This isn't well understood by many homeowners who continue to add humidity to their homes. They are frustrated by condensation on windows and sliding glass doors. This often leads to replacement of windows and doors with better insulated, more tightly sealed, units in an effort to eliminate condensation. In most cases, the problem isn't the quality of the windows and doors – it's the elevated moisture level created inside the house.

Old Windows Indicated Humidity Levels	Old windows were great. Moisture control in homes used to be easier. Many people ignored their humidistat on a furnace humidifier, but paid attention to the condensation on windows. As windows began to get condensation, people turned down the humidifier. They weren't trying to save their wall cavities; they were trying to prevent water damage to window sills and the walls below. Older, single-glazed windows typically had a very cold surface. The cold glass would cool the air adjacent to the window, often to the dew point, resulting in condensation.
Loose-Fitting Storms Are Good	Loose-fitting storm windows cut down on drafts and raised the temperature of the inner glass slightly, reducing condensation. Any warm, moist air that leaked past the first window would have no difficulty getting past the loose-fitting storm window, so there was no condensation problem with the storm window.
Double And Triple Glazing	Modern windows are more energy efficient than old windows. The inner surface of the glass is warmer as a result of carefully spaced double or triple glazing. Insulated sashes help raise the temperature of the inner surface of the windows. Newer windows and doors are also tighter than old ones. They allow less air leakage.
A Good Thing Or A Bad Thing?	Modern windows don't lose heat as quickly as older windows (although their R-values are still quite low). We don't get condensation on windows as quickly, and, therefore, don't make the same efforts to reduce the humidity in houses that we used to. The reduced air leakage through and around new doors and windows leads to higher indoor humidity levels. While all this is nice for the homeowner and great for windowsills, it is hard on wall and roof cavities. If the house humidity levels are higher, occupants will be happier, but the building suffers. The warmer glass surfaces reduce one symptom, but don't do anything to stop moisture problems in walls and roofs.
Exhausting The Moisture	Many people have realized that with fewer air changes we are elevating the moisture levels inside the home. Because people and houses generate moisture all the time, the longer any given bundle of air stays in the house, the more moisture it will contain. One strategy is to dump the warm, moist air directly outside. Exhaust fans close to sources of moisture, such as kitchens and bathrooms, make sense.

There are two advantages:

Two Advantages	• Kitchen and bathroom fans throw out moisture with that air – moisture that could damage the building envelope. • Exhaust fans lower the air pressure inside, relative to the outdoors. In cold climates, we've said that warm, dry air leaking in through the walls isn't a problem. The added benefit of exhausting warm, moist air is increased movement of dry air through walls and roofs.

There are also two disadvantages:

Two Disadvantages	• Exhausting air from the house wastes energy. We are throwing out air we have heated to a comfortable temperature. • The lower indoor air pressure can cause backdraft of combustion appliances. Exhaust products from gas and oil burners and smoke from wood stoves and fireplaces may get into living spaces.

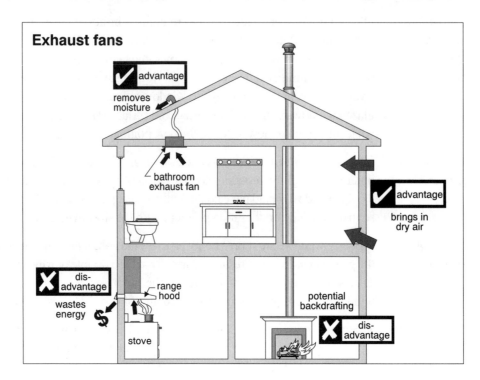

Exhaust fans

✔ advantage
removes moisture

bathroom exhaust fan

✔ advantage
brings in dry air

✗ dis-advantage
wastes energy

range hood

potential backdrafting

✗ dis-advantage

stove

What About The Furnace?   In a traditional home, it's very risky to create lower air pressure inside the house than outside. This leads to backdraft of combustion appliances. Furnaces and water heaters, for example, use house air for combustion and draft, and vent all that air outside. Natural-draft appliances worked well historically, because of stack effect and the higher air pressures in houses caused by warmer air. If we reduce the air pressure inside the house, we risk backdraft chimneys. Air moving down through chimneys to neutralize the pressure difference between indoors and outdoors can prevent exhaust gases going up chimneys.

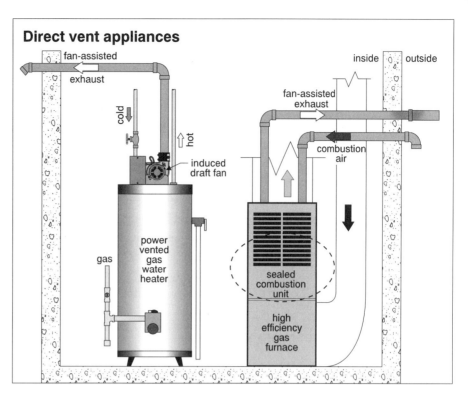

Direct vent appliances

Direct-Vent Appliances Combustion appliances have changed significantly in recent years. There is less reliance on natural-draft chimneys. Induced-draft fans are common and sealed combustion units using outdoor air for combustion are becoming more popular. Induced-draft fans push air outside through chimneys and vents. They don't rely on natural draft. Sealed combustion units don't care if the air pressure inside is low. Combustion air is drawn directly from outside.

Negative Indoor Air Pressure Okay The need to maintain a positive indoor air pressure is diminishing. (We're talking about a cold climate problem here. In hot climates we want higher indoor air pressure!)

Controlled Ventilation Controlled exhaust is a common strategy in controlling house moisture in North America in the 1990s. In some areas, energy conservation efforts have been coupled with the recognized need to exhaust warm, moist air. This has led to the development of a new home appliance.

Heat Recovery Ventilators **The heat recovery ventilator (HRV), energy recovery ventilator (ERV), heat exchanger or air-to-air heat exchanger** has been around for several years. These devices are popular in some areas and, falling from favor in others. They use two simple concepts:

• Since we have to exhaust air anyway, let's remove some of the heat from the air before we exhaust it.
• Let's control the location and amount of fresh air drawn into the house.

How They Work

HRVs use one duct to throw warm, moist air outside and another duct to draw cool, dry air in. On the way through, some of the heat from the warm, moist exhaust air is transferred to the cool, dry air coming in through the heat exchanger. If we set up the pressures correctly, we can minimize the amount of air leakage through roofs and walls in either direction. This should solve all our problems!

What's In An HRV?

• An HRV contains two duct systems. One collects exhaust air from the house and pushes it outside. The second draws fresh air into the house. Fans move the air into and out of the house at the desired rate.

• A heat exchanger captures some of the heat from the exhaust air and transfers it to the incoming air.

• A defrost mechanism removes ice from the heat exchanger.

• Balancing dampers and flow collars allow the system to be adjusted.

• A drain carries away condensed humidity and melted frost.

• Controls determine the speed of fans and the volume of air moved, depending on manual controls, humidistats and/or timers.

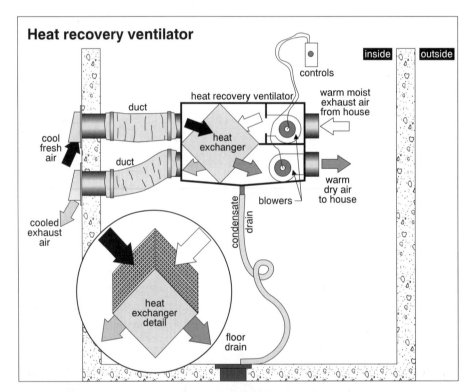

Thousands of homes have been provided with HRVs. Have they worked? Yes and no. The theory and the mechanics of these systems are fine. In practice, there have been many problems. As with any new technology, installation methods don't always match the designers' intent. Many of these units were installed incorrectly. Not surprisingly, with more installations, we've learned more about how these units should be installed and operated. The technology has changed somewhat over the years.

Do They Work Now? The outdoor environment is changeable. Temperatures, moisture levels, wind direction and intensity vary. Heat recovery ventilators are typically set up with a single operating parameter. In many cases, they are not set up as intended by the designer, or not maintained by the occupant. As a result, many heat recovery ventilators do not provide the optimal interior air conditions. In some studies, the majority of HRVs were found to be set up incorrectly. Many were shut off by homeowners who either didn't understand them or found they were not operating properly.

Efficiency Question Many systems don't deliver the desired interior conditions. Perhaps because of enthusiastic marketing, homeowner neglect or both, many units do not deliver expected efficiencies. The value of the heat reclaimed from the exhaust air can often be largely offset by the electricity cost of the continuous operation of the HRV and the furnace fan.

Time will tell whether heat recovery ventilators become more or less popular. Home inspectors should understand the operating principles and installation issues for these units. We'll talk about them in more detail later.

Two Types Of Ventilation We've been looking at ventilation systems as a way to control moisture damage in buildings caused by condensation from air leakage. But there is another reason ventilation is important. There are two types of house ventilation:

1. Venting of roof and wall spaces to flush warm, moist air out of the building components.
2. Exhausting stale air and supplying fresh air to the living space in a home. This type of ventilation can eliminate excess moisture before it has a chance to get into wall systems. However, it has other functions. Since we've reduced the number of air changes, it's not just moisture that lingers in our houses. Odors from cooking and bathrooms, for example, stay in the home longer. Other indoor air pollutants also accumulate in the home.

Can't Filter The Air There are no common filter systems that allow us to remove all the pollutants from the indoor air and keep the same air in the home.

Indoor Air Quality Indoor air quality (IAQ) has become a big issue in the 1980s and '90s. A great deal of attention has been paid to this issue, and much research is ongoing. Given the sensitivity of the typical North American to health concerns, we expect more issues like radon, asbestos, urea formaldehyde, PCBs, electromagnetic radiation and dangerous molds will drive efforts to improve house ventilation systems.

Home Inspectors' Approaches Home inspectors have a responsibility to be sensitive to what's happening in homes and to the perceptions of our clients. A whole new profession is springing up around indoor air quality. Some inspectors are embracing this as a business opportunity. Others are dismissing it as a fad. Many inspectors are sitting on the sidelines, referring concerned clients to specialists in indoor air quality.

*Separate
Ventilation
Issues*

It's important to understand the two types of ventilation. It's difficult because there is a little bit of overlap. We vent wall and roof spaces to get rid of moisture that can cause damage to the structure. We vent indoor living spaces to improve the air quality. This includes removing excess moisture in the air.

Summary

We've spent a lot of time talking about controlling heat and moisture flow. This lays the foundation for our discussion of the insulation, air/vapor barriers and ventilation systems you'll see in the field.

Before we move on to look at our inspection strategies for insulation, air/vapor barriers and ventilation systems in various house components, let's revisit houses in hot climates and recap what should be done differently when air conditioning is more common than heating.

Moisture control in hot climates

*Moisture Is
Outdoors*

We've talked about the source of moisture being on the outside of the house, rather than the inside, in cooling climates. In these houses, it's the warm, moist air outside leaking in through wall systems toward cool air-conditioned space that causes the problem. Air leaking out through the building is not a problem. We want to avoid depressurizing these houses. Depressurization may occur as a result of –

• leaking ductwork on air conditioning systems,
• clothes dryers venting outside,
• central vacuum systems,
• kitchen exhaust fans, and
• bathroom exhaust fans.

Kitchen and bathroom fans should be operated to remove high levels of moisture and odor from the house; however, they should be turned off as soon as they have performed their tasks. Timers on these fans may be a good idea.

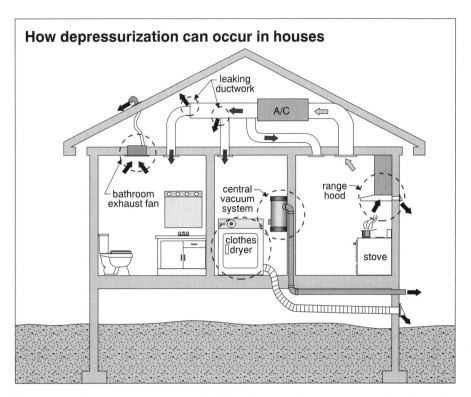

How depressurization can occur in houses

Moisture Through The Floors	In summer climates, houses are typically slab-on-grade construction. Good practice includes six-mil polyethylene film below the slabs and a couple of inches of sand below the poly to provide good drainage. We don't want moisture to accumulate under the slab and move up through the concrete. The poly stops moisture migration up through cracks and voids in the concrete.
Crawlspaces	Unfinished crawlspace floors can be a significant source of moisture in homes, which can cause damage in both heating and cooling climates. We discussed earlier that poly applied over an unfinished floor and sealed at joints in the perimeter is very effective. It's common to apply gravel over the poly to keep it in place. Home inspectors sometimes have to dig through gravel to determine if there's poly on the floor.
Don't Vent Crawlspaces	The conventional wisdom has held that crawlspaces should be vented to get moisture out. That wisdom is being questioned. Current thinking is that the warm, moist outdoor air leaking into the cool space under the house actually brings moisture into the crawlspace and causes problems. Many now say that crawlspaces should not be ventilated at all.

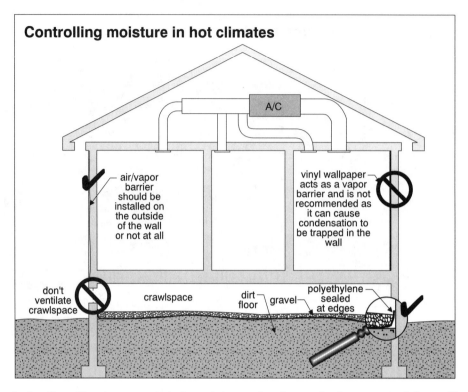

Controlling moisture in hot climates

A/C

✓ air/vapor barrier should be installed on the outside of the wall or not at all

⊘ vinyl wallpaper acts as a vapor barrier and is not recommended as it can cause condensation to be trapped in the wall

⊘ don't ventilate crawlspace

crawlspace

dirt floor gravel

✓ polyethylene sealed at edges

Walls	In heating climates, we wanted an air/vapor barrier on the inner part of the wall. Inner air/vapor barriers cause problems in cooling climates. The warm, moist air moving indoors from outside may condense on the outer face of the air/vapor barrier, resulting in mold, mildew and rot within the wall cavity. Air/vapor barriers should be installed on the outside of the wall, if at all, in cooling climates.
Avoid Vinyl Wallpaper	Vinyl wallpaper is a good air/vapor barrier. It should be avoided in cooling climates because it's an interior air/vapor barrier.
Poor Drying Potential	Many modern homes in cooling climates have poor drying potential. This is particularly true with face-sealed exterior siding systems such as stucco and EIFS. This can contribute to deterioration in walls since trapped moisture cannot easily escape.
Seal Exterior Of Walls	In hot climates, tightly fit OSB and plywood sheathing help keep warm, moist air out. Tightly fit building paper, housewrap and heat-reflecting foils behind siding also help prevent air moving into the walls and condensing as it cools. Providing a drainage plane between the siding and the sheathing papers helps prevent moisture getting into wall systems.

Roof Systems   Many homes in the hottest climates use radiant barriers to reflect heat. These foils are typically placed on the underside of sheathings. This foil is much more effective on the underside of the roof than it is on the attic floor, for example. We want to avoid dust and condensation, which can reduce the reflective performance of radiant barriers. Good venting of the attic space is also important to eliminate as much heat as possible. A combination of soffit and high-level vents is required.

Those of you working in hot climates will have to be sensitive to the fact that many of the building science principles conventionally accepted in North America do not apply. Use common sense to think about the source of moisture and how to control or eliminate it.

Now let's look at some specific building details, and find out what home inspectors can do during an inspection of insulation, air/vapor barriers and ventilation systems – after the Quiz, of course!

Insulation & Interiors
M O D U L E

QUICK QUIZ 4

☑ INSTRUCTIONS

- You should finish Study Session 4 before doing this Quiz.
- Write your answers in the spaces provided.
- Check your answers against ours at the end of this Section.
- If you have trouble with the Quiz, re-read the Study Session and try the Quiz again.
- If you did well, it's time for Study Session 5.

1. Is vapor diffusion or air leakage more important with respect to moisture problems in homes?

2. List two common air/vapor barrier materials. Which one is more commonly used today?

3. Why are roofs vented in the winter in cold climates?

4. How have modern wall and roof sheathings affected air leakage in modern homes?

5. How is housewrap different from building paper?

_____

_____

6. How do modern windows affect moisture levels and moisture control in homes?

_____

_____

7. List two advantages of exhaust fans in homes.

_____

_____

8. List two disadvantages of exhaust fans in homes.

_____

_____

9. How have induced-draft fans and direct-vent furnaces affected moisture levels in homes?

_____

_____

10. What is the purpose of a heat recovery ventilator?

_____

_____

_____

_____

_____

11. List six components of a heat recovery ventilator.

_____

_____

_____

_____

_____

_____

12. The term _ventilation_ might mean two very different things with respect to homes. Explain the two different types of ventilation.

_____

_____

13. Describe the source of moisture that can cause condensation problems in walls in hot climates.

_____

_____

14. Indicate where the air/vapor barrier should be located in the wall assembly in a hot, humid climate.

_____

If you had no trouble with this Quiz, you are ready for Study Session 5.

Key Words:
- **_Air/vapor barriers_**
- **_Sheathing_**
- **_Housewrap_**
- **_Building paper_**
- **_Windows_**
- **_Exhaust fans_**
- **_Direct-vent appliances_**
- **_Heat recovery ventilators (HRVs)_**
- **_Venting roofs_**
- **_Venting living spaces_**
- **_Indoor air quality (IAQ)_**
- **_Depressurization_**

Insulation & Interiors

MODULE

STUDY SESSION 5

1. You should have completed Study Session 4 and Quick Quiz 4 before starting this Session.

2. This Session covers common insulation materials and their characteristics.

3. At the end of this Session, you should be able to –
   • describe in one sentence the function of insulation.
   • list eight characteristics of an ideal insulation material.
   • define in one sentence **thermal bridges**.
   • explain in two sentences the effect of convective loops and voids in insulation on R-values.
   • describe in one sentence **wind washing**.
   • list four common forms of insulation materials.
   • list eight common insulation materials.
   • list the forms that each of the eight materials are commonly found in.
   • list one, two or three advantages and disadvantages of each of the various insulation materials.
   • describe in five sentences the issues surrounding urea formaldehyde foam insulation.

4. Before you start this Session, read sections 1.0 to 9.0 in the Insulation Chapter of **The Home Reference Book™**.

5. This Session may take you roughly one hour.

6. Quick Quiz 5 awaits you at the end of this section.

Key Words:

- *Heat transfer*
- *Thermal bridges*
- *Voids*
- *Convective loops*
- *Wind washing*
- *Housewraps*
- *Loose fill*
- *Batts or blankets*
- *Rigid board*
- *Foamed-in-place*
- *Fiberglass*
- *Cellulose*
- *Mineral wool*
- *Vermiculite*
- *Perlite*
- *Expanded polystyrene*
- *Extruded polystyrene*
- *Phenolic*
- *Polyurethane*
- *Polyisocyanurate*
- *Isocyanate*
- *Urea formaldehyde foam insulation (UFFI)*
- *Radiant barrier*

▶ 4.0 INSULATION MATERIALS

Function

The purpose of insulation is to slow the rate of heat transfer. This is true in both hot and cold climates. In cold climates, we are trying to stop the flow of heat out of the building. In hot climates, we are trying to slow the movement of heat into the building. Fortunately, insulation works in both directions. For the many climates in North America that need heating and cooling at different times of the year, insulation serves both functions.

Insulation Doesn't Stop Heat

We should emphasize again that insulation only slows the movement of heat. It does not stop it altogether. We should also remind ourselves that it really isn't the insulation material that stops most of the heat. The trapped air pockets within the insulation do most of the work. The solid material of the insulation does reduce direct radiation. Conduction and convection are controlled by the still air.

Characteristics

The ideal insulation material would have the following characteristics:

- High resistance to heat flow (a high R-value)
- Inexpensive
- Durable (lasts the life of the home)
- Completely fills cavities
- Air barrier (stops air leaks)
- Vapor barrier (stops vapor diffusion)
- Moisture and rot resistant (because all houses eventually leak)
- Non-combustible
- Chemically inert

Not surprisingly, no insulation meets all these criteria.

Measuring Insulation

We have already discussed that the resistance to thermal conductivity is referred to as an R-value. The metric equivalent to this is **RSI**. To convert from R to RSI, multiply the R by 0.1761. To convert from RSI to R, multiply the RSI by 5.679.

- RSI = R x 0.1761
- R = RSI x 5.679

All Materials Add To R-Value

The insulation value of walls is more than the value of the insulation itself. All materials have some insulating value, including the plaster or drywall, polyethylene air/vapor barrier, wood sheathing, building paper, siding and the air films. However, the majority of the R-value of the wall is attributable to the insulation.

Thermal Bridges

Thermal bridges occur at studs, top plates, sill plates, etc. These are points where the insulation is not continuous. Solid, high-density materials like wood studs conduct heat readily though walls. That's why these are called thermal bridges.

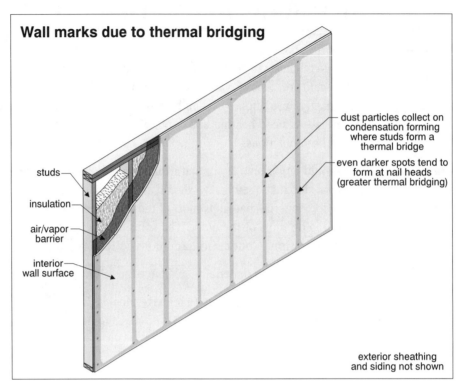

Wall marks due to thermal bridging

studs

insulation

air/vapor barrier

interior wall surface

dust particles collect on condensation forming where studs form a thermal bridge

even darker spots tend to form at nail heads (greater thermal bridging)

exterior sheathing and siding not shown

Wall Marks Due To Bridging

Thermal bridges not only cause heat loss, but can lead to localized cool temperatures on interior surfaces. Cool interior surfaces can result in surface condensation. The moisture on the surface attracts dust particles and, over a period of years, will create a permanent pattern on the wall. You may have seen dirt marks on walls in older houses that have not been painted for a long time. Very often, these dirt marks follow the pattern of the wall studs. In other cases, they may follow the wood lath. The dark spots are typically the areas where thermal bridging has resulted in cool interior surfaces and localized condensation on the plaster surface.

Insulation Voids And Convective Loops

Gaps in insulation can dramatically reduce the R-value of a wall assembly. If insulation batts do not completely fill stud cavities side to side and top to bottom, **convective loops** can be established as the air moves freely through the stud cavity. Voids can allow a great deal of heat through the wall. It is important that voids are filled. Mineral wool and fiberglass batts can lose up to 30 percent of their R-value if there is 4 percent gap in the insulation, measured over the exposed, insulated surface.

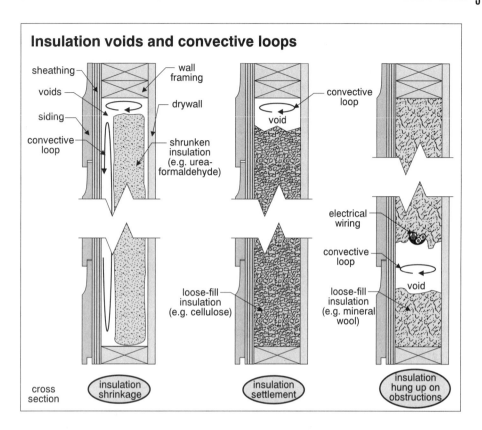

Insulation voids and convective loops

sheathing — wall framing — voids — drywall — siding — convective loop — shrunken insulation (e.g. urea-formaldehyde)

convective loop — void

convective loop

electrical wiring — convective loop — void

loose-fill insulation (e.g. cellulose)

loose-fill insulation (e.g. mineral wool)

cross section

insulation shrinkage

insulation settlement

insulation hung up on obstructions

Loose-fill insulation that is blown or poured in can also settle, creating voids at tops of wall cavities, for example. Where insulation is poured or blown into wall cavities, voids may develop if the insulation gets hung up on obstructions partway down the cavity. Voids may also develop in insulations that shrink. Urea-formaldehyde foam insulation sometimes shrank dramatically after it was installed, reducing its R-value.

Heat Loss Through Air Movement

Insulation is not typically designed to prevent heat loss from air leakage. Air sealing, including caulking and weather-stripping, reduces air leakage.

Wind Washing-Roofs

We've touched on a type of air movement that can dramatically reduce the R-value of insulations such as loose-fill fiberglass and mineral wool. Depending on the density of the insulation, it may be relatively easy for air to blow through the insulation. On a windy day, considerable air movement may be experienced in an attic, for example. If air blows through the insulation on the attic floor, the R-value will be dramatically reduced. Remember that the goal of insulation is to trap air pockets and keep them still.

Wind Washing-Walls

Similarly, the windward side of a wall may allow enough air movement into the wall system to dramatically reduce the performance of fiberglass batts in the wall. Dense insulation materials, including cellulose fiber and most of the rigid board insulations, are not as susceptible to wind washing as fiberglass and mineral wool.

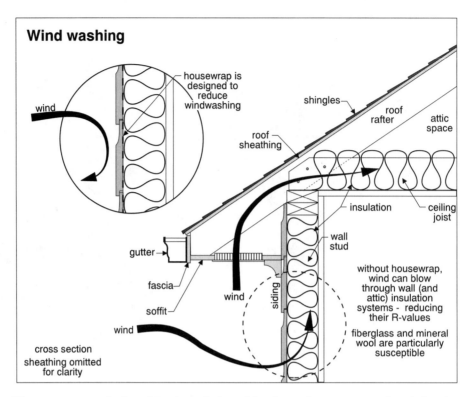

Wind washing

housewrap is designed to reduce windwashing

wind

shingles

roof rafter

attic space

roof sheathing

insulation

ceiling joist

gutter

wall stud

fascia

without housewrap, wind can blow through wall (and attic) insulation systems - reducing their R-values

wind

soffit

siding

fiberglass and mineral wool are particularly susceptible

wind

cross section sheathing omitted for clarity

Housewraps   Housewraps are designed to stop wind washing by acting as an exterior air barrier. These breathable fabrics form a tighter windbreaker or outer jacket on buildings than does traditional building paper. While some debate the merit of housewrap over building paper, reduced wind washing is one of the advantages of house-wraps, according to their supporters.

4.1 FORMS OF INSULATION

Insulation typically takes one of four forms:

1. **Loose fill**. Loose fill, which can be blown or poured, is common in roofs and walls. Materials such as cellulose fiber, glass fiber, mineral wool, vermiculite and perlite have all been commonly used. Other materials may also be found, including sawdust, wood shavings, shaved leather, asbestos and gypsum slag.
2. **Batts or blankets**. Fiberglass and mineral wool commonly come in batts or blankets. Batts are typically 12 to 24 inches wide and fit between ceiling joists, rafters and wall studs. Blankets can be up to six feet wide and are typically laid over ceiling joists.
3. **Rigid boards**. Rigid board insulation can be fit between studs or on the outer face of studs, replacing exterior sheathing. Common board insulations include fiberglass, expanded polystyrene, extruded polystyrene, polyurethane and polyisocyanurate. These materials are often faced with housewrap-type products, kraft paper, asphalt-impregnated kraft paper or aluminum foil. Some of these materials have very high R-values.
4. **Foamed-in-place insulations**. Foamed-in-place insulations are becoming more popular. They include polyurethane and **polyisocyanate**. Notice it's similar to the word **polyisocyanurate**, but we left out the **"ur"** on purpose.

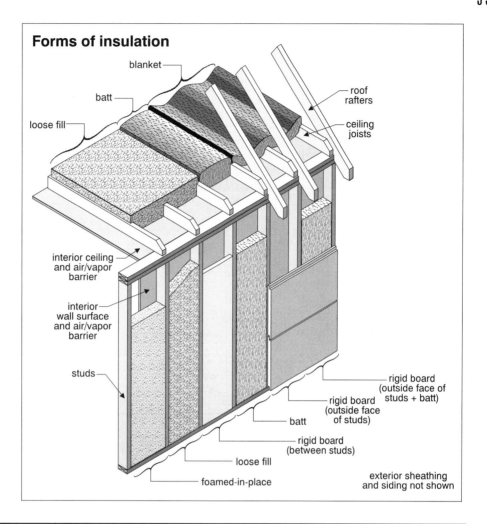

Forms of insulation

4.2 COMMON INSULATION MATERIALS AND
    THEIR PROPERTIES

We've talked about some of the desirable features of insulation materials. The chart sets out the properties of various insulating materials. Some of the information in the chart has been approximated and there are exceptions to these numbers. We understand there are variances with each material, depending on density and other factors, but this is a broad guideline.

PROPERTIES OF COMMON INSULATION MATERIALS						
Material (and form)	Approx.R-value per inch	Air barrier?	Vapor barrier?	Combustible?	Resistant to sunlight?	Resistant to moisture?
Fiberglass (loose fill)	3.0	No	No	No, or slightly	Yes	Somewhat
Fiberglass (batt)	3.2	No	No	No, or slightly	Yes	Somewhat
Fiberglass (board)	4.3	Some have air barrier attached	No	No, or slightly	Yes	Somewhat
Cellulose fiber (loose fill)	3.5	Limited extent	No	Yes, although treated with fire retardants	Yes	No
Mineral wool (loose fill)	3.1	No	No	No	Yes	Yes
Mineral wool (batt)	3.3	No	No	Mostly no	Yes	Yes
Mineral wool (board)	4.3	No	No	No	Yes	Yes
Vermiculite/ Perlite (loose fill – poured only)	2.4	No	No	No	Yes	Absorbs moisture if untreated
Expanded polystyrene (board)	4.0	Yes, if joints are sealed	Almost (thicker boards retard more)	Yes	No	Somewhat
Extruded polystyrene (board)	5.0	Yes, if joints are sealed	Almost (thicker boards retard more)	Yes	No	Yes
Closed-cell phenolic plastic (board)	8.0	Yes, if joints are sealed	No	Very slightly	No	Yes
Polyisocyanurate (board)	7.0	Yes, if joints are sealed	Yes	Yes	No	No
Polyurethane (board)	6.3	Yes, if joints are sealed	Yes	Yes	No	No
Polyurethane (foamed-in-place)	6.0	Yes	Sometimes	Yes	No	Somewhat
Isocyanate or polyisocyanate (foamed-in-place)	4.3	Yes thicknesses	Yes, in typical	Yes	Yes	Somewhat

Now let's look at some of the insulation materials.

Fiberglass

Fiberglass is a commonly used, versatile, inexpensive insulation. The glass itself is not combustible, although the binders used in it may be. Fiberglass may have an air/vapor retarder fixed to the surface. This may be kraft paper, foil, vinyl or a housewrap.

Skin And Lung Irritant

We encourage home inspectors to wear masks and goggles in attic areas because many of the insulations are dusty, and touching and inhaling the fibers is not good for you. HEPA (High Efficiency Particulate Arresting) or P100-type filters are appropriate for masks. We also encourage wearing gloves and long sleeves when moving through insulation. Fiberglass is an irritant to both skin and lungs. There have been some questions raised about the long-term health effects of exposure to fiberglass. There isn't much evidence that occupants of a typical home will have a problem.

R-Values Vary

Fiberglass is often blown in as a loose-fill insulation. There are a couple of ways that the blown-in insulation is made, and the densities can vary dramatically. Consequently, the R-values of loose-fill fiberglass insulation can range from the low 2s to the mid 3s per inch.

Rigid Boards

There are at least two types of semi-rigid fiberglass insulation boards commonly used on houses. There is one type designed for above-grade use. It typically has an air barrier-type housewrap bonded to the surface. The second type is intended for use below grade and does not have a facing. The below-grade material is an oriented strand product, designed to be installed with the strands aligned vertically. This material is designed to act as both a thermal insulation and a drainage layer. The theory is that water accumulating outside the foundation wall will be able to fall through the insulation material to the perimeter drainage tile. This avoids hydrostatic pressure on the walls. A third advantage of the rigid fiberglass insulation on the exterior of the foundation wall is the cushioning effect of this semi-rigid material during back filling. There is some debate as to how well this works.

Wet Fiberglass Poor Insulator

When fiberglass insulation gets wet, its R-value drops dramatically. When the insulation dries out, it will tend to regain its R-value, although if the insulation is compressed as a result of the weight of the moisture, it won't regain all of its insulating capabilities. Loose fill or batts that are compressed by weight or by squeezing oversized batts into cavities that are too small will under-perform.

Mineral Wool

Mineral wool insulation is very similar to fiberglass. It is made with slag or rock spun into a wool. It is typically treated with oil to reduce the dust, and with binders to hold it together. It is non-combustible, water-resistant, rot resistant and inexpensive.

Cellulose

Cellulose is shredded newspaper, chemically treated to resist fire, rot and fungus and to prevent corrosion when wet. Cellulose is more dense and is a better air barrier than fiberglass. Cellulose is less susceptible to wind washing and convective air movement through the insulation itself.

Absorbs Water   Cellulose insulation absorbs water readily. This reduces its insulation value in the short term and tends to compress the insulation, reducing its R-value over the long term, after it has dried.

Combustibility   Because cellulose is made from wood fibers, it is inherently combustible. The fire-retardant chemicals are generally considered effective, although there is some recent evidence these chemicals are applied with varying levels of success. The effectiveness of the chemicals may diminish over time. There is also some evidence that water in the insulation can eliminate the fire retardants.

Sprayed-On   Some cellulose fiber can be sprayed in place. Binders are added to the cellulose to
Cellulose   hold the cellulose together and to bond it to the substrate.

Vermiculite or Perlite

Vermiculite is expanded mica, and perlite is a rock similar to granite. Perlite, when expanded, takes on a popcorn-like appearance. Vermiculite, when expanded, becomes small, light cubes.

Heavy And   As insulation materials, vermiculite and perlite are relatively heavy. They are also
Expensive   relatively expensive. Vermiculite will absorb water unless it is treated with an asphalt coating to prevent moisture absorption.

Non-   Vermiculite and perlite are not combustible unless treated. They are typically only
combustible   used as a poured loose-fill insulation. They aren't common in new construction.

Polystyrene

Polystyrene insulation is a combustible plastic. It comes in two common forms:

• **Expanded** polystyrene or **beadboard** is made by compressing small beads of plastic together to form a board. It is typically white and has a makeup similar to foam coffee cups.
• **Extruded** polystyrene is a closed-cell insulation material. The beadboard has balls of foam surrounded by air. Extruded polystyrene has balls of air surrounded by foam.

Extruded Is   Extruded polystyrene traps the air better, and therefore is a slightly more effective
Better   insulation. Its R-value per inch is roughly 5.0. The R-value of expanded polystyrene is typically 4.0 per inch. There are different densities of polystyrene available, and insulation levels vary.

Damaged By Sunlight And Chemicals

Polystyrene insulation is very susceptible to deterioration by sunlight. It's also very susceptible to the solvents in some adhesives and asphalts.

Boards

These insulations typically come in boards. Boards may be 2ft by 4ft, for example, with thicknesses ranging from 1/4 inch to 4 inches. There are many variations, including interior finishes bonded to the insulation panels, and concrete surfaces bonded to insulation pads used as rooftop walking surfaces.

Highly Combustible

Like many plastics, polystyrene is very combustible, although it may be difficult to ignite. It produces thick, black, highly toxic smoke.

Below-Grade Use

Extruded polystyrene may be suitable for use below grade. Expanded polystyrene is typically not suited for use below grade level. Both materials come in different densities, which yield different strengths. When used as an exterior sheathing material, these insulations can add some rigidity to the structure. This is usually incidental.

Drainage Layers

Extruded polystyrene is sometimes used as an exterior insulation and drainage layer. Grooves or slots cut into the insulation allow water adjacent to the building foundation to flow freely down into the perimeter drainage tile. This prevents hydrostatic pressure and leakage of water through the foundation walls.

Stress Skin Panels

Polystyrene is common in prefabricated wall panels that may include plywood or OSB panels sandwiched onto both sides of a block of insulation.

Concrete Forms

Polystyrene may also be used as permanent forms for poured concrete, both above and below grade. This provides a concrete wall with good insulating value.

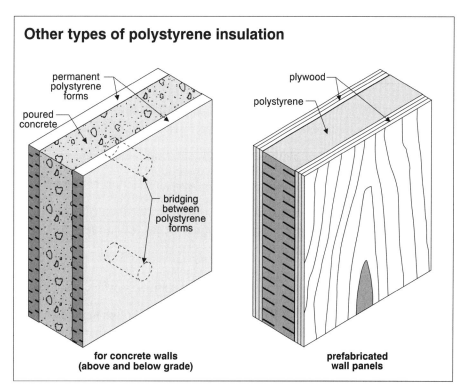

Other types of polystyrene insulation

permanent polystyrene forms

poured concrete

bridging between polystyrene forms

plywood

polystyrene

for concrete walls (above and below grade)

prefabricated wall panels

Phenolic Board

Phenolic board insulation is an expensive material with a very high R-value per inch. The surface is often coated with kraft paper or foil. Phenolic board insulation is vapor permeable, so it isn't a vapor barrier. It does prevent air movement if the joints are taped.

Only Slightly Combustible

Phenolic board insulation is less combustible than other plastic insulations such as polyurethane and polyisocyanurate. When phenolic insulation is exposed to water, there is some suspicion that corrosive acids may form, which can corrode metal. This is a topic of some controversy in commercial roofing, where phenolic board insulation was used extensively on metal roof decks.

Polyurethane and Isocyanurate Boards

These are closed-cell foams with refrigerant gasses (fluorocarbons) used in the bubbles instead of air. These are typically boards and are often foil-finished on one or both sides. These are expensive insulations, usually used where a high R-value is important and space is limited.

Foamed-in-place Polyurethane

Polyurethane can be foamed in place. This material expands dramatically when foamed, and cannot be used in closed cavities. It is susceptible to deterioration by sunlight and is very combustible. The material has to be covered. The R-value of foamed in-place insulations is often said to deteriorate over time.

Foamed-in-place Polyisocyanate or Polyicynene

These foamed-in-place plastic insulations have carbon dioxide as the blowing agent. Again, they can't be used in closed cavities, although they are often used in new construction in wall and floor cavities, before interior or exterior finishes are applied.

Air Barriers

All of the foamed-in-place insulations make good air barriers, and, in adequate thicknesses, also perform as vapor barriers. These materials are generally combustible and, up to a point, resist moisture damage.

Urea-Formaldehyde Foam Insulation (UFFI)

This type of insulation deserves special attention, because it has been controversial.

What is UFFI?

Urea-formaldehyde foam insulation, or UFFI (pronounced "you-fee"), was a common retrofit insulation in the 1970s. It was typically injected as a mixture of urea-formaldehyde resin, an acidic foaming agent, and a propellant such as air. It was typically used in existing houses, most often injected into wood frame wall cavities, or other areas where installation of conventional insulation was impractical.

Insulating Value

UFFI has a good R-value (roughly 5.0), although in many cases the foam has been found to shrink over time and its R-value is dramatically reduced. It is very weak and can be crumbled into a dusty powder in a person's hand. It absorbs moisture readily and cannot act as a vapor barrier. It is vapor permeable.

Formaldehyde Gas	Some formaldehyde gas was released during the on-site mixing and curing of the foam. Formaldehyde is colorless, but has a strong odor that can be generally be detected by nose at concentrations of more than one part per million. It is this by-product of the curing of the foam that became a controversial issue.
Formaldehyde Is Common	Formaldehyde is both a naturally occurring chemical and an industrial chemical. It is found in dry cleaning chemicals, paper products, no-iron fabrics, diapers, glue in particleboard and plywood, cosmetics, paints, cigarette smoke, automobile exhaust, gas appliances, fireplaces and wood stoves. It occurs naturally in forests and is a necessary metabolite in our body cells.
Amount Of Formaldehyde in Homes	Ambient formaldehyde levels in houses are typically 0.03 to 0.04 parts per million (ppm). The smoking section of a cafeteria may have formaldehyde levels of 0.16 ppm. Houses with new carpeting may have similar formaldehyde levels
Dependent On Temperature And Humidity	The rate at which formaldehyde gasses are released from materials depends on temperature and humidity. The higher the humidity levels and the temperature, the more quickly gas will be released.
UFFI In The U.S. And Europe	UFFI was used in the United States during the 1970s and has been used in Europe over the last 35 years. UFFI is still used in Europe, where it was never banned and is considered one of the better retrofit insulations.
	In 1982 a law prohibiting the sale of urea-formaldehyde insulation was passed. In April 1983 the U.S. Court of Appeals struck down the law because there was no substantial evidence linking UFFI to health complaints. UFFI is not widely used in the United States today.
UFFI In Canada	The insulation was used in Canada in the 1970s, most extensively from 1975 to 1978. Financial incentives, such as the Canadian Home Insulation Program (CHIP), stimulated the sale of retrofit insulation materials, including UFFI. UFFI was banned in December 1980 in Canada. It's believed that more than 100,000 homes in Canada were insulated with UFFI.
Installation	UFFI was not a do-it-yourself product. The foam was machine-mixed on-site and injected into wall cavities, where it expanded to fill the cavities. Workmanship and quality control were often less than perfect, as is typical in any new and fast-growing industry.
The Controversy	One of the first problem cases involving formaldehyde was in the United States. This involved an extremely air-tight and poorly ventilated mobile home. The home had been insulated with an apparently poorly mixed, half-formed UFFI. This started to raise government suspicions about the insulation. Interestingly, in subsequent mobile home studies, elevated levels of formaldehyde were traced to paneling or carpets, rather than UFFI.

Cancer In Rats

A laboratory study that produced nasal cancer in rats exposed to high levels of formaldehyde increased the concern. After some media releases and cautioning by authorities, a number of homeowners reported problems that included respiratory difficulties, eye irritation, running noses, nose bleeds, headaches and fatigue. Fear and suspicion quickly led to the conclusion that a problem must exist. The foam was banned as a precautionary measure, although there were no substantiated problems attributed to the foam. Research was initiated to evaluate the problem and determine a course of action.

Tough To Identify UFFI Homes

There are no specific records of which homes in North America had UFFI. It's often difficult to identify UFFI since the product was injected into cavities that are typically not visible. Although the product was designed for wood-frame walls, it was sometimes used inappropriately in solid masonry homes, in open attic areas, and as acoustic insulation in party walls in row houses and in the ceilings between floors.

Real Estate Values

Fears of cancer and other health problems led to a reduction in the value of homes with UFFI in many communities. The costly remedial measures and long-term stigma attached to UFFI houses became a market reality because of suspected health problems.

Formaldehyde Level Guidelines

Governments set guidelines for formaldehyde levels in houses and in some cases specified techniques for removing UFFI. The initial threshold levels set for formaldehyde gas was 1.0 ppm. As testing methods improved, the level was reduced to 0.5 ppm, and eventually to 0.1 ppm. This threshold level was very conservative.

Where's the Problem?

Researchers charged with the task of designing and refining remedial measures looked for worst-case homes to try out their theories, but they met an obstacle: They couldn't find any UFFI-insulated homes with formaldehyde gas levels above 0.1 ppm. Even in the few houses that tested at levels approaching this, these levels were rarely duplicated in subsequent testing.

UFFI Levels Fall With Time

We found out that levels of formaldehyde decrease rapidly after the foam has been installed. Within several days of the application, formaldehyde levels typically return to ambient levels.

Tests of several thousand homes did not reveal any with formaldehyde levels that remained above 0.1 ppm. The highest levels were found in homes with new carpeting, tested on a hot summer day. The same houses tested two weeks later showed typical ambient levels.

UFFI Does Not Add To House Formaldehyde

The conclusion was that UFFI does not affect the amount of formaldehyde in indoor air.

People Who Work With Formaldehyde

In a study in Great Britain, people who worked in environments with high formaldehyde levels, such as morticians and laboratory technicians, were examined for possible health effects. These people were found to have a less than average number of respiratory diseases and actually lived slightly longer than the population average.

Symptoms Appear After Ban

A number of studies have examined the health effects of UFFI. Studies using random samples of both UFFI and non-UFFI homes done before the ban showed no impact of UFFI on health. Studies after the ban showed increasing reports of symptoms, including such things as constipation and deafness, which have no basis for a biological connection to UFFI.

Other House Problems

Investigation into mold, fungi, dust mites and other gases found no correlation between any of these and UFFI. Further, no damage was found to house framing or building materials attributable to UFFI.

Summary

UFFI has been studied extensively and there is no evidence of any health issues. From a real estate value standpoint, UFFI may have an impact in your working area; however, our research has not convinced us of any reason a homeowner should have any UFFI-related health concerns.

Radiant Barriers

Aluminum foil may be used as an insulating material called a **radiant barrier**. Radiant barriers –

• reflect heat back to the source (the house in winter and outdoors in summer), and
• slow heat radiation through the barrier.

Warm Climates Only

Radiant barriers are used more in hot climates. Some authorities do not allow them in cold climates, especially if they are made of multiple layers of aluminum foil separated by an air space. In some tests, these systems used in walls have a show a negative R-value! Heat loss was actually greater with these systems in place.

Forms

The foil can be –

• single-sided with a kraft paper or plastic backing;
• double-sided, sandwiched around a backing material; attached to one or both faces of a rigid insulation board; or
• a multi-layered foil system.

Shiny And Dead Air Space

The foil needs to remain clean and shiny to do its job. It also needs a minimum of half an inch of air space in front of it. Burying the foil in insulation will dramatically reduce its effectiveness. The air space has to be still for the system to be effective. Convective loops eliminate the insulating value and may pump heat out of the house. This is the suspected mechanism that resulted in the negative R-value in tests.

Where Used   Radian barriers may be used in walls; if so, you won't usually find them. You may see them attached to the underside of roof sheathing or rafters in attics.

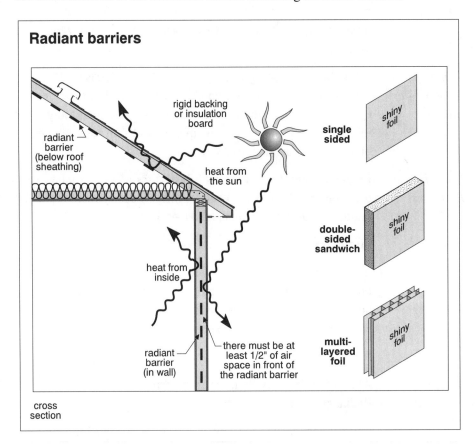

Radiant barriers

rigid backing
or insulation
board

radiant
barrier
(below roof
sheathing)

heat from
the sun

heat from
inside

radiant
barrier
(in wall)

there must be at
least 1/2" of air
space in front of
the radiant barrier

single
sided

*shiny
foil*

double-
sided
sandwich

*shiny
foil*

multi-
layered
foil

*shiny
foil*

cross
section

What To Say   These systems may be more effective helping to cool houses in summer than helping to control winter heat loss. There is no downside to their presence, although there may be no big upside either, especially if the foil is dust-covered or if there is no dead air space.

Wrong Side   Radiant barriers can be a problem if they are installed on the cold side of insulation.

Vapor Barrier   There they may act as a vapor barrier on the wrong side.

Inspection Tip   You'll want to learn to identify the various insulation materials so you can—

• properly describe them during your inspections, and
• estimate the total insulation value in attics, for example.

You can find samples of most common insulations at building supply houses. You may want to keep small samples with you, to help identify these materials in the field.

Insulation
& Interiors
M O D U L E

QUICK QUIZ 5

☑ INSTRUCTIONS

- You should finish Study Session 5 before doing this Quiz.
- Write your answers in the spaces provided.
- Check your answers against ours at the end of this Section.
- If you have trouble with the Quiz, reread the Study Session and try the Quiz again.
- If you did well, it's time for Study Session 6.

1. What is the function of insulation?

_____

2. List at least eight characteristics of an ideal insulation. (You may come up with nine.)

_____

_____

_____

_____

_____

_____

_____

_____

_____

3. Define **thermal bridges**.

   _____

   _____

4. How do convective loops and insulation voids affect R-values?

   _____

   _____

5. What is **wind washing**?

   _____

   _____

6. List four common forms that insulation materials may be found in.

   _____

   _____

   _____

   _____

7. List 10 common insulation materials. (Hint: vermiculite and perlite only count as one. There are two types of polystyrene.)

   1. _____

   2. _____

   3. _____

   4. _____

   5. _____

   6. _____

   7. _____

   8. _____

   9. _____

   10. _____

8. What forms does each of these materials come in?

1. _____

2. _____

3. _____

4. _____

5. _____

6. _____

7. _____

8. _____

9. _____

10. _____

9. List, in point form, as many of the advantages and disadvantages you can think of for each material.

1. _____

Advantages: Disadvantages:

_____   _____

_____   _____

_____   _____

2. _____

Advantages: Disadvantages:

_____   _____

_____   _____

_____   _____

3. _____

Advantages: Disadvantages:

_____   _____

_____   _____

_____   _____

4. _____

Advantages: Disadvantages:

_____   _____

_____   _____

_____   _____

5. _____

Advantages: Disadvantages:

_____   _____

_____   _____

_____   _____

6. _____

Advantages: Disadvantages:

_____   _____

_____   _____

_____   _____

7. _____

Advantages: Disadvantages:

_____   _____

_____   _____

_____   _____

8. _____

Advantages: Disadvantages:

_____   _____

_____   _____

_____   _____

9. _____

Advantages: Disadvantages:

_____   _____

_____   _____

_____   _____

10. _____

Advantages: Disadvantages:

_____   _____

_____   _____

_____   _____

10. How is urea formaldehyde foam insulation installed? Why is it
controversial?

_____

_____

_____

_____

If you had no trouble with this Quiz, you are ready for Study Session 6.

Key Words:

- **Heat transfer**
- **Thermal bridges**
- **Voids**
- **Convective loops**
- **Wind washing**
- **Housewraps**
- **Loose fill**
- **Batts or blankets**
- **Rigid board**
- **Foamed-in-place**
- **Fiberglass**
- **Cellulose**

- **Mineral wool**
- **Vermiculite**
- **Perlite**
- **Expanded polystyrene**
- **Extruded polystyrene**
- **Phenolic**
- **Polyurethane**
- **Polyisocyanurate**
- **Isocyanate**
- **Urea formaldehyde foam insulation (uffi)**
- **Radiant barrier**

Insulation & Interiors

M O D U L E

STUDY SESSION 6

1. You should have completed Study Session 5 before starting this Session.

2. This Session covers the materials and characteristics of air/vapor barriers.

3. At the end of this Session, you should be able to –
   • describe in one sentence the function of an air barrier.
   • give two reasons it is important to control air movement through building walls and roofs.
   • describe in one sentence the function of a vapor barrier.
   • list two other names for vapor barriers.
   • list six qualities of a good air barrier.
   • list ten building materials that may act as an air barrier.
   • list five qualities of a good vapor barrier.
   • indicate whether it is more important for an air barrier or a vapor barrier to be continuous.
   • indicate whether vapor barriers should be on the warm or cold side of walls.
   • indicate whether the same is true in hot, humid climates.
   • define perm in one sentence.
   • list seven common building materials that may act as vapor barriers.
   • explain in two sentences why a vapor barrier should be laid on an earth floor in a crawlspace.

4. Before you start this Session, read sections 13.0 and 14.0 of the Insulation chapter of **The Home Reference Book™.**

5. This Session may take you roughly one hour.

6. Quick Quiz 6 is at the end of this Study Session.

Key Words:

• *Air barrier*

• *Vapor barrier*

• *Vapor diffusion retarder (VDR)*

• *Polyethylene*

• *Perm*

• *Crawlspace floors*

► 5.0 AIR/VAPOR BARRIER MATERIALS (VAPOR RETARDERS)

The title of this section suggests that air barriers and vapor barriers are the same things. While we often use the same material to do both, the functions are different.

Assume a heating situation

In climates like Florida where cooling is more important, keep in mind the warm, moist air we're trying to stop is on the outside, rather than the inside. The function of air barriers is the same; the only thing that changes from heating to cooling climates is the direction of movement that causes the problem. In a heating climate, we don't want the warm, moist air to leak into and through the walls to the outside. In a cooling climate, we don't want the warm, moist outdoor air to leak into and through the walls to the inside. For our discussions, we're going to assume climates where heating is a bigger issue than cooling.

Function Of Air Barriers

Air barriers are designed to **stop air movement** through the building walls and roof. There are two reasons this is important:

- Air carries heat. We want to minimize the flow of building heat to the outdoors.
- Air carries moisture. This moisture may be deposited in the building structure as it cools and condenses. This can cause damage to the building.

Function Of Vapor Barriers

A **vapor barrier**, **vapor retarder** or **vapor diffusion retarder (VDR)** is designed to protect the building from moisture damage. A VDR minimizes (but does not completely stop) the diffusion of vapor from inside the house to the wall or roof cavity. Remember that no air movement is necessary for vapor diffusion to take place. We also said that air leakage is roughly one hundred times more important than vapor diffusion, with respect to moisture damage to buildings.

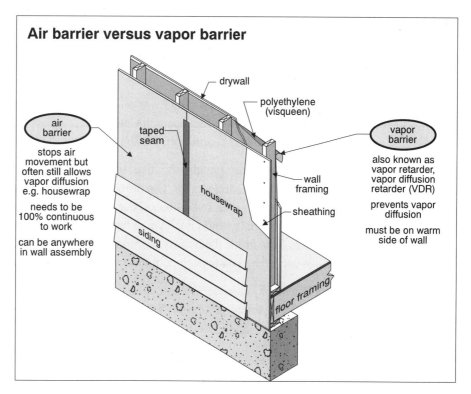

Air barrier versus vapor barrier

drywall

polyethylene
(visqueen)

air
barrier

taped
seam

stops air
movement but
often still allows
vapor diffusion
e.g. housewrap

needs to be
100% continuous
to work

can be anywhere
in wall assembly

housewrap

wall
framing

sheathing

vapor
barrier

also known as
vapor retarder,
vapor diffusion
retarder (VDR)

prevents vapor
diffusion

must be on warm
side of wall

siding

floor framing

As you can see, a vapor diffusion retarder has fewer functions and is less critical than an air barrier. Let's look at air barriers first.

5.1 AIR BARRIERS

Qualities Of A Good Air Barrier

These are qualities of a good air barrier:

• Stops air movement.
• Durable, ideally lasting the life of the building.
• Strong, and either rigid or well enough supported to stay in place.
• Continuous. A bucket that is missing one percent of its bottom cannot do its job. Similarly, an air barrier that is 99 percent intact cannot do its job.
• Inexpensive.
• Resistant to moisture, rot and chemicals.

What Makes Up The Air Barrier?

In a typical home, the air barrier may be thought of as a system, rather than a single component. Many people think only of polyethylene film as an air barrier. People don't usually think of windows as air barriers, but they are. So are drywall, plaster, doors, caulking, weather stripping and many other building materials.

Polyethylene

Polyethylene film is often installed as an air barrier because it is more continuous than other building components. However, polyethylene film is often not continuous, because of a number of factors, including –

• poorly sealed joints;
• discontinuities in the film at partitions, wall/ceiling intersections, wall/floor intersections, door and window openings, etc.; and
• plumbing, electrical and heating penetrations.

How Do We Know Air Barriers Are Not Effective?

If the air barrier were completely effective, there would be less need to ventilate roof spaces. Attic ventilation wouldn't be needed in heating climates because there would be no warm, moist air to be flushed out of the building before it condensed. Exterior siding materials would not have to breathe to allow moist air leaking into the wall spaces from inside the house to escape outdoors. We'd have no rot damage to structures caused by condensation, at least in heating climates.

Never Perfect

Even people who pay considerable attention to effective air sealing of homes have been less than completely successful. Wherever high indoor air pressure pushes air against wall and roof cavities, the air will find a way in.

Insulation As An Air Barrier!

Some insulations work very well as air barriers. Most of the rigid board insulations, for example, are effective air barriers, if their joints are taped, caulked or protected by gaskets. Foamed-in-place insulations are good air barriers. Some loose-fill insulations act as a decent air barrier. Cellulose fiber, in particular, installed at the appropriate densities, can greatly restrict air movement through walls and roofs.

Insulation Better Than Caulking!

Surprisingly, some studies have shown that, while fairly extensive air sealing efforts with caulking did not dramatically reduce the air leakage of a building, blowing loose-fill cellulose fiber insulation into wall cavities did dramatically reduce air leakage. That's because air sealing can only address the problems that are accessible. There are many gaps in buildings through walls and roofs that are not visible or accessible.

Common materials

Common air barrier materials include polyethylene film and housewraps. We'll look at those first, and then look at several other materials that act as part of the air barrier system in a house, sometimes by accident.

1. Polyethylene

Polyethylene (also known as **visqueen**) is typically used as an air barrier in a six-mil thickness. This means it is 6/1000 of an inch thick. (Don't confuse **mils** with **millimeters**. A millimeter (mm) is 1/1000 of a meter.) Polyethylene sheets have traditionally been provided on the inside face of wall studs and on the underside of ceilings, immediately behind the plaster or drywall. Polyethylene film is light and inexpensive to work with. It is also a vapor barrier.

Damaged By
Sun

Polyethylene film doesn't stand up very long if it's exposed to the ultraviolet light of the sun. It must be covered. It doesn't have a lot of mechanical strength and is easily punctured. There is some suggestion that over a period of years the material becomes brittle and may break down because of the pumping action on walls exerted by wind pressures.

Despite some limitations, polyethylene remains a very common air barrier.

Kraft Paper

We've left kraft paper off this list because it was usually attached to insulation batts or blankets, and is almost never continuous. Kraft paper could be an air barrier, but in practice has not been installed as an effective air barrier.

2. Housewraps

Housewraps are typically spun-bonded polyolefin or polypropylene fabrics. These are good air barriers but are not vapor barriers. You can think of them as a wind-breaker. They will allow vapor diffusion readily, but will not allow wind to blow through them.

Unlike polyethylene film, housewraps are typically installed on the exterior of a building, on the outside of the sheathing. Housewraps typically come in wide rolls so that the entire height of a floor level can be covered without seams. Some insulating sheathings include a housewrap bonded to the surface of the insulation. The insulating sheathing is put up and the joints are taped to create an air barrier.

Why On The
Outside?

It's confusing that polyethylene film is installed near the inside of the wall assembly and housewraps are usually installed on the outside. Remember that polyethylene film is also a vapor barrier—housewraps are not. We don't want a vapor barrier on the outside part of the wall assembly because it is cold. A vapor barrier prevents moisture from getting into the wall, where it will condense. Putting polyethylene film on the outside of a building, allowing vapor diffusion into the wall cavity from indoors, would not be effective.

Continuous
Air Barrier Is
Fine

We said an air barrier can be placed anywhere in a wall assembly and still do its job. As long as it prevents air from moving from the inside of the house to the outside, there won't be a supply of warm, moist air from the house into the wall assembly. An outside air barrier works just fine.

Easier To Be
Continuous
On Outside

Housewraps could go on the inside of the wall, in the same place as polyethylene films. However, it's easier to install them on the outside because there are typically fewer wall penetrations. You don't have to worry about wall/wall, wall/floor or wall/ceiling intersections. There are far fewer electrical and plumbing penetrations in the exterior wall assembly. In short, it's easier to make the air barrier continuous if it's installed on the outside of the building.

Reduced
Wind Washing

The other advantage to an exterior air barrier is reduced wind washing. We spoke earlier of how wind blowing through a wall assembly can dramatically reduce the insulating value of fiberglass, for example. The housewrap prevents wind washing.

Didn't Building Paper Do The Same Thing?

Building paper was less effective as an air barrier because it was typically installed in three-foot-wide sheets with loose overlaps. Air could move readily through a typical building paper installation.

What About Plywood Or OSB Sheathing?

These large panel-type sheathing materials are, in fact, excellent air barriers. They also approach being vapor barriers. We don't want a vapor barrier on the outer part of the wall. As a result of their near vapor-tight nature, these panel-type exterior sheathings are usually not installed tightly together. Tongue and groove exterior sheathing is not common, for example. Air moves through the sheathing along the joints between panels. Remember we said earlier that an air barrier that isn't continuous is not very effective.

Expansion And Contraction

Another reason for not butting panel-type sheathing tightly together is the risk of buckling on expansion. When a plywood sheet gets wet, it tries to expand. If there are no gaps around the perimeter, we can get buckling. We see this regularly on roof sheathing, for example, where adjacent panels have not been appropriately spaced.

Can Housewrap Replace Building Paper?

Housewrap does stop air leakage better than building paper. That's one of the functions of building paper. Building paper is vapor permeable. So is housewrap. Perhaps the most important function of building paper is as a water barrier, acting as a backup to the siding material. Housewrap is water repellent (although some questions have been raised as to its effectiveness in this area). It replaces this function of building paper, as well.

Performance Depends On Installation

Proponents of housewrap say it does a better job because it comes in larger rolls and there are fewer seams. The construction reality is that if either building paper or housewrap is not carefully installed, it will not perform as a backup water barrier, particularly around windows and doors.

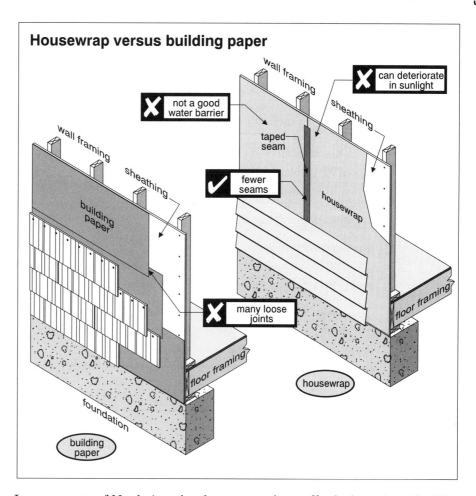

Housewrap versus building paper

wall framing

not a good
water barrier

taped
seam

fewer
seams

wall framing

sheathing

building
paper

many loose
joints

can deteriorate
in sunlight

sheathing

housewrap

floor framing

floor framing

housewrap

foundation

building
paper

Housewraps Have Replaced Building Paper

In many parts of North America, housewraps have effectively replaced building paper on walls. Their long-term performance remains to be seen.

Sun Attacks Housewraps

Like most plastics, housewraps cannot afford to be exposed to the sun over the long term. They will deteriorate. Siding should be installed soon after the housewrap.

Hard To See

All of this may not seem to be important to the home inspector who, in most cases, can't see into the wall assembly. You may find clues around electrical boxes, at tops and bottoms of walls, or where walls have been damaged. It is important to know the function and location of these components, so that if you see them, you can determine if they have been installed correctly and are likely to perform their intended function. It's also important to know what they do and where they belong so that if you see moisture damage to a wall, and the air barrier, for example, is missing, you will have a clue to the cause.

3. Foam insulation boards

Most of the foam insulation boards, including polystyrene, polyurethane, isocyanurate and phenolic board, are good air barriers as long as their seams are sealed with tape, caulking or gaskets. These are typically installed on the outside of a building.

Vapor Barrier In Wrong Location

An interesting question arises when one considers that these materials can also approach vapor barrier performance. In some situations, the board provides enough insulating value to keep the entire wall cavity warm. A vapor barrier at the warm side of the insulation is not a problem. In these cases, the joints may be taped. This depends on climate, other insulation in the wall, and local building practices.

4. Drywall, plaster and wood paneling

Most interior wall and ceiling finishes are effective air barriers, but there are many gaps at the edges and penetrations through the finishes. Conventional wisdom has suggested we cannot rely on these systems as air barriers.

Air Drywall Approach

One school of thought says if we seal the drywall to other building materials, we can create an effective air barrier between the house air and the wall assembly with drywall. The **Air Drywall Approach (ADA)** relies heavily on attention to detail during installation, and the long-term performance of inaccessible caulking materials. This is not a common approach.

5. Sheathing

We've talked about **plywood** and OSB sheathings. We've explained how they are usually not effective air barriers because of the intentional gaps at the joints. **Lumber** sheathing is not an effective air or vapor barrier because of the large number of unsealed joints.

Fiberboard

Asphalt-impregnated fiberboard sheathing is popular in some areas. It is a good air barrier, although again there are openings at joints. Fiberboard is vapor permeable, and joints could be sealed without risk of creating an exterior vapor barrier. We have not found many cases of fiberboard sheathing being relied upon as an air barrier. **Gypsum board** can be an air barrier, again if joints are sealed. It is not a vapor barrier.

6. Building paper

Building paper does not perform well as an air barrier because of the many loose joints in a typical building paper application.

7. Sill gaskets

Polyethylene foam sill gaskets, a quarter of an inch thick and six to eight inches wide, are laid on foundation walls under the sill plates. These sill gaskets significantly reduce air leakage between the top of the foundation and the sill plate. These are air barriers specific to this part of the home.

Sill gaskets and electrical box enclosures

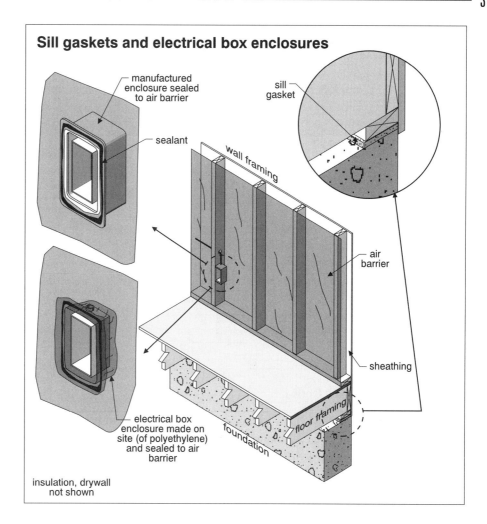

manufactured enclosure sealed to air barrier

sealant

sill gasket

wall framing

air barrier

sheathing

floor framing

foundation

electrical box enclosure made on site (of polyethylene) and sealed to air barrier

insulation, drywall not shown

8. Gaskets for electrical boxes and plastic enclosures around electrical boxes

Electrical boxes often defeat air barriers. Gaskets for electrical boxes and plastic enclosures around electrical boxes help maintain the continuity of the air barrier at the interior wall surface.

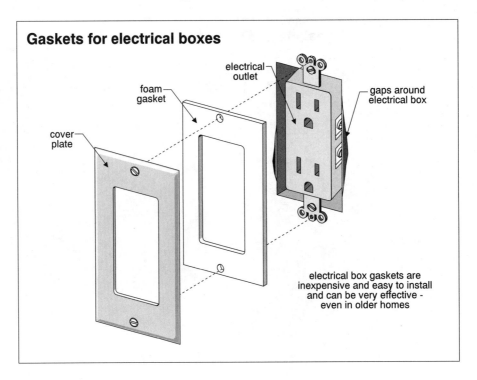

Gaskets for electrical boxes

cover plate

foam gasket

electrical outlet

gaps around electrical box

electrical box gaskets are inexpensive and easy to install and can be very effective - even in older homes

9. Backer Rods

Polyethylene foam backer rods are often used to bridge gaps between building materials that are too wide to simply be caulked. Like sill gaskets, backer rods are effective air barriers.

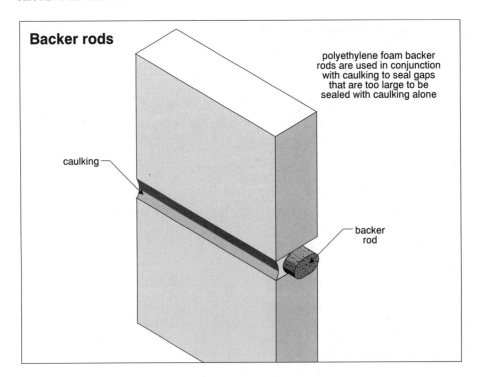

Backer rods

polyethylene foam backer rods are used in conjunction with caulking to seal gaps that are too large to be sealed with caulking alone

caulking

backer rod

10. Caulking and Weather Stripping

Caulking and weather stripping are also air barrier materials that are used as part of the air barrier system in every house.

11. Polyurethane foams

Single-component polyurethane foams in convenient spray cans are commonly used as air barriers around door, window and other opening in walls and ceilings. Both high-expansion and low-expansion foams are available. These materials both insulate and air seal.

12. Duct Tape and Duct Mastic

Duct tape and duct mastic are used to control duct leaks. Air leakage into and out of duct systems for heating or air conditioning can be important contributors to heat transfer and moisture movement. This is particularly true where duct systems are in unconditioned spaces, such as attics or crawlspaces. Some basements are effectively treated as unconditioned spaces and in these, too, duct leakage can cause considerable problems. We want to stop the leakage of warm, moist air into cool spaces. There are energy efficiency and comfort reasons for controlling air leakage at ducts, as well as moisture-damage prevention reasons.

Caulking and Weather Stripping

Home inspectors don't spend a lot of time looking at caulking and weather stripping. We should understand, however, that these can have a significant effect both on heating costs and air and moisture movement into and out of buildings. Let's focus on heating climates, and look at caulking for a moment.

Caulk Indoors   People caulk buildings on both the inside and the outside. Caulking on the outside

Or Out   is done to keep rain out of the wall systems. Caulking from the inside prevents air leakage into the wall system. When we're trying to preserve energy and prevent condensation in building components, we should be caulking from the inside.

Caulking - indoors or out?

exterior caulking is done to
keep rain out of wall systems

interior caulking is intended to prevent
air leakage into the wall system

The Ideal
Caulking

The ideal caulking –

• has a life that is the same as that of the building;
• stays flexible indefinitely;
• has great adhesion – it sticks to everything;
• spans large gaps without support behind;
• is paintable; and
• is mechanically resilient – able to stand up to foot traffic, for example.

There are no perfect caulkings. There are many ways to look at caulking, but we
can look at four basic types:

• Silicone
• Latex acrylic
• Polyurethane
• Synthetic rubber (block copolymers)

Silicones

Silicone caulking is used as a caulking material and as a glue. It is strong, weather-
resistant and resists ultraviolet light. Most silicone is not paintable and does not
adhere well to wood, concrete or masonry. It bonds well to bare aluminum, but
bonds better to other metals when primed. Silicone can stain porous stones. High-
temperature silicone caulking is available for metal chimneys and vents. Silicone
caulking is often used on EIFS (Exterior Insulation and Finishing Systems) houses.

Latex acrylic

Latex acrylic caulking is a good indoor caulking. It has good flexibility and is paintable. It's easy to work with and does not have the strong odor of many silicone caulkings. Cleanup is easy. Some latex acrylic caulks perform well outdoors. Latex acrylics are not as flexible as silicone or polyurethane. There is a wide range of qualities of latex acrylic caulking, from inexpensive to premium levels.

Polyurethane

Polyurethane caulking sticks well to most surfaces and is paintable. It is available in both water- and air-cured types.

Synthetic rubber or block copolymers

This new family of caulkings does a very good job of sticking to most surfaces. It will, however, attack foamed plastics and should not be used with them. It is a clear, unobtrusive caulking. Application results in strong odors, so the product is best used outside. These caulks shrink considerably after installation. They are paintable.

Weather Stripping

There are a whole family of weather stripping materials and, as you might suspect, the quality ranges are wide. Weather stripping is used on movable joints in buildings, typically at doors and windows. Repairing and replacing weather stripping is normal maintenance; the same is true of most caulkings. The illustration shows some common types of weather stripping.

Air Sealing Is New Trade

Air sealing of homes has become a new industry. Many companies specialize in reducing air leakage. Obviously caulking and weather stripping are part of their arsenal. Gaskets and products that seal around light fixtures, electrical receptacles and other wall penetrations are also essential components of an air-sealing program.

Pot Lights

Recessed light fixtures in ceilings can cause considerable heat loss through air movement. Considerable moisture can also be dumped into roof spaces through light fixtures. Recessed light fixtures in flat and cathedral ceilings are particularly bad as the moisture lingers in these typically poorly ventilated spaces.

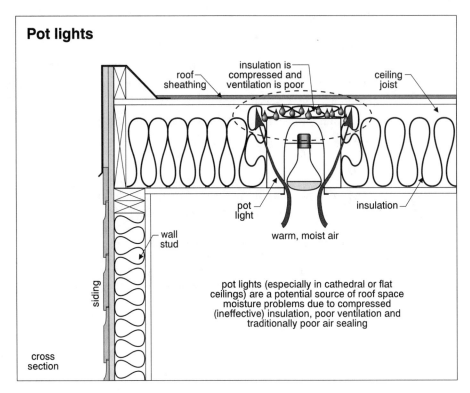

Pot lights

roof sheathing

insulation is compressed and ventilation is poor

ceiling joist

pot light

insulation

wall stud

warm, moist air

siding

pot lights (especially in cathedral or flat ceilings) are a potential source of roof space moisture problems due to compressed (ineffective) insulation, poor ventilation and traditionally poor air sealing

cross section

Air Sealing Saves Money

Some studies have shown that improved insulation did not significantly reduce heating costs, while improved air sealing did make substantial differences in heating costs.

Double-Edged Sword

The air sealing work that is often done reduces heat loss and moisture migration into building components. On the other hand, it reduces the natural ventilation levels in houses. Indoor air quality may deteriorate after an air-sealing project. House humidity levels and indoor air pollutant levels rise as the number of air changes is reduced. The house may be more comfortable and less expensive to heat, but may become less pleasant in other respects.

5.2 VAPOR BARRIERS

Vapor barriers, vapor retarders or vapor diffusion retarders (VDRs), have a different function than air barriers, although the same materials are sometimes used for both. Let's look at the properties of a good vapor barrier:

• Vapor-diffusion resistant
• Durable
• Moisture and rot resistant
• Chemically inert
• Inexpensive

Vapor Barrier
Defined

How do we know if a material is a vapor barrier? Vapor barriers are described by their **permeance**. The unit of permeance is the **perm**. The lower the perms, the more effective the vapor barrier. Unfortunately, there are metric and imperial perms. An imperial perm is the number of grains of water that will move through one square foot of material in one hour, under a pressure difference of one inch of mercury. One grain is 0.002285 ounces.

Metric Perm

A metric perm is the number of nanograms of water that will pass through one square meter of material per second, under a pressure difference of one pascal. A nanogram is one one-billionth of a gram.

Vapor Barrier
Has Perm Less
Than 1.0

Generally speaking, a material is considered a vapor barrier or vapor diffusion retarder if its perm rating is less than 1.0 (imperial) or 57.5 (metric).

Continuity
Not Critical

Vapor barriers do not have to be continuous to be effective. A vapor barrier is not like a bucket with one percent of its bottom missing; a vapor barrier with one percent of its material missing will still be 99 percent effective. Vapor diffusion is a function of the surface area across which the water molecules can move. If we block most of the surface area, we'll stop most of the vapor diffusion. Again, this is different from an air barrier.

Must Be On
The Warm Side

Vapor barriers must be on the warm side of the wall to perform their function. We said that air barriers could be on the warm side, in the middle, or on the cool side of a wall assembly. A vapor retarder must be on the warm side. If the water molecules are allowed to move into a cool space, they are likely to condense. A vapor retarder on the outside will not protect the wall from moisture damage due to vapor diffusion.

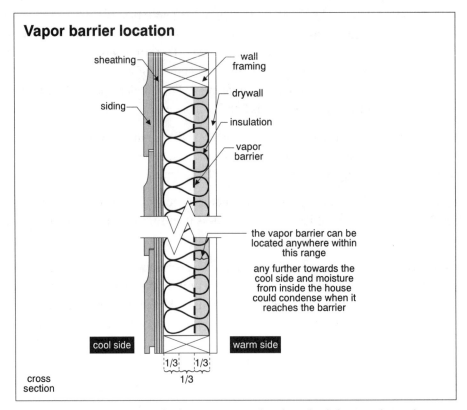

Vapor barrier location

sheathing

wall framing

siding

drywall

insulation

vapor barrier

the vapor barrier can be located anywhere within this range

any further towards the cool side and moisture from inside the house could condense when it reaches the barrier

cool side

warm side

1/3   1/3
1/3

cross section

Within The Warm Third

There is a widely quoted rule that says vapor barriers don't have to be at the warm face of the insulation. They can be a third of the way through the insulation, closer to the warm side. The research that this rule is based on is a 1950s study of questionable authority, but the rule has been relied on without serious trouble for a long time. It seems to work.

Drying Potential

There's another reason we don't want a vapor barrier on the exterior of a wall in a cold climate. Most wall systems are less than perfect at keeping rain and snow from getting into the wall from outside, and keeping warm, moist air from getting into the wall from indoors. Building materials such as wood will store a certain amount of moisture without causing damage. When the outdoor temperatures rise and relative humidities drop, building materials typically dry to the outside. If there were an exterior vapor barrier, this drying potential would be greatly reduced. Breathable exterior components on wall assemblies prevent water damage by allowing good drying to the outdoors.

Don't Want Two Vapor Barriers

Consider a wall with a vapor barrier on the inside surface. Can we have a second vapor barrier on the exterior surface? No. Vapor barriers are not likely to be 100 percent effective. We don't want to allow moisture to get into the wall cavity, and then trap it there. We want to allow that moisture to move out through the wall to the outdoors. Any vapor that leaks past the internal vapor barrier should be allowed to flush itself out to the exterior.

Building Paper Vapor Permeable

That's why building paper has traditionally been used on the exterior of buildings. It is very good at stopping external water from getting into the wall. It does not, however, stop vapor diffusion moving out through the wall. Housewraps are also vapor permeable.

Cooling Climate

Where air conditioning is an issue, we still want the vapor barrier on the **warm** side of the wall. In this case, it's the **outside**. We don't want the high vapor pressure from the outdoors pushing water vapor into the cooler wall cavities, where it may condense.

What About Cooling Climates?

In hot climates, the walls store moisture and dry to the **inside**. Again, in Florida for example, we wouldn't want a vapor barrier on the inside, since that would inhibit drying of the walls.

Thank Goodness We're Just Looking For Damage

Let's stop before things get too complicated. What do we need to know? Home inspectors should understand the mechanisms of moisture damage in walls and roofs. We should be able to recognize it in the field. A complete diagnosis is often not possible during a one-time visual home inspection.

Common Vapor Barrier Materials

Let's look at some of the materials that are commonly used as vapor barriers.

1. **Polyethylene film (visqueen)**. This is probably the most common material used as a vapor barrier. As we discussed, it's typically also used as an air barrier, immediately behind the drywall in wall and ceiling assemblies.

2. **Kraft paper**. Old fiberglass and mineral wool insulation batts were often faced with brown kraft paper, which is a vapor barrier.

3. **Aluminum foil**. This vapor barrier may also be used as a radiant barrier to reflect heat.

4. **Oil-based paints and vapor-retardant paints**. Many paints act as vapor barriers. Latex paints generally do not, unless they are specially formulated to act this way. Varnishes and shellacs also act as vapor barriers.

5. **Insulations**. Some insulation materials act as vapor barriers. This includes polyethylene and polyisocyanurate boards. Expanded and extruded polystyrene boards can also act as vapor barriers if they are thick enough. The same is true of foamed-in-place polyurethane and isocyanates.

6. **Vinyl wallpaper**. Vinyl wallpapers make quite good vapor barriers. This is unfortunate for people in hot climates who don't want vapor barriers on the interior of their wall assemblies.

7. **Plywood and OSB sheathings**. While these materials may or may not be quite vapor barriers in the true sense of the word, they do have fairly low perm ratings. They are almost vapor barriers.

Let's look at some common building materials and their permeance. We'll deal with imperial units only. Anything with a perm rating of less than 1.0 can be considered a vapor barrier.

PERMEANCE OF COMMON BUILDING MATERIALS	
Material	**Perm Rating (Imperial)**
6 mil Polyethylene film	0.06
0.35 mil Aluminum foil	0.05
Exterior-grade plywood (¼ inch)	0.7
Drywall	50
Plaster on gypsum lath	20
Plaster on wood lath	11
Brick (4 inches thick)	0.8
Concrete block (8 inches thick)	2.4
Asbestos cement board (1/8 inch)	4 to 8
Asphalt-impregnated kraft paper used as facing on old fiberglass batts	0.3
Two coats of oil-based paint on plaster	1.5 to 3.0
One coat of latex vapor diffusion retarder paint	0.45
Two coats of aluminum paint	0.3 to 0.5
Fiberglass insulation, cellulose or mineral wool insulation (4 inches)	29
Expanded polystyrene insulation board (1 inch)	2.0 to 5.8
Extruded polystyrene (1 inch)	0.4 to 1.6
Polyurethane insulation board (1 inch)	1.2
Fiberboard insulation (½ inch)	20 to 50
Spun-bonded polyolefin housewrap	63
Spun-bonded polypropylene housewrap	15

Vapor Barriers on Crawlspace Floors

We've been talking about vapor barriers that separate the living space from the outdoors. There's another place where vapor barriers are extremely important. Houses that have crawlspaces with earth floors can suffer considerable moisture damage. A typical earth floor in a crawlspace can add much more water to a home than an average family. Several gallons of moisture vapor can be added to the house air every day!

Cover The Floor

If the sub-grade area has a concrete floor, there is very little problem. If the floor is unfinished earth, we recommend a vapor barrier such as 6 mil polyethylene. This makes a tremendous difference if it is sealed at joints and around the edges and covered with gravel to keep it in place. The several gallons of water that find their way from the soil into the house air every day will be eliminated. Home inspectors commonly find serious rot in wood framing members in crawlspaces with earth floors. It's one good reason to make the effort to get into crawlspaces.

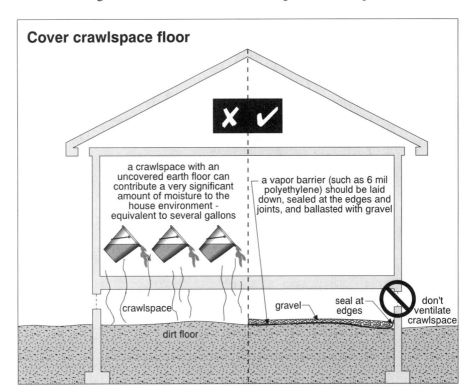

Cover crawlspace floor

a crawlspace with an uncovered earth floor can contribute a very significant amount of moisture to the house environment - equivalent to several gallons

a vapor barrier (such as 6 mil polyethylene) should be laid down, sealed at the edges and joints, and ballasted with gravel

crawlspace

gravel

seal at edges

don't ventilate crawlspace

dirt floor

Inspection Strategy

If you are inspecting a house with a crawlspace you can't get into, this should raise a yellow flag. Make sure your client understands there may be concealed damage. This is a significant limitation to your inspection.

Summary

So far we've talked about insulation materials and air/vapor barriers designed to control air and moisture flow in the home. Now let's assume that our efforts to control airflow and moisture have failed. This is a good assumption. What do we do now? Let's look at some ventilation systems.

Insulation
& Interiors
MODULE

QUICK QUIZ 6

☑ INSTRUCTIONS

•You should finish Study Session 6 before doing this Quiz.

•Write your answers in the spaces provided.

• Check your answers against ours at the end of this Section.

• If you have trouble with the Quiz, re-read the Study Session and try the Quiz again.

• If you did well, it's time for Study Session 7.

1. What is the function of an air barrier?

    _____

2. Indicate two reasons we are interested in controlling air movement through building wall and roofs.

    _____

    _____

3. Describe the function of a vapor barrier.

    _____

    _____

4. List two other names for vapor barriers.

    _____

    _____

5. List six qualities of a good air barrier.

_____

_____

_____

_____

_____

_____

6. List at least 10 building materials that may act as an air barrier. (You may come up with 12.)

_____

_____

_____

_____

_____

_____

_____

_____

_____

_____

_____

_____

7. List five qualities of a good vapor barrier.

_____

_____

_____

_____

_____

8. Is it more important for the air barrier or vapor barrier to be continuous?

_____

9. Should the vapor barrier be on the warm side or cold side of the wall assembly?

_____

10. Should the vapor barrier be on the warm side or cold side of the wall assembly in a hot, humid climate?

_____

11. Define **perm** using Imperial units.

_____

_____

12. List seven common building materials that may act as vapor barriers.

_____

_____

_____

_____

_____

_____

_____

13. Why would a vapor barrier be laid on an earth crawlspace floor?

_____

_____

If you had no trouble with this Quiz, you are ready for Study Session 7.

Key Words:
- **Air barrier**
- **Vapor barrier**
- **Vapor diffusion retarder (VDR)**
- **Polyethylene**
- **Perm**
- **Crawlspace floors**

Insulation
& Interiors
MODULE

STUDY SESSION 7

1. You should have completed Study Session 6 and Quick Quiz 6 before starting this Session.

2. This Session covers venting of roof spaces.

3. At the end of this Session, you should be able to –
   - list three functions of roof vents.
   - list four types of roof vents.
   - describe the percentage of venting that should be at the soffits and near the peak of the roof.
   - describe in two sentences the pros and cons of turbine type vents.
   - describe in one sentence the function of soffit vent baffles.
   - define how much ventilation is needed for attics.
   - define how much ventilation is needed for low-slope, flat and cathedral roofs.
   - describe in two sentences how mansard and gambrel roofs should be vented.
   - describe in one sentence whether over venting is possible.
   - describe in one sentence why power roof vents should not operate in the winter.

4. Before you start this Session, read section 15.0 of the Insulation Chapter of **The Home Reference Book.**

5. This is a short Study Session. It may take you roughly one half hour.

6. Quick Quiz 7 is at the end of this Study Session.

Key Words:

- *Ice dams*
- *Heat*
- *Moisture*
- *Soffit vents*
- *Ridge vents*
- *Roof vents*
- *Gable vents*
- *Turbine vents*
- *Power vents*
- *Soffit vent baffles*
- *Depressurization*

► 6.0 VENTING ROOFS

Venting Living Spaces

We've talked about two different kinds of ventilation – one is for living spaces and the other is for unconditioned spaces. Ventilation of living spaces is designed to –

- Remove excess moisture from sources such as kitchens, bathrooms and clothes dryers. This helps protect the structure from moisture damage by exhausting the moisture directly rather than allowing it to leak into wall and roof systems.
- Replace stale, polluted air with fresh air. This helps to maintain good indoor air quality.

Ventilating Unconditioned Spaces

The second type of ventilation is designed to flush out the warm, moist air that has escaped from the living space (despite our best efforts) before it has a chance to deposit its moisture into the building structure where it can do some damage. This ventilation is most visible on roofs, but is also inherent in walls and may be built into sub-grade areas such as crawlspaces. Our discussion here will focus on venting roofs.

Ventilation To Eliminate Heat

There's another good reason to ventilate unconditioned spaces, especially in climates with hot summers. The air in a roof space can become very hot. The sun beating down on a roof surface can drive heat into the roof space. Summer attic temperatures of 130°F are not unusual. If the outdoor temperature is 100°F, it's tough enough to keep a building cool. However, if the air immediately above the building is 130°F, it's even harder to cool. The ventilation system allows the 130°F air to be replaced with 100°F air. Roof ventilation systems flush out the super-heated air and help keep the building cooler.

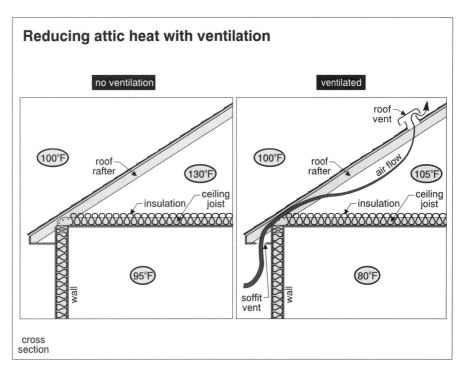

Reducing attic heat with ventilation

121

Prolonged Life Of Roof Shingles

Some people believe the high temperatures found on poorly ventilated roofs in warm climates contribute to a shortened life expectancy of roofing products such as asphalt shingles. Incidentally, some studies have shown attic temperatures to be higher on homes with dark color roofs, than with light color roofs.

FRT Plywood Sheathing Problems

Some also believe these high attic temperatures may have contributed to the problems associated with fire retardant treated (FRT) plywood roof sheathing. This is discussed in the Structure Module. This special plywood deteriorates due to charring at low temperatures as a result of the chemical treatment designed to reduce the combustibility of the plywood.

Prevent Ice Dams

In northern climates, there is a third good reason to ventilate unconditioned roof spaces. Good roof venting helps prevent ice dams. Ice dams (discussed in the Roofing Module) form at the lower edges of roofs when the warm air in the attic melts the snow on the roof above the attic. The melted snow runs down to the edge of the roof, which is colder because there is no attic beneath it. The water refreezes before it drops off the edge of the roof. This freezing water builds a dam along the eaves. Subsequent melted snow running down the roof will run into the dam and back up under roof shingles, leaking into the building.

Venting Keeps Roof Cold

A well-ventilated roof will be cold. The colder the roof, the less likely snow is to melt above the attic. Less melted snow means less chance of ice dams.

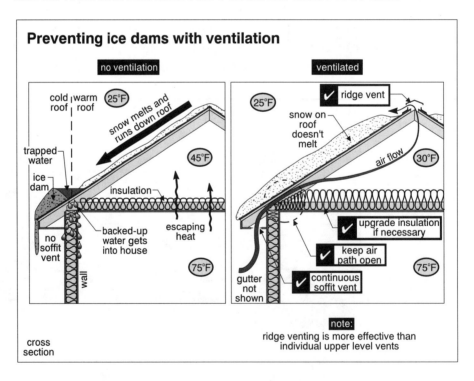

6.1 VENTING ROOF SPACES

Purpose   We've touched on three functions of roof venting. Let's review them:

• Venting allows warm, moist air out of the attic before the moisture condenses on structural members.
• Venting reduces attic temperatures in the summer by allowing hot air to escape.
• Venting helps prevent ice dams by keeping the attic cold in winter.

There are four common types of vents:

Types And   • Soffit vents
Locations Of • Ridge vents
Vents   • Roof vents
               • Gable vents

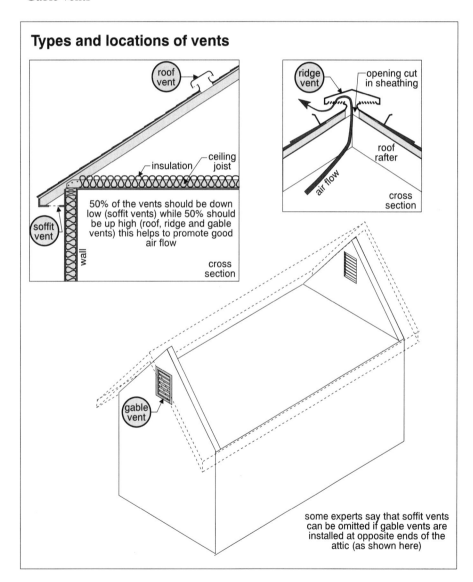

Types and locations of vents

roof vent

ridge vent — opening cut in sheathing

insulation — ceiling joist

air flow

roof rafter

cross section

soffit vent

wall

50% of the vents should be down low (soffit vents) while 50% should be up high (roof, ridge and gable vents) this helps to promote good air flow

cross section

gable vent

some experts say that soffit vents can be omitted if gable vents are installed at opposite ends of the attic (as shown here)

High And Low Vents   Soffit vents typically comprise at least fifty percent of the ventilation. Ridge, roof or gable vents make up the balance. The purpose of low and high vents is to encourage convective air movement through the attic. Air is drawn in through the soffit vents and leaves through the ridge, roof or gable vents.

Gable Vents At Opposing Ends   Some experts say that soffit vents can be omitted where there are gable vents at opposing ends of attic areas. While we prefer to see soffit vents too, we have seen many houses vented this way with no evidence of problems.

Vent Both Roof Surfaces   In many locations, roof vents are required on both sides of a gable roof, for example. This is good practice because wind comes from different directions on different days. If the roof vents are on the downwind side of the roof, air will tend to be drawn out of the roof vents as intended. However, if the roof the vents are on the windward side of a roof, air may be forced into the attic through the roof vents. If there are vents on both sides of the roof, we are assured that air will be drawn out of at least some of the roof vents.

Soffit Vents At Least 50% Of Total   We've said that soffit vents should make up at least half of the total venting. Because the air tends to be drawn in though soffit vents, soffit vents tend to put the attic under positive pressure. This is a good thing in one sense. We don't want to create negative pressure in the attic because this will draw more house air out of the living space and up into the attic.

Roof And Ridge Vents Depressurize Attics   If we have more roof or ridge venting than soffit venting, we can create low pressure in the attic. This causes more warm, moist air from the house to be sucked out through the ceilings into the attic. This increases attic moisture levels, which is what we are trying to avoid.

Turbine Vents   Turbine-type vents may be found on the roof surface. We don't recommend these for several reasons:

- Wind-driven vents don't work on calm days, but we still need ventilation on calm days.
- They are often noisy or seized.
- They can depressurize the attic on windy days.
- We often find these covered with garbage bags to prevent water leakage through the vents.

Turbine vents

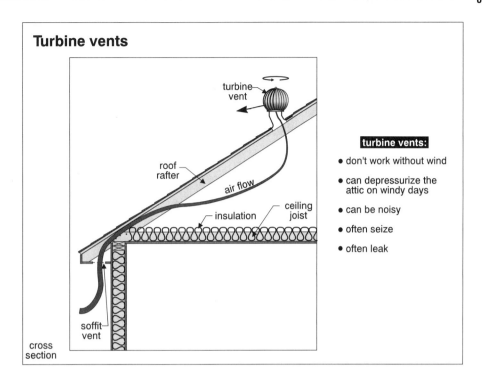

turbine
vent

roof
rafter

air flow

insulation

ceiling
joist

soffit
vent

cross
section

turbine vents:

- don't work without wind
- can depressurize the attic on windy days
- can be noisy
- often seize
- often leak

Baffles For Soffit Vents   It's common to find soffit vents blocked by insulation. Good installations include cardboard, plywood or expanded polystyrene baffles. These baffles prevent insulation from covering the roof vents and allow air to flow up through the soffit vents into the roof space. Baffles also direct the air away from the insulation and help reduce wind washing (unwanted air movement through the insulation, which reduces its R-value).

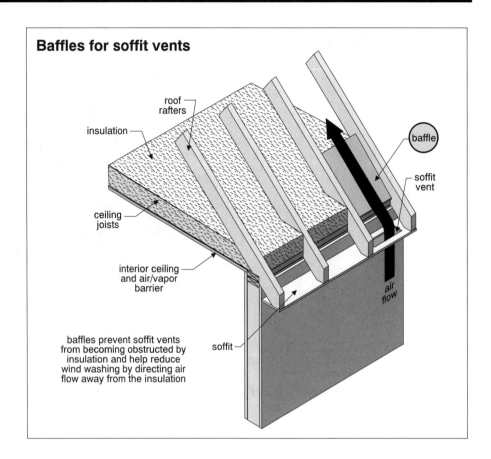

Baffles for soffit vents

roof rafters

insulation

baffle

soffit vent

ceiling joists

interior ceiling and air/vapor barrier

air flow

baffles prevent soffit vents from becoming obstructed by insulation and help reduce wind washing by directing air flow away from the insulation

soffit

Recommended Amounts Of Attic Ventilation

The total vent area is often recommended to be 1/300 of the floor space of the attic. If the attic floor is 600 square feet, we would look for 2 square feet of unobstructed or free vent area. At least one square foot of this vent area would be at the soffits and the other square foot would be at the ridge, roof or gable vents. The actual vent size has to be larger because the vent area is reduced by louvers or screens to keep out insects, rain and snow.

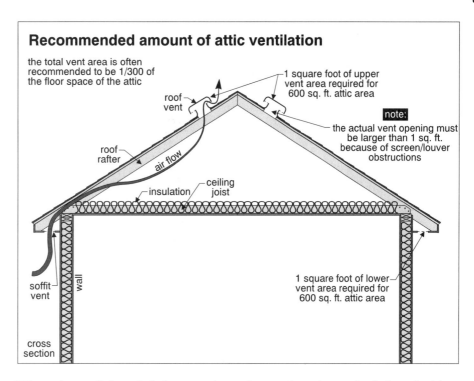

Recommended amount of attic ventilation

the total vent area is often recommended to be 1/300 of the floor space of the attic

1 square foot of upper vent area required for 600 sq. ft. attic area

roof vent

note: the actual vent opening must be larger than 1 sq. ft. because of screen/louver obstructions

roof rafter

air flow

insulation

ceiling joist

soffit vent

wall

1 square foot of lower vent area required for 600 sq. ft. attic area

cross section

Low-Slope, Flat And Cathedral Roofs   Where the roof slope is below two in twelve, or there is a cathedral roof with no attic space, the recommended vent area is often increased to 1/150 of the roof area.

Flat Roofs   Flat roofs typically have ventilation on opposing sides to promote cross ventilation. Since the roof surface is virtually flat, convective airflow does not help much. These roofs are vulnerable to condensation damage.

Vent Spaces When There Is No Attic   Roofs with a slope of less than two in twelve are often ventilated by maintaining 2½ inches between the top of the insulation and the underside of the roof sheathing. Cross ventilation may be promoted with 2-inch by 2-inch purlins on top of, and perpendicular to the roof joists. The insulation level is kept one inch below the top of the joist. The purlin provides another 1½ inch of air space. Air can move in any direction with this configuration.

Venting cathedral roofs

the recommended vent area for cathedral roofs is often increased to 1 square foot for every 150 square feet of roof area

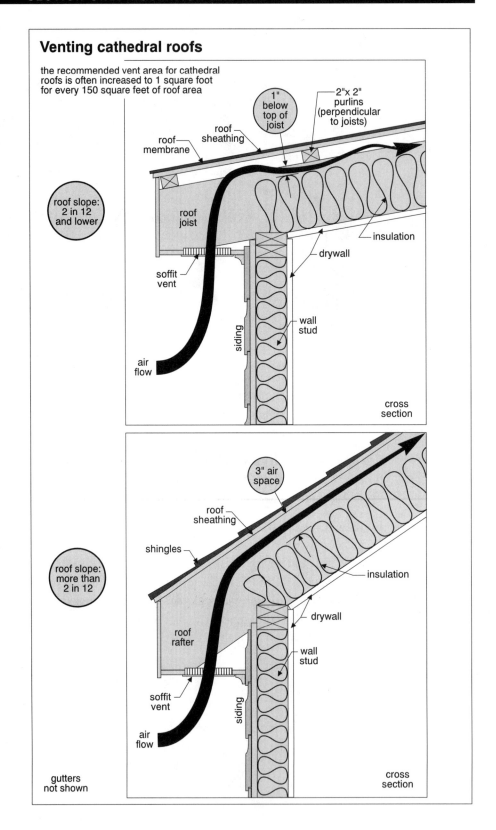

roof slope: 2 in 12 and lower

roof membrane

roof sheathing

1" below top of joist

2"x 2" purlins (perpendicular to joists)

roof joist

insulation

drywall

soffit vent

wall stud

siding

air flow

cross section

roof slope: more than 2 in 12

3" air space

roof sheathing

shingles

insulation

roof rafter

drywall

wall stud

soffit vent

siding

air flow

gutters not shown

cross section

Vent Spaces On Cathedral Roofs

Cathedral roofs with a slope of more than two in twelve are often ventilated without purlins. A three-inch air space is often recommended between the top of the insulation and the underside of the sheathing. Vents at the soffits and the ridge allow air to move up the roof space between rafter pairs.

Mansard And Gambrel Roofs

Mansard and gambrel roofs have an upper and lower section with different slopes. Most jurisdictions do not require the steep, lower slope of the roof to be vented. Vents are required at the ridge and at the bottom of the upper section. This is a complicated roof detail, and many mansard and gambrel roofs do not enjoy this type of venting.

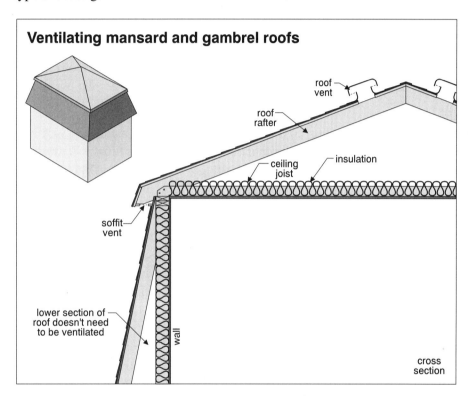

Ventilating mansard and gambrel roofs

roof vent

roof rafter

insulation

ceiling joist

soffit vent

lower section of roof doesn't need to be ventilated

wall

cross section

Can You Have Too Much Venting?

Some people believe there can be too much roof venting, particularly if the ceiling (attic floor) is not well sealed. Too much venting may create negative air pressure in the attic, which will tend to draw warm, moist air out of the living space. As we've discussed, this can promote, rather than prevent, condensation and damage structural members.

No Power Vents

We recommend against the use of exhaust fans or power vents in attics during the heating season. While these can help reduce attic heat during summer months, they tend to depressurize the attic and draw warm, moist air up into the roof space, again possibly causing condensation and damage. These fans require electricity and controls. They are also mechanical devices in a hostile environment and are prone to mechanical failure. They may be controlled with manual switches, humidistats or thermostats.

A Roof Venting Counterpoint

There is one school of thought that says venting well-insulated roofs can make things worse, rather than better. The colder the climate, the riskier venting attics may be.

The reasoning goes like this:

Wood Sees Relative Humidity

1. Wood structural members respond to relative humidity, rather than absolute humidity. Wood absorbs moisture from 20F winter air at 80% R.H., even though there is very little moisture in the air.

Poorly Insulated Attics

Poorly insulated attics have lots of heat loss. The attic air is warm. The heat raises the temperature of the outdoor air moving through the vents. This lowers the relative humidity of the air, and the air helps keep the wood dry.

Well Insulated Attics

As attics become better insulated, the air temperature drops and the relative humidity rises. This higher relative humidity does not dry the wood, and can increase the moisture level.

Attics Colder Than Outdoors

2. On clear, cold winter nights, the air temperature in a well-insulated attic can be lower than the outdoor temperature, due to **night sky radiation** – heat transfer by warm bodies (houses) to cold bodies (the sky) by radiation. Dropping the air temperature raises the relative humidity, often to the saturation point. Again, wood is getting wetter, rather than drier.

Venting Pulls

Air Out Of Houses

3. We've talked about this one already. Attic vents, especially turbine type or power vents, can depressurize the attic, drawing warm, moist air out of the home. This warm, moist air cools and condenses in the attic, again contributing to moisture problems.

Summary

While there is some validity to these points, and there are some counterpoints we don't discuss, even the proponents of these ideas agree that in most parts of North America, most homes benefit from attic ventilation. These arguments open the door to designing unvented systems, but many variables have to be considered. We believe unvented roof spaces are a risky business.

Insulation
& Interiors
M O D U L E

QUICK QUIZ 7

⚓ INSTRUCTIONS

• You should finish Study Session 7 before doing this Quiz.

• Write your answers in the spaces provided.

• Check your answers against ours at the end of this Section.

• If you have trouble with the Quiz, re-read the Study Session and try the Quiz again.

• If you did well, it's time for Study Session 8.

1. List three functions of roof vents.

   _____

   _____

   _____

2. List four types of roof vents.

   _____

   _____

   _____

   _____

3. What percentage of the total venting should be at the soffits?

   _____

4. What percentage of the total venting should be high on the roof?

   _____

5. When can soffit venting be omitted safely?

_____

6. What problems are commonly found with turbine vents?

_____

7. What do soffit vent baffles do?

_____

_____

8. What amount of ventilation is typically recommended for attics?

_____

9. What amount of ventilation is typically recommended for low-slope, flat or cathedral roofs?

_____

10. How are mansard or gambrel roofs ideally vented?

_____

11. Is it possible to over-ventilate an attic? If so, why is this an issue?

_____

_____

12. Why should power roof vents not be operated in the winter?

_____

_____

If you had no trouble with this Quiz, you are ready for Study Session 8.

Key Words:

- **Ice dams**
- **Heat**
- **Moisture**
- **Soffit vents**
- **Ridge vents**
- **Roof vents**
- **Gable vents**
- **Turbine vents**
- **Power vents**
- **Soffit vent baffles**
- **Depressurization**

Insulation & Interiors

M O D U L E

STUDY SESSION 8

1. You should have completed Study Session 7 before starting this Session.

2. This Session covers the venting of living spaces, including heat recovery ventilators.

3. At the end of this Session, you should be able to—
   • gve two reasons for venting house air.
   • explain in one sentence why venting is more important now than it used to be.
   • list three general approaches to ventilation.
   • explain in two sentences what is meant by a **balanced ventilation** system.
   • list nine components of a heat recovery ventilator.
   • explain in two sentences how heat is recovered in an HRV.
   • explain in two sentences which ducts on an HRV should be insulated and why.
   • explain in two sentences the function and location of flow measuring stations and balancing dampers.
   • explain in two sentences where fresh air from an HRV may be introduced to a home.
   • explain in two sentences where exhaust air for an HRV may be drawn from the home.
   • describe six characteristics of good air intake points for an HRV.
   • describe four characteristics of good air exhaust points for an HRV.
   • List three control methods for an HRV.
   • List four defrost methods for an HRV.
   • Describe in two sentences the common relationship between an HRV and forced-air furnace.

4. This Study Session may take you roughly one hour.

5. Quick Quiz 8 is lurking at the end of the session.

Key Words:

- *Moisture*
- *Odors*
- *Indoor air pollutants*
- *Fresh air*
- *Air sealing*
- *Exhaust ventilation*
- *Supply ventilation*
- *Balanced ventilation*
- *Heat recovery ventilators (HRVS)*
- *Energy recovery ventilators (ERVS)*
- *Heat exchangers*
- *Air exchangers*
- *Flow measuring stations*
- *Balancing dampers*
- *Condensate systems*
- *Defrost systems*
- *Duct insulation*
- *Air filters*

► 7.0 VENTING LIVING SPACES

Purposes

Living spaces are ventilated primarily to accomplish two things:

• Eliminate moisture, odors, and other indoor air pollutants from the home
• Bring fresh air into the home

Through the 1980s and '90s, more and more attention has been paid to this issue.

Indoor Air Pollutants

Cigarette smoke, dust, animal dander, radon, soil gases, household chemicals and solvents, formaldehyde from furniture, draperies, rugs, glues, particleboard and waferboard, are all indoor air polutants.

Ozone

Ozone is an indoor air pollutant. It makes the respiratory system sensitive. It can trigger symptoms in asthmatics. Ozone levels may be high if electronic air cleaners are not working properly. Low humidity can cause more ozone production by electronic air cleaners.

Carbon Monoxide

CO levels of 100 parts per million can result in a coma after ten to sixteen hours' exposure. Levels of 200 ppm cause a coma in one to two hours. Detectors are often set at 50 ppm.

Controlling Pollutants

Indoor air quality depends on controlling pollutants as well as ventilation. Pollutant control includes –

• removing source,
• replacing pollutants,
• isolating pollutants, and
• filtering, humidifying or dehumidifying air.

How Did We Ventilate Living Spaces In The Past?

Houses used to be so leaky we didn't need to worry about venting the living spaces. We typically had more than one air change per hour in our homes. This provided good air quality and flushed out moisture.

Houses Tighter

Since the oil crisis of the early 1970s, we've started paying more and more attention to heating costs and to air leakage. Our houses have become tighter. We have reduced the number of air changes per hour and raised the moisture and pollutant levels inside our homes. Many homes have also eliminated chimneys, which were a natural exhaust device that removed stale air and allowed fresh air to be drawn in through cracks, holes and other imperfections in the house envelope.

Improving Air Quality Through Ventilation

We've spent so much time and effort tightening houses up, it's ironic that we now find our indoor air quality has deteriorated. It's too simple (or is it?) to suggest that people crack their windows open to provide fresh air and improve air quality. The downsides of this include the cool drafts and wasted energy that our parents lived with.

Three
Approaches

There are three general approaches we can take to improve indoor air quality:

Exhaust

1. We can build an exhaust system that will remove moist, polluted air from the house. We can rely on natural infiltration and leakage to bring fresh air in.

Supply

2. We can force air into the house with a fan and push the stale air out of the house through cracks, holes and other incidental openings.

Balanced

3. We can create a balanced ventilation system controlling both the exhaust air and the fresh air supply.

Let's look at these.

1. Exhaust-only Venting Systems

We've been using exhaust-only systems for a number of years without describing them this way. Kitchen and bathroom exhaust fans, for example, are exhaust-only ventilation systems. So are clothes dryers and central vacuum systems that throw a considerable amount of air outdoors. Let's look at some pros and cons.

- An exhaust-only venting system can depressurize the house, leading to back-draft of combustion appliances. It can pull smoke out of fireplaces and generate unhealthy conditions inside the house. (If there are no natural-draft appliances, this may not be a problem.)
- The air leaking in through the unintentional openings in the building may create drafts.
- The exhaust-only system does not include any heat recovery, so there is a waste of heat.
- There's an energy consumption issue because electricity is required to operate exhaust fans.
- An advantage to exhaust-only systems is that the cold dry air we pull in through wall and roof spaces has a drying effect on the structure (in a heating climate).

2. Supply-only Ventilation Systems

These systems use a duct and usually a fan to bring fresh air into the house. The fresh air is usually dumped into the furnace return duct, where it can be heated before it is distributed through the home.

- This can be uncomfortable, creating cool drafts, unless we ensure the furnace is heating the air every time fresh air moves through it.
- We may pressurize the house with this system. We don't want to do this, because we are likely to drive moisture into the structure where it may cool, condense, and damage the building.
- A supply-only venting system is not very energy efficient since there is no heat recovery.
- There's also electricity involved if there is a fan to bring the air into the house. This may be a dedicated fan, or it may be the furnace blower.

3. Balanced Ventilation Systems

These systems use both supply and exhaust air fans to maintain a neutral pressure inside the house. These generally create fewer drafts and provide better comfort.

• There are no pressurization or depressurization problems if the system is set up correctly.
• Again, if there is no energy recovery, there is some wasted heat.
• Electricity is consumed by the fans.

As you can imagine, the balanced ventilation systems are popular among people who give this a lot of thought. In a moment, we'll look at a way to improve the energy efficiency of balanced ventilation systems, but first let's take a quick look at some of the components of the exhaust and supply systems.

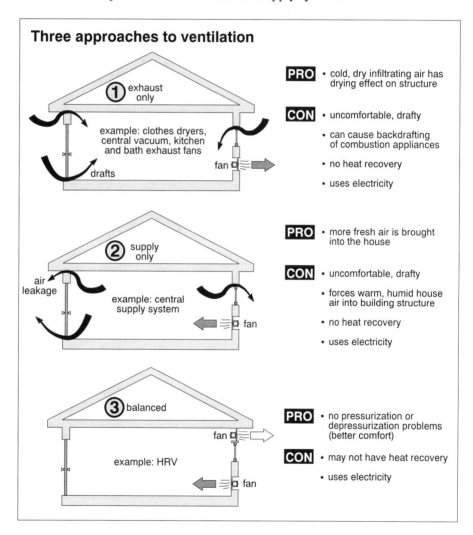

Exhaust System Components

These are typically quiet fans, often located in bathrooms or kitchens, since these are the sources of concentrated pollutants, including moisture and odors. The fans are typically designed to operate continuously and are sometimes two-speed or multi-speed fans. There is usually a manual control, and sometimes an automatic control such as a dehumidistat. There may also be a timer.

Discharge
Outside

The fans are connected to ductwork and discharge outside the building. The outlets are typically four to eight inches above grade, and about six feet away from other mechanical inlets for the building. We don't want to exhaust air and then pull it right back into the house.

Termination
Points

The discharge point of the exhaust system should not be under decks or in garages, crawlspaces or other confined areas.

The duct system is sealed at the wall or ceiling penetration to prevent air leakage and keep moisture out of the building. The terminations are hooded to keep snow, wind and rain out of the duct. There's typically a flap to keep out drafts as well as animals, birds and insects when the system is not operating.

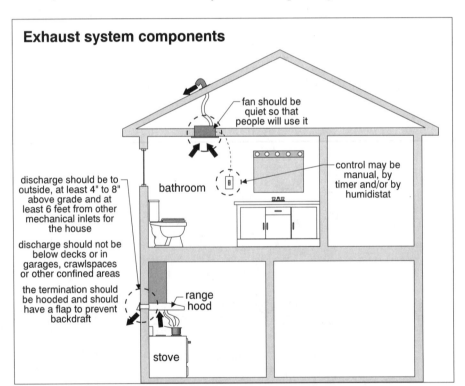

Supply Systems

No Fan

There are two types of supply systems. The first has no fan, relying on the forced-air furnace blower. There's an outdoor air intake and an insulated duct connected to a furnace return duct. Air is drawn in when the furnace fan runs.

Inlet

The intake point on the outside of the wall is hooded, with a screen or grille, to keep out animals, birds and insects. The inlet is typically eighteen inches above grade and separated several feet from building and automobile exhausts. We don't want to pull these back into the living space.

Supply Ventilation Systems With Fans

This system has a fan in the intake ductwork. This system does not rely on a furnace fan. In this type of system, the air may be run through dedicated ducts to the living space. Fresh air is typically delivered into rooms at or near the ceiling level where it mixes with the warm air. In some cases, preheaters are used to temper the air before releasing it into the rooms. This approach is slightly more complex than simply using the furnace blower.

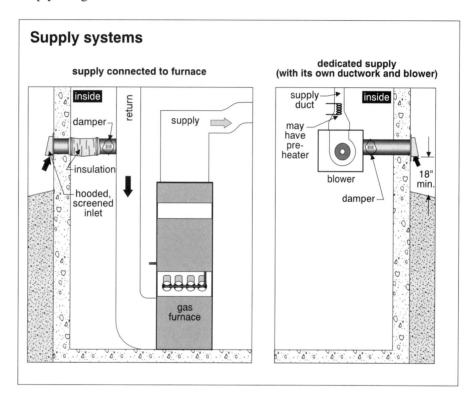

Heat Recovery Ventilators (HRVs)

The Principle

Heat recovery ventilators, energy recovery ventilators (ERVS), heat exchangers or air exchangers as they are sometimes called, improve the energy efficiency in a balanced ventilation system. Inlet and exhaust airflows are controlled by fans so that house pressures are balanced. Heat is recovered from the exhaust air by transferring it to the cool inlet air. Energy recovery ventilators transfer house moisture as well as heat from the exhaust air into the fresh air coming into the home. For simplicity, we'll call them all heat recovery ventilators or HRVs.

Components   The components of the heat recovery ventilator include

1. A cabinet
2. A heat exchanger
3. Inlet and exhaust fans
4. A duct system
5. Flow measuring stations
6. Controls
7. Air filters
8. A condensate system
9. A defrost system

Cabinet   The heat recovery ventilator is typically a sheet metal cabinet in a basement or closet area.

How Big Are They?   Heat recovery ventilators are typically the size of small furnaces. They are often hung from ceilings and are usually longer than they are tall. They might be 30 inches long, 18 inches tall and 18 inches wide, for example.

Location   Heat recovery ventilators are usually located in conditioned spaces, often in basements. Putting them in garages or attics is considered poor practice as this reduces the energy efficiency of the systems.

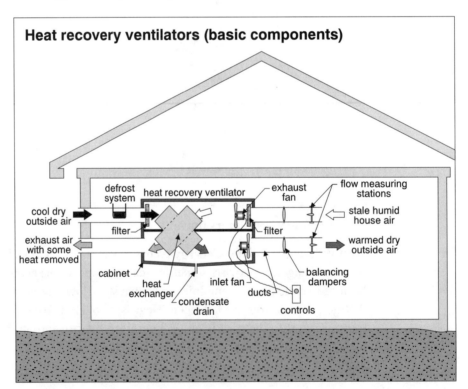

Heat recovery ventilators (basic components)

Heat Exchangers

There are six types of heat exchangers used in heat recovery ventilators:

1. Flat plate
2. Rotary wheel
3. Concentric tube
4. Heat pipe
5. Capillary blower
6. Heat pump

Flat-plate heat exchangers are made of metal, aluminum or plastics such as polypropylene. Fresh air and exhaust air usually flow in opposite directions. Exhaust and supply air are kept separate. There is no moisture transfer in most units. Some have paper cores with limited moisture transfer.

Rotary wheel. This type has a slowly rotating wheel. Fresh air moves in one direction; exhaust air moves in the other. A stationary baffle separates the warm side from the cold side. Heat collects on the warm side and as the core rotates to the cold side, heat is released to the cold air. The wheel turns at about 30 rpm. There is some cross leakage between exhaust and indoor air. Some moisture is transferred.

Concentric tube-type ventilator. This is a pipe within a pipe. The fresh air flows in opposite direction to exhaust air. Fresh air may flow through the core and second ring; exhaust air may flow through the first and third rings, for example. Heat is transferred from the exhaust air to the fresh air.

Heat pipe ventilator. This unit has a core with aluminum or copper fins on tubes filled with a refrigerant such as Freon™. A baffle separates the warm side and the cold side. The cold side is slightly higher than the warm side of each tube. Warm air flowing over the outside of the tubes boils the refrigerant from a liquid to a gas. The gas migrates to the high side. Cool air passing over the high side takes the heat from the refrigerant. The refrigerant condenses and flows back to the low side.

Capillary blower-type. This system has a foam ring that rotates at high speed. It's like a doughnut with a stationary baffle across the middle of the hole. Cold air comes in to the center of the doughnut on one side of the baffle and exhaust air comes into the hole on the other side. The rotating ring picks up heat and moisture from the warm exhaust air and transfers it to the cold air exhaust outlet. Heat is transferred to the cool air before it is discharged through the indoor supply duct. The ring is permeable and some moisture transfers from the exhaust air to the intake air.

Heat pump-type. The heat pump system has a compressor and an evaporator and condenser coil. Heat from the warm exhaust boils a refrigerant in the evaporator, giving off its heat to the refrigerant. The refrigerant goes to the compressor where it is heated before being released into the condenser. The cool inlet air passing over the warm condenser picks up heat. The refrigerant in the condenser cools and condenses to a liquid from a gas. The liquid refrigerant moves to the evaporator and the cycle repeats.

The illustration shows how these systems work.

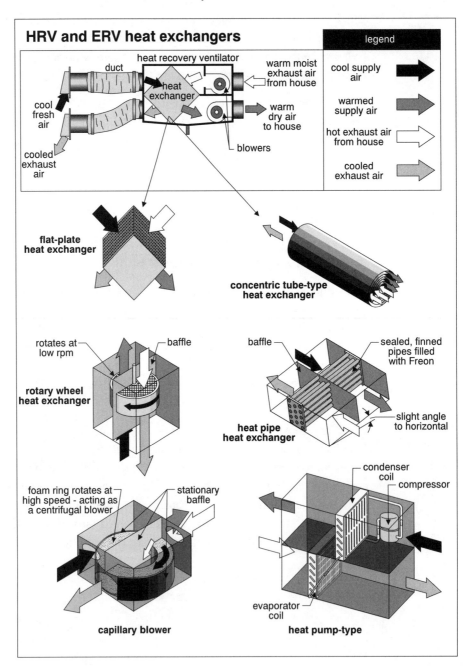

The Goal   The different types of heat exchangers are trying to accomplish the same thing. As the warm air is exhausted from the house, cold air is brought in. The warm air is moved through one side of the heat exchanger and the cool, outdoor air is moved through the other side. Heat moves from the exhaust air into the intake air. This cools the exhaust air before it is dumped outside, and warms the fresh intake air before it is distributed through the house.

Fans   Fans are used to move the exhaust air through the heat recovery ventilator, out of the house, and to move the fresh air from outside through the HRV and into the home. There may be two fans and two motors, or two fans driven by a single motor, in a typical heat recovery ventilator.

Fans are typically 120-volt, but may be 240 volts if an electric preheater is used for the fresh air.

Efficiency

HRVS are typically 55 to 85 percent efficient, with an average of about 70 percent. Efficiency varies with outdoor temperature and the design and performance of the unit.

The amount of ventilation

There are a number of ways to look at how much ventilation a house should have. Common recommendations include –

• One-third air change per hour (0.3 to 0.35 ACH). Note: Levels should be higher if people smoke or if there is a hot tub or swimming pool in the home. Or,—
• There should be 15 cfm (cubic feet per minute) per person. This is adequate for a person at rest to avoid excess carbon dioxide levels. Or,
• There should be 65 cfm total for the kitchen, dining room, living room, one bathroom and a master bedroom. Add 10 cfm for each additional room, and 20 cfm for an unfinished basement. We need 65 cfm for the base group of rooms, and 10 cfm each for the second bathroom and three other bedrooms, and 20 cfm for the basement. This totals (65+10+30+20) 125 cfm.
• There is a second set of criteria to calculate. The total number should include 50 cfm for each bathroom and 100 cfm for the kitchen. For example, a house with four bedrooms, a family room, two bathrooms and an unfinished basement will need 100 cfm for the kitchen and (50x2) 100 cfm for the bathrooms. This totals 200 cfm. The first criteria called for 125 cfm. We use the larger, so we require 200 cfm.

This total ventilation is typically achieved with exhaust fans and an HRV. The HRV typically would handle about 50 percent of the total demand. An average design for a house might by 200 cfm. The HRV might operate in the 70 to 100 cfm range on normal speed, and have a maximum capacity of 200 cfm on high speed. High speed numbers in the area of 200 cfm may only be required if the system is also used as a kitchen or bathroom exhaust fan.

Over-Ventilating and Under-Ventilating.

We want to provide enough ventilation, but not too much. If we don't provide enough, moisture levels in the house will rise and indoor air quality will suffer, as the levels of other pollutants rise. Over-ventilating reduces moisture levels in the house to an uncomfortable level and may use too much energy. Exhausting air does cost some energy, even if the HRV is operating at eighty percent efficiency.

Balanced Ventilation

It's important the HRV be balanced so the house will not be pressurized or depressurized. Pressurizing the house drives moisture into the structure. Depressurizing the house may cause combustion appliances to backdraft.

Carbon Monoxide Sensors For Wood Stoves And Fireplaces

In some jurisdictions, houses with mechanical ventilation systems require carbon monoxide sensors in rooms with wood stoves or fireplaces. The risk of depressurizing the house and causing backdraft of these appliances warrants the installation of these sensors. Obviously, there's considerable indoor air pollution and a risk to health if wood-burning appliances backdraft.

Insulate Ducts

Ducts running through unconditioned spaces to or from heat recovery ventilators should be insulated to prevent heat loss and condensation. The ducts on the cold side of the HRV (the exhaust duct running from the HRV out through the wall and the fresh air intake running from the outdoors to the HRV) should be insulated. Flex duct with one inch of fiberglass insulation (R-value of 3.0) is usually considered adequate. Where ducts are longer than ten feet, some recommend additional insulation.

Cold-Side Ducts Short And Straight

The ducts on the cold side of the HRV should be kept as short and straight as possible. Flex duct often sags, which creates additional friction loss and reduces flow rates. The sags may also be spots where condensation collects in the exhaust duct. Flex ducts should be supported every three feet, approximately.

Exhaust Should Slope

Some experts recommend the cold-side ducts slope slightly down, toward outdoors, from the HRV so any condensation will drain outside.

Warm-Side Ducts

The warm-side ducts are usually uninsulated rigid metal. Combustible ducts on an HRV system are allowed under some circumstances, but most are metal.

Balancing Dampers

Balancing dampers are usually provided on warm-side ducts, close to the HRV. These are needed to balance the system from time to time, so we don't over-pressurize or depressurize the home.

HRV (heat recovery ventilator) components

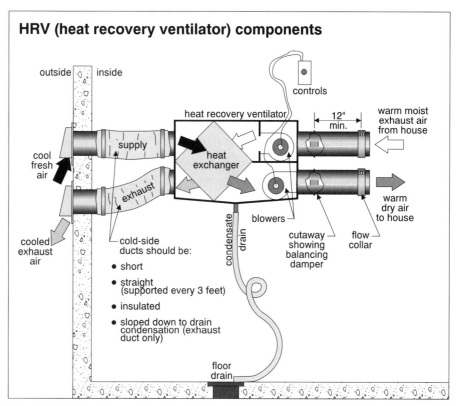

Flow Measuring Stations

Flow measuring stations, flow stations, or **flow collars** may be installed on the warm-side ductwork to allow balancing of the system. These collars are typically installed at least twelve inches away from the balancing dampers. They can usually be identified by the two pins protruding from the collar. These pins are connected to the gauges. Where flow collars are provided, there is usually one on the fresh air intake and one on the exhaust duct. Balancing should be checked and adjusted annually.

Exhaust Ducts The exhaust air from the house can be taken from –

- Kitchens and bathrooms, sources of high humidity and pollution. If exhaust air is drawn from the kitchen, it should not be close to the stove. The grease may clog the system. An exhaust grille in the kitchen should have a grease filter, even if it is well away from the cooktop. There may be a separate range hood vent.
- Any other rooms in the home.
- The return air plenum of a forced-air heating system.

Fresh Air
Supplies
The fresh air supply coming from the HRV into the home can go –

- through dedicated ducts to various rooms of the house. In this arrangement, the supply registers are usually at the ceiling level, or on walls within twelve inches of the ceiling.
- into the cold air return plenum for the furnace.

If the exhaust air is also drawn from the return plenum, the supply connection should be at least three feet downstream of the exhaust connection. The fresh air connection to the plenum may be a direct or indirect connection. In an indirect connection, an open T may be provided on the top of the fresh air duct, near the return plenum. Alternatively, the fresh air duct may also terminate four to twelve inches in front of a grille on the side of the return plenum. Indirect connections should not be made inside a furnace room. The furnace may draw room air in through the opening in the return duct, depressurizing the furnace room. This may lead to backdraft of the furnace.

Outdoor terminations for exhaust and fresh air intakes

Fresh Air
Locations Intake
The fresh air intake should be –

- located six feet away from the exhaust;
- 18 inches above grade;
- 40 inches away from the corners of buildings, to avoid turbulent air;
- three feet from gas meters;
- well away from driveways and garages (to avoid bringing car exhaust fumes into the house); and
- three feet from clothes dryer vents, exhaust fan vents, boilers vents, furnace vents, water heater vents, and fuel oil fill and vent lines.

The fresh air intake should not be –

- in crawlspaces or attics, or
- in areas where snow may accumulate.

Hooded And
Screened
The fresh air intake should be arranged so that rain, snow and wind won't enter the duct system or the home. Screening should be provided to keep out animals, birds and insects.

Exhaust
Terminations
Exhaust terminations should be away from fresh air intakes and away from attics, garages and crawlspaces. The exhaust outlet should be a minimum of four to eight inches above grade, and should be protected from rain and snow with a hood similar to the fresh air intake. Screening may or may not be provided on the exhaust, but it should have a damper that opens readily when the system is operating and closes tightly when the system is at rest.

Labeling
In some jurisdictions, the fresh air intake and exhaust outlets must be labeled on the building exterior.

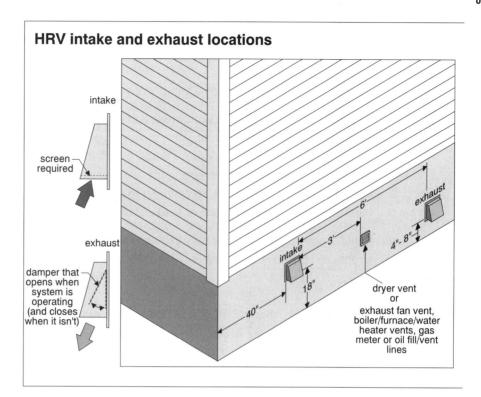

HRV intake and exhaust locations

intake

screen required

exhaust

damper that opens when system is operating (and closes when it isn't)

intake

exhaust

6'

3'

4"- 8"

18"

40"

dryer vent or exhaust fan vent, boiler/furnace/water heater vents, gas meter or oil fill/vent lines

Controls

HRVs are controlled a number of ways:

• Thermostat
• Humidistat
• Dehumidistat
• Manual switches
• Timers
• Fan speed controls

A manual switch near the center of the house and labeled "Ventilation Fan" is a common way to control the system. In modern installations, where the HRV is interconnected with furnace ductwork, the furnace fan is often interlocked with the HRV, so that when the HRV fan is operating, the furnace fan operates as well. Some older systems were not set up this way. Sometimes the ventilation fan would be labeled and an adjacent switch labeled "Circulation Fan" would control the furnace fan.

Many new systems have the HRV and furnace operate continuously at low speed. The "Ventilation Fan" switch operates both fans at high speed.

Furnace Fans Use Lots Of Electricity HRVs use some electricity to move 70 to 200 cfm air. Furnace fans move considerably more air, and consume substantial amounts of electricity, especially if they run continuously to support a ventilation system. There are modern, highly efficient motors (ECM motors, for example) that use less electricity. These are rare in residential applications.

Exhaust Fan Controls

Many ventilation systems depend on separate exhaust fans for bathroom and kitchen areas. These fans can be controlled manually, with dehumidistats, or with timers.

HRV Controls

The HRV can also be controlled a number of ways:

- Manual operation, as we have discussed above.
- Automatic operation. These may use timers to operate the HRV a specified number of hours per day at various times, or dehumidistats that activate the system when humidity levels rise.
- Continuous operation. Some HRVs are set to operate continuously at low speed, and move to high speed in response to a manual switch or a dehumidistat.

Fancy Systems

Sophisticated control systems may include occupancy sensors that detect people in a room and operate the system only when the room is occupied.

Air Filters

Most HRVs have integral air filters that require regular inspection, cleaning or replacement. These are similar to furnace filters and are usually accessed by opening a cover on the HRV itself.

Condensate Drains

Size And Traps

Many HRVs have condensate pans and drainpipes. The drain pipe is typically half an inch in diameter and includes a two-inch trap. The trap is often created by simply looping the pipe.

Drain Termination

The drain should slope to carry water by gravity to a safe location. An outdoor termination may be susceptible to freezing. An air gap is required where the drain dumps into a plumbing fixture or drain. In many jurisdictions, hard connections to plumbing stacks, for example, are not allowed.

Filters and condensate drains

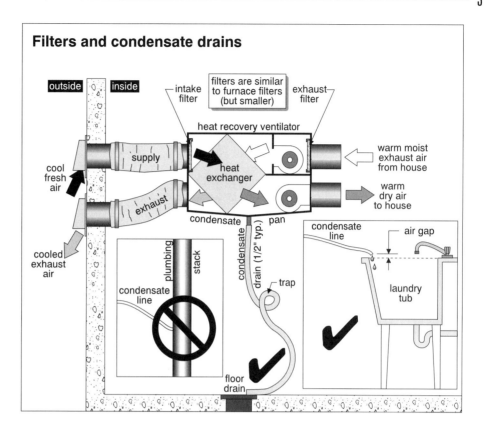

Energy Recovery Wheels

Some HRVs, including energy recovery wheels or ERVs, allow some moisture to move from the exhaust air into the supply air. Manufacturers argue that this improves the quality of the fresh, indoor air, and maintain the level of humidity can be controlled. These systems typically do not have condensate drains since a good deal of the moisture is transferred into the incoming fresh air.

Defrost Control

Very cold fresh air coming in can cause condensation and frost in the heat exchanger. This can ice up the unit. Defrost cycles are often automatic, coming on for a few minutes every hour when the temperature is below a given set point (e.g., 20°F), There are four ways the defrost is typically arranged:

1. The incoming supply air is preheated with an electric duct heater.
2. The exhaust air leaving the HRV is recirculated through the fresh air inlet. The HRV sees no fresh air during this cycle.
3. The exhaust fan stops and the fresh air intake is blocked with the damper. Warm house air is drawn through the fresh air side of the heat exchanger.
4. The fresh air fan is shut off and the exhaust fan continues to move warm air through the HRV.

Defrost control

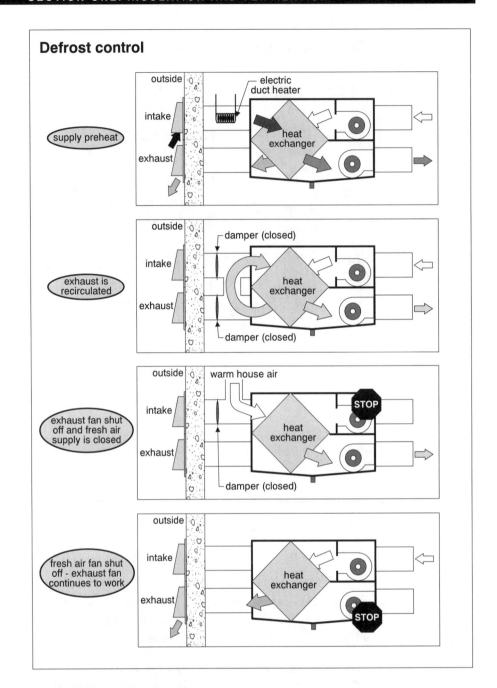

Regular maintenance includes –

Regular

HRV

Maintenance

• clean or replace air filters monthly.
• clean fresh air inlet and exhaust air hoods.
• clean the heat exchanger core
• clean the condensate drain.
• lubricate the fan motors as necessary.
• clean the ductwork.
• clean the dehumidistat (if used).
• check and balance flow rates.

Summary

It's probably fair to say that house ventilation systems are evolving. Many people believe the heat recovery ventilator is the best approach to solving indoor air quality problems, controlling house moisture and minimizing heat loss.

The Counterpoints

• Others feel these systems are too expensive to install and operate.
• There are many examples of poorly installed and operating systems.
• An unbalanced ventilation system can be worse than no ventilation system.
• Ventilation systems are not sophisticated and do not respond automatically to changing conditions in houses.
• Many homeowners do not understand the purpose and operation of ventilation systems, and often shut the system down or operate it without proper maintenance.

Energy Costs

While the HRV does capture some of the heat that would otherwise be exhausted, there is a price to pay. In addition to the original installation costs, there is the electrical cost to operate the fans in the HRV, and often a larger cost to operate the furnace fan.

Complexity

Some maintain that houses are already too complicated for the average homeowner to maintain, and devices such as HRVs are more trouble than worth. We believe we will see more changes in house ventilation and we hope solutions become simpler, less expensive and more user-friendly. For the time being, home inspectors should be able to recognize and test these systems. Fully testing a ventilation system is beyond the scope of a standard home inspection, but is often recommended.

Balanced Ventilation without Heat Recovery

There are many types of ventilation systems. This one is not as fancy as an HRV, but is more sophisticated than an exhaust-only or supply-only system.

Makeup Air

A balanced ventilation system can be established with kitchen and bathroom exhaust fans and **makeup** air. Supply air is brought in from the outside to balance the air exhausted by kitchen and bathroom fans, clothes dryers, central vacuums, fireplaces and anything else that tends to depressurize a house.

Makeup Air Supplies

A makeup air system is typically an inlet on the exterior wall connected to a duct that discharges into the furnace return plenum. The duct is typically at least ten feet away from the furnace, so the cold air can be mixed with the warm return air from the house. We don't want to subject the heat exchanger in the furnace to low temperatures.

Balancing Damper   Many makeup air systems have a manual damper to allow adjustment of the amount of makeup air. Some of these dampers are automatic. They may be activated when exhaust fans are turned on. There are other operational systems that can be set up for these.

Tough To Balance   Houses are dynamic. The conditions inside change as people come and go, and use the various appliances. The house is also in a dynamic environment. Outdoor temperatures and pressures change regularly; wind velocity and direction can have a significant effect on the pressure differences between the outdoor air and the house itself.

No Perfect Ventilation System   All ventilation systems, including supply-only, exhaust-only, and any of the balanced ventilation systems we've looked at, fall short of being perfect in preventing air movement through the building structure. There are too many variables to achieve constant pressure balance between indoors and out, over the entire volume of the home. The best ventilation systems minimize the effect of pressure and the movement of air into and out of houses, but none prevent that movement.

Insulation & Interiors

MODULE

QUICK QUIZ 8

☑ INSTRUCTIONS

- You should finish Study Session 8 before doing this Quiz.
- Write your answers in the spaces provided.
- Check your answers against ours at the end of this Section.
- If you have trouble with the Quiz, re-read the Study Session and try the Quiz again.
- If you did well, it's time for Study Session 9.

1. Why do we want to change the air in houses? (two reasons)

   _____

   _____

2. Why is venting of house air more important now than it used to be?

   _____

   _____

3. List three general approaches to venting homes.

   _____

   _____

   _____

4. What is meant by a balanced ventilation system?

   _____

5. What is the principle of a heat recovery ventilator?

    _____

6. List nine components of an HRV.

    _____

    _____

    _____

    _____

    _____

    _____

    _____

    _____

    _____

7. Explain how heat is recovered in an HRV.

    _____

    _____

8. Explain which ducts should be insulated on an HRV and why.

    _____

    _____

    _____

9. Explain the function and location of flow measuring stations on an HRV.

    _____

    _____

    _____

10. Explain the function and location of balancing dampers on an HRV.

    _____

    _____

    _____

11. Where is fresh air from an HRV typically introduced to a home?

_____

_____

12. Where is exhaust air typically drawn from within the home?

_____

_____

13. What things would you be looking for on the fresh air intake for an HRV? (list six)

_____

_____

_____

_____

_____

_____

14. What things would you be looking for on the outdoor exhaust outlet for an HRV? (list four)

_____

_____

_____

_____

15. List at least three different control methods for an HRV.

_____

_____

_____

16. List four different defrosting methods for an HRV.

_____

_____

_____

_____

17. How is the operation of the HRV fan and the furnace fan related in many installations?

_____

_____

18. Why is a condensate drain necessary on an HRV? Is it always needed?

_____

_____

Bonus Question:

19. Would an HRV be more important in a cold climate or a warm climate?

_____

_____

If you had no trouble with this Quiz, you are ready for Study Session 9.

Key Words:

- *Moisture*
- *Odors*
- *Indoor air pollutants*
- *Fresh air*
- *Air sealing*
- *Exhaust ventilation*
- *Supply ventilation*
- *Balanced ventilation*
- *Heat recovery ventilators (HRVS)*
- *Energy recovery ventilators (ERVS)*
- *Heat exchangers*
- *Air exchangers*
- *Flow measuring stations*
- *Balancing dampers*
- *Condensate systems*
- *Defrost systems*
- *Duct insulation*
- *Air filters*

Insulation
& Interiors
M O D U L E

STUDY SESSION 9

1. You should have completed Study Sessions 1 to 8 before starting this Session.

2. This Session covers the inspection of attics (including safety issues), the inspection of other house systems and common insulation and ventilation problems in attics.

3. At the end of this Session, you should be able to—
   • State at least three precautions you should take when inspecting attics.
   • List six other house system components you'll be typically looking at when in the attic.
   • List twenty-six insulation and ventilation related problems (and their implications) you may find in attics.
   • Describe in one sentence the inspection strategy for each of these problems

4. Before you start this Session, read sections A to O, sections 10 to 12, and sections 16 to 19 in the Insulation chapter of **The Home Reference Book**.

5. This Study Session may take you roughly one hour and a half.

6. Quick Quiz 9 is at the end of the session.

Key Words:

- *Mask*
- *Goggles*
- *Gloves*
- *Access hatch*
- *Staircase*
- *Pull-down stairs*
- *Insulation*
- *Dropped ceilings*
- *Recessed lights*
- *Knee walls*
- *Skylights*
- *Duct insulation*
- *Air/vapor barrier*
- *Chimneys*
- *Soffit vents*
- *Roof vents*
- *Ridge vents*
- *Gable vents*
- *Turbine vents*
- *Power vents*
- *Rot, mold and mildew*
- *Delaminated or buckled sheathing*
- *Whole house fans*

► 8.0 ATTIC INSPECTIONS

8.1 SAFETY FIRST

Fall, Shock Hazard And Irritant
The attic inspection is dangerous. You may fall through a ceiling, damage finishes, get an electric shock, or irritate your lungs, eyes or skin with insulation materials.

Hot, Cold And Infested
Attics can be hot in the summer and cold in the winter. They are often very difficult to move through and may be the home for birds and animals you'd rather not meet. Watch for animal droppings, which may pose a health hazard. You may be startled by bats, raccoons, squirrels, mice and other animals, causing you to lose footing or step carelessly, putting a foot through a ceiling. Bees, hornets and other stinging insects can also spoil your inspection.

Mask And Goggles
We recommend wearing an appropriate mask with proper filters (HEPA High Efficiency Particulate Arresting or P100-type) and goggles to keep irritants out of your eyes. We also suggest gloves and long sleeves with tight cuffs when working in an attic.

How Far To Go
When the attic insulation completely covers the ceiling joists, we do not walk through the attic. We simply look at the attic from the access hatch. When you can't see the ceiling joists, it's difficult to know where to step. You may be able to find the ceiling joists with your foot by feeling around through the insulation. However, you may step on a wire or open electrical junction box, or on a joist that has been cut or is cracked. We recommend you do not move through attics where there is risk of damaging the property or injuring yourself. We are also cautious about planks that have been laid as walkways across attic areas. Unless they are clearly well traveled, we move carefully.

Note: Some inspectors take more chances in attics. We wish them well.

Describe Your Inspection
Whether you moved through the attic or looked at it from the attic access hatch, you should describe in your report how you performed your attic inspection. If you leave clients with the impression your inspection was thorough, you'll have trouble explaining why you didn't identify problems that are discovered later.

Attic Access
It is good practice to provide access into any attic. As a general rule, if the attic is larger than 100 square feet and has at least 24 inches of headroom, an access hatch should be provided.

Access Hatches
Attic access hatches are ideally at least 20 inches by 28 inches, although many are smaller. Access hatches should be insulated and weather-stripped to minimize the air leakage into the attic.

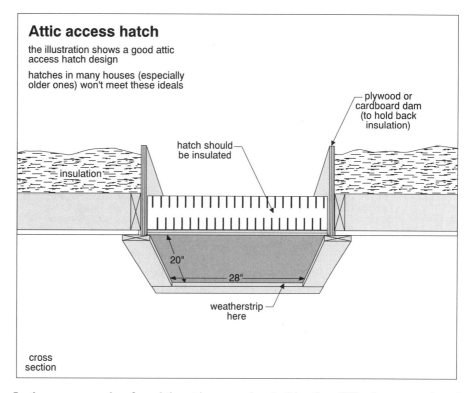

Attic access hatch

the illustration shows a good attic
access hatch design

hatches in many houses (especially
older ones) won't meet these ideals

plywood or
cardboard dam
(to hold back
insulation)

hatch should
be insulated

insulation

20"

28"

weatherstrip
here

cross
section

Opening *Access Hatches*	Let's assume you've found the attic access hatch. It's often difficult to get to it and it may be difficult to open. You'll have to use common sense here. We'll remove clothes from a closet and clear stored items from shelves so we can get into an attic. We arrange any materials we move so we are able to replace them in exactly the same order and location.
Hatches Screwed, Nailed Or Painted Shut	Most attic hatches do not require tools to open. Where the hatch has been sealed, you have to decide whether you're prepared to open it. Understand there is some risk of damaging cosmetic finishes.
Use Gloves To Avoid Fingerprints	Bare hands often leave dirty fingerprints on attic hatches. We recommend using gloves to avoid this. You should wipe off any marks you leave on the access hatch and trim. Many inspectors push access hatches open with flashlights to avoid touching them with their fingers.
Don't Dismantle Shelves	In some cases, shelving units have been added, making it impossible to get through the access hatch without removing them. If shelves can be removed without tools we will take them out. If the shelves are screwed or nailed into place, we will not usually disassemble them to gain access.
Drop Cloths And Vacuum Cleaners	Many inspectors use drop cloths below the access hatch and carry battery-powered vacuum cleaners to clean up. It's not unusual for a small amount of insulation to fall through the hatch when removing or replacing it.
Careful With The Hatches	We open all access hatches very carefully.

Insulation is often blown into an attic through a roof or gable vent. Twelve inches of loose-fill insulation over an access hatch can create quite a mess in the home when the hatch is opened. Insulation can be an irritant. If you dump insulation on people's clothes, perhaps you should offer to dry-clean them.

Stairwells To Attics

Some homes have full staircases leading to attics. There is usually a door at the bottom, and the staircase may be open to the attic. The stairwell walls and underside of the staircase should be insulated. The door should be weather-stripped. If the stairs have a ceiling, it should be insulated.

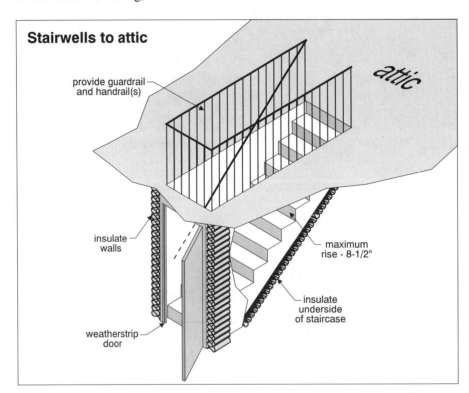

Stairwells to attic

attic

provide guardrail and handrail(s)

insulate walls

maximum rise - 8-1/2"

weatherstrip door

insulate underside of staircase

Safety

The stairs should be uniform with adequate tread width and a maximum rise of 8½ inches. Headroom should be adequate, ideally six and a half feet. There should be handrails on the stairs and a guardrail around the top. These are often missing.

Lighting

There should be electric lighting for the stairwell.

Pull-Down Stairs

Pull-down stairs can be dangerous. Be careful when pulling the stairs down. They may come down very quickly if mechanical components are loose or broken. When climbing the stairs, be careful. Treads or stringers may be loose or broken. Bolts may be loose or missing. Some home inspectors have been injured by pull-down stairs.

Insulating Cover

Pull-down stairs allow considerable heat loss and air leakage into attics. An insulated, weather-stripped, box can be placed over the stairs to help reduce heat loss and air leakage.

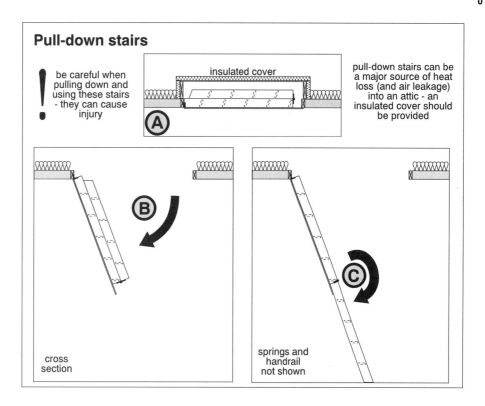

Pull-down stairs

be careful when pulling down and using these stairs - they can cause injury

insulated cover

(A)

pull-down stairs can be a major source of heat loss (and air leakage) into an attic - an insulated cover should be provided

(B)

cross section

(C)

springs and handrail not shown

8.2 INSPECTING OTHER SYSTEMS

Before we talk about insulation and air/vapor barriers, we should remind you that you may be inspecting components of several other house systems in the attic:

- **Roofing.** You may have seen some suspect flashing details or damaged roofing material on the outside of the home. When you are in the attic, look for evidence of leakage on the sheathing and rafters below.
- **Structure.** You'll be looking at the roof framing, including the sheathing, ridge boards or beams, rafters, collar ties, knee walls, struts, purlins and ceiling joists while in the attic. We talk about these in the Structure Module. You may have noticed a sag or dish in the roof when outside. The attic is the place to find the source of that problem.
- **Electrical.** Most houses have wiring running through the attic. Where it is visible, you'll want to inspect the wiring condition and connections.
- **Heating**. It's common to find chimneys and vents running through attic areas. You may have seen conditions on the outside of the home or in the floors below that make you suspicious about chimneys and vents. Look at the visible portions of these systems in the attic. We have occasionally found chimneys that stop in the attic and discharge their exhaust gases here rather than outside!

Hot water heating systems may have expansion tanks in the attic. Electric radiant heating systems may only be visible from the attic. Attic ductwork is common, particularly in warmer climates. There may be combustion air supplies for gas- and oil-fired appliances in the house below.

- **Air conditioning** and **heat pump** systems are often in the attic. You may find the fan coil and ductwork in the attic, for example. You'll be looking at the equipment, electrical supplies and condensate handling system for this equipment while in the attic.
- **Plumbing**. It's common to find plumbing stacks extending through the attic to the outdoors. In warm climates, supply piping may be run through the attic.

Attics Are Important

It's apparent that aside from insulation, air/vapor barriers and ventilation, there are many reasons to look in the attic. Many home inspectors go to considerable lengths to get into attics, since they are so important.

Let's turn our attention to the insulation-related issues in an attic. We'll start by listing the common problems.

8.3 CONDITIONS

Common attic problems include the following:

1. Access hatch problems
2. Attic staircase problems
3. Pull-down stair problems
4. Insulation amount – too little
5. Insulation – wet
6. Insulation – compressed
7. Insulation – gaps or voids
8. Insulation – missing at dropped ceilings
9. Insulation covering recessed lights
10. Insulation inadequate in knee wall areas.
11. Insulation inadequate at skylights and light wells.
12. Ducts leaking or disconnected, or insulation missing or loose
13. Duct air/vapor barrier missing or damaged
14. Insulation too close to chimneys
15. Air/vapor barrier missing, incomplete or wrong location
16. Air leakage excessive
17. Venting missing or inadequate
18. Venting obstructed at soffits or roof vents
19. Snow or wet spots below roof vents
20. Turbine vents noisy or seized
21. Power vents operating in winter
22. Power vents – poor wiring
23. Power vents inoperative in summer
24. Rafters and sheathing – rot, mold or mildew
25. Plywood sheathing delaminating or buckling
26. Whole-house fan with no insulated cover

8.3.1 ACCESS HATCH PROBLEMS

Problems include –

• not insulated or weather-stripped
• missing or inaccessible

The attic access hatch should be insulated, ideally to the same level as the rest of the attic. Rigid or batt insulation can be glued to the top of the hatch. Air leakage around the access hatch can be controlled if the hatch cover is weather-stripped. Sometimes there is no access into the attic.

Cause   These are installation issues.

Implications   Heat loss is an issue where the access hatch is not insulated. Missing hatches should be written up as a limitation to your inspection. Air leaking from the house into the attic is both a heat loss and moisture damage issue. Warm, moist air leaking into the attic can condense on structural members, causing damage.

Strategy   As you remove the access hatch, it's easy to see whether it's insulated and weather-stripped. It's a minor recommendation to advise that insulation and weather stripping be added.

Access To Secondary Attics   Home inspectors frequently fail to report the fact they could not get into secondary attics. As you look for the attic access hatch, keep in mind what the house looks like from the exterior. Is there only one attic? Many houses have more than one attic or roof space. If some of the roof space is inaccessible, make sure you let your client know, in writing, about this limitation.

Examples Of Secondary Attics   Two-story houses with one-story additions should have two attics. Split-level homes typically have two attics and two access hatches. Access to the lower attic is often through a sidewall on the upper floor. This may be in the back of a clothes closet, for example.

1½ And 2½- Story Houses   Houses with one and a half or two and a half stories often have three attic areas. There are typically at least two knee wall areas and a small attic above an upper floor. There may or may not be access into all of these. If there are dormers, there may be even more attics. A dormer can separate a knee wall attic into two separate areas. Again, there may or may not be access into each. Be sure to let you client know what parts of the house you could not get to.

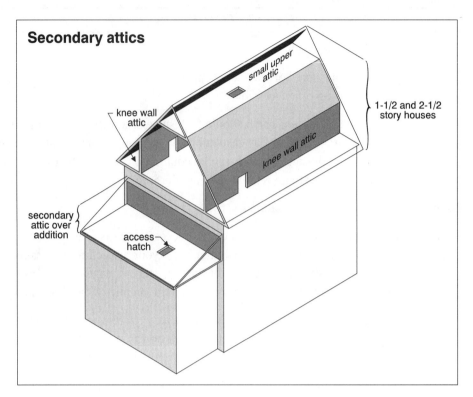

Secondary attics

small upper attic

knee wall attic

1-1/2 and 2-1/2 story houses

knee wall attic

secondary attic over addition

access hatch

Inaccessible   If the hatch is blocked by shelving, for example, or the cover is nailed on, many inspectors will not get into the attic. Report that you didn't gain access to the attic, and why.

8.3.2 ATTIC STAIRCASE PROBLEMS

These include –

• Inadequate insulation and weather stripping
• Stair rise, run and tread problems
• Handrail and guardrail problems
• Lighting problems
• Headroom problems

Causes   These are installation issues.

Implications   Heat loss and air leakage are the implications of inadequate insulation and weather stripping. Heat gain and air leakage result in cooling climates. Rise, run and tread problems; railing problems; and lighting problems are safety concerns.

Strategy   Check for insulation on the walls and underside of the stairwell. Look for weather stripping on the door.

Check the safety of stair treads. Is the run at least 8¼ inches? Is the rise 8½ inches, maximum? Are the treads uniform and level? Is there a handrail on the stairs that is easy to grab? Is there a guardrail around the top of the stair opening? Is there adequate headroom? Many standards call for 6-feet 8-inches clearance.

8.3.3 PULL-DOWN STAIR PROBLEMS

These include –

- stairs coming down too fast,
- stairs not solid and stable, and
- stairs not insulated.

Causes

This may be the result of poor installation, wear and tear, or mechanical damage.

Implications

Pull-down stairs are dangerous if they are not properly installed and maintained.

Stairs that are not insulated and weather-stripped are a source of heat loss and air leakage into the attic. Houses with air conditioning will suffer from heat gain and air leakage around pull-down stairs.

Strategy

Be very careful when opening pull-down stairs. Be ready to slow down a set that is coming too quickly, or to step out of the way if it is out of control. Look at the stairs once they are down for broken, loose or missing components.

Climb Carefully

Be careful, too, when you climb the stairs. Spread your weight over both hands and feet, each supported by different components. If in doubt, don't use the stairs. Many inspectors use their ladders even when there are pull-down stairs. Don't allow anyone else to be on the stairs at the same time you are. They aren't designed for two people.

8.3.4 INSULATION AMOUNT – TOO LITTLE

Many older homes will not have as much insulation as is currently recommended. You'll need to know the recommendations for new construction in your area as a benchmark. In some areas, the recommended insulation levels are higher if you have electric heat (because it's more expensive). If this is the case, you should be aware of these requirements and the heating type before you evaluate the insulation.

Cause

Inadequate insulation levels are an original construction issue.

Implications

The obvious implication of inadequate insulation is heat loss. In areas where snow accumulation and ice dams are an issue, low attic insulation levels promote ice dams.

Strategy

To determine the R-value of the attic insulation, you'll need to identify the insulation material and check its R-value per inch. You'll then need to measure the average depth of insulation in the attic. For example, if there are four inches of mineral wool insulation, the R-value will be 13 (3.2 x 4). If the recommended attic insulation level is R-25, we have only roughly half the insulation level that would be ideal. We would recommend that a client make this improvement, although we would not rank it as a priority measure.

Improvement Rather Than Repair

Clients should understand that adding insulation to an attic is an improvement, rather than a repair. The home has survived with the existing insulation level to this point and would continue to be habitable if no improvements were made. Increased attic insulation will reduce heating costs and may help prevent ice dams. The house may be more comfortable as a result, but this is not necessarily so.

The Risks Of More Insulation

Adding insulation improperly can cause problems. For example, increasing the attic insulation level will make the attic colder, but it won't reduce the amount of warm, moist air that leaks from the house into the attic. Now that the attic is colder, there may be more condensation as the warm, moist air cools faster and condenses before it can be flushed out. This can cause rot damage.

Inadequate Ventilation

Higher insulation levels without adequate attic ventilation may result in rot. When insulation levels were low, ventilation was not so important because the attic was relatively warm and moisture didn't condense quickly. With improved insulation and a colder attic, it's more important to flush the warm, moist air out quickly.

Adding Insulation Can Reduce Ventilation

The careless addition of attic insulation can obstruct soffit vents and dramatically reduce the attic ventilation. This is likely to result in moisture damage.

Burying Recessed Light Fixtures

Insulation added to an attic can cover recessed light fixtures that should be kept clear of insulation.

Too Close To Chimneys

Insulation may also be installed too close to metal chimneys, and if the insulation is combustible, it may be inappropriately installed against masonry chimneys.

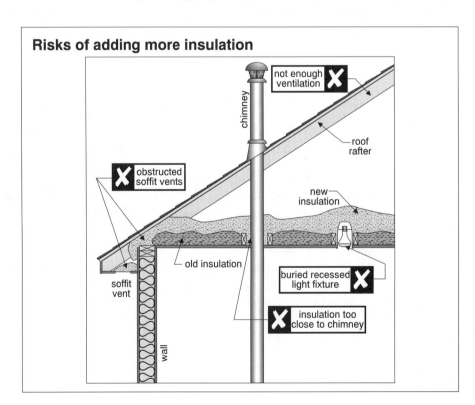

8.3.5 INSULATION – WET

Wet insulation does not insulate very well. Some materials recover their insulating properties as they dry. Others do not.

Causes   Insulation may be wet as a result of –
• a roof leak,
• condensation, or
• rain or snow blowing in through roof vents.

Implications   Wet insulation won't perform its function. If the amount of moisture is significant, damage to ceiling joists and finishes can result.

Strategy   As you look at the attic insulation look for evidence of moisture. In some cases this will appear as a dark patch. In some cases it appears as matted or compressed area of insulation. Cellulose fiber insulation, for example, will be compressed when it's wet and will not rebound. Fiberglass insulation tends to rebound and recover its insulating value as it dries.

Vulnerable Areas Of Roof   Check at vulnerable points of the roof, such as chimney flashings and below valleys.

Below Vents   We occasionally find piles of snow in the attic under roof vents. This is common where soffit vents have been blocked, and some roof vents act as inlets while others act as outlets for air. Roof vents on windward sides may allow blowing rain or snow in through the vents.

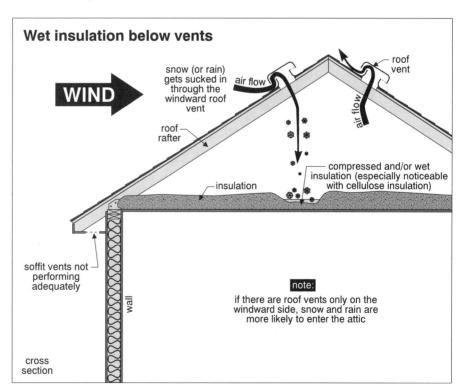

Wet insulation below vents

WIND

snow (or rain) gets sucked in through the windward roof vent

air flow

roof vent

air flow

roof rafter

insulation

compressed and/or wet insulation (especially noticeable with cellulose insulation)

soffit vents not performing adequately

wall

note:
if there are roof vents only on the windward side, snow and rain are more likely to enter the attic

cross section

8.3.6 INSULATION – COMPRESSED

Insulation works by trapping pockets of air. When insulation is compressed, it can no longer insulate.

Causes We've talked about wet insulation being compressed. Insulation can also be compressed by foot traffic in the attic or storage on top of the insulation. Insulation can also be compressed by poor installation practices, if the insulation is jammed and forced into enclosed areas. This is more common in walls than in attics.

Implications The implications of compressed insulation are reduced insulating values.

Strategy Look for areas where insulation has obviously been compressed. Point out the reduced insulating abilities, and if the area is significant, recommend improvements. Adding insulation over the compressed insulation may be the easiest approach.

8.3.7 INSULATION – GAPS OR VOIDS

Attic insulation is not perfectly uniform. It's not unusual to find areas where there is little or no insulation.

Causes Gaps or voids in the insulation may be an original installation issue. More commonly, it's the result of work being done in the attic. People working in other trades often remove insulation to facilitate their work. The insulation may not have been replaced when their work was finished.

Wind Although it's rare, it's possible that wind may move insulation around in an attic. This is most likely with a low-density blown insulation and large gable vents at either end of an attic.

Implication The implication of gaps and voids in insulation is increased localized heat loss.

Strategy Check the insulation for uniform depth across the attic

8.3.8 INSULATION – MISSING AT DROPPED CEILINGS

Some houses have ceilings that are dropped over bathtub enclosures or kitchen cabinets, for example. Split-level homes have two attic levels. Wherever the attic level changes, it's possible that insulation is omitted on the vertical surfaces separating the upper and lower attic levels.

Cause This is an installation issue in most cases. Insulation may have been put on vertical wall surfaces, but subsequently fell off.

Implications There may be considerable heat loss through the wall sections between the upper and lower attic levels.

Strategy As you move through the attic, you may see a recess in the insulation. If you can get over to that section, look down for insulation below. Make sure there is insulation on the floor and wall surfaces.

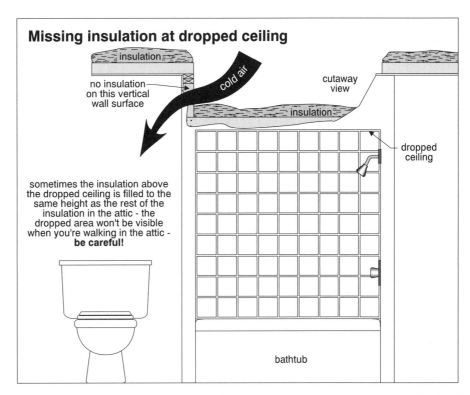

Missing insulation at dropped ceiling

Safety Issue   In some cases, the lowered ceilings for bathtub enclosures are completely filled with insulation. If the insulation depth over the majority of the attic is six inches, a twelve-inch lower attic floor above a bathtub enclosure may have eighteen inches of insulation. As you look across the attic, you won't notice the recessed ceilings. If you walk through the attic you may discover the recess in a dramatic and unpleasant way. This is another reason to be very careful when walking through attics where you cannot see the ceiling joists.

8.3.9 INSULATION COVERING RECESSED LIGHTS

Most people who are serious about energy efficiency do not like **recessed lights** (also called **pot lights**, or **high hat lights**). Light fixtures above the ceiling get hot, and may be covered with insulation. Some types are designed to be covered with insulation; these typically have a double shell with an air space between to help cool the fixture

Air Leakage   We've talked about trying to restrict air leakage into the attic. Recessed light fixtures can be the source of considerable air leakage. They are difficult to seal.

Causes   Recessed light fixtures inappropriately covered with insulation are an insulation installation issue.

Implications   This poses a fire hazard unless the fixture is designed to be covered with insulation.

Strategy   Where you see recessed light fixtures surrounded by insulation, look for a double shell on the fixture. You may be able to see a designation that includes the letters **IC** indicating appropriate for use in an **insulated ceiling**.

Boxed-In
Fixtures

Some recessed light fixtures have plywood or drywall boxes built around them to keep the insulation away. Insulation was often applied over the box to maintain continuity of the attic insulation while separating the light fixture from the insulation. This arrangement is not common but is acceptable.

8.3.10 INSULATION INADEQUATE IN KNEE WALL AREAS

Knee wall areas form triangles. There are two insulation strategies that can be used.

Floor And
Wall

The most common approach is to insulate the floor and the wall of the attic. If loose-fill or batt insulation is used on the attic floor, solid blocking should be provided at the point where the floor meets the wall. This blocking could be rigid polystyrene insulation or wood, for example. It should be caulked to prevent air leakage from the living space ceiling into the attic. In many cases, this detail is not well done.

The wall should be insulated, very often with conventional batts. The air/vapor barrier should be placed on the warm side of the wall, as usual.

Insulating
Sloped Roofs
And End Walls

The less common approach to insulating knee walls is to add insulation to the underside roof sheathing between the rafters, and to insulate the end walls. This arrangement is less desirable because it does not allow for venting of the roof space and it involves heating a knee wall area that is not living space. In some cases, the knee wall area is used as storage, which may make it desirable to heat and insulate the knee wall areas.

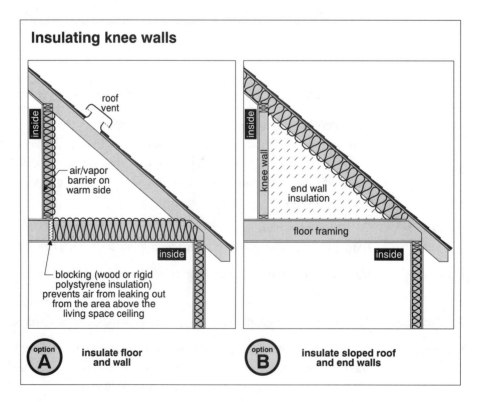

Insulating knee walls

roof vent

inside

air/vapor barrier on warm side

blocking (wood or rigid polystyrene insulation) prevents air from leaking out from the area above the living space ceiling

inside

option A — **insulate floor and wall**

inside

knee wall

end wall insulation

floor framing

inside

option B — **insulate sloped roof and end walls**

Causes   Inadequate knee wall insulation is an installation issue. In many cases, people do not understand the insulation principles. We commonly find people have insulated the floor and the sloped ceiling, or the wall and the sloped ceiling, only. The goal is to create a continuous separation between indoor and outdoor environments.

Implication   The implication of inadequate knee wall insulation is increased heating cost.

Strategy   Check for insulation on the wall separating the knee wall from the living space, the floor, the rafters and the end walls. Determine which approach has been followed. If possible, look at the floor/wall intersection for an air barrier between the ceiling joists where the attic joins the living space.

Air/Vapor Barriers And Ventilation   Look for continuous air/vapor barriers on the warm side of the insulation, no matter which approach is taken. Look also for ventilation of knee wall areas. In many cases it will be less than optimum. Look for evidence of mold, mildew or rot as a result of restricted air/vapor barrier or venting problems.

8.3.11 INSULATION INADEQUATE AT SKYLIGHTS AND LIGHT WELLS

Skylights on roofs typically have a light well that extends through the attic. This light well should be insulated to prevent heat loss into the attic. The air/vapor barrier should be on the warm side, which is the side closest to the interior finish. The light wells can be thought of as conventional walls, exposed to the attic. These are similar to knee walls.

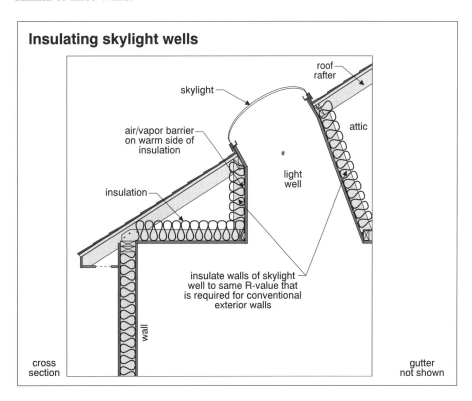

Insulating skylight wells

roof rafter

skylight

attic

air/vapor barrier on warm side of insulation

light well

insulation

insulate walls of skylight well to same R-value that is required for conventional exterior walls

wall

cross section

gutter not shown

173

Causes   Many light wells are not properly insulated. This is most common where the skylight has been added to the home.

Implications   Considerable heat loss can take place through the skylight well. Condensation is another possible implication. Skylights are often in kitchens and bathrooms, both high humidity areas. Warm, moist air encountering the cold drywall surface of a light well is likely to condense. Moisture condensing and running down the skylight well surface inside the home is often mistaken for skylight leakage. You can identify the water as condensation by –

• a lack of insulation;
• a uniform water accumulation around the perimeter of the skylight (not typical of a leak);
• the problem occurring only in cold weather; or
• the problem not appearing during or after rains.

Strategy   When you're in the attic, look for insulation and air/vapor barriers on skylight wells. The insulation level would ideally be the same as any wall (R-12 to R-20 for example).

8.3.12 DUCTS LEAKING OR DISCONNECTED, OR INSULATION MISSING OR LOOSE

Heating and air conditioning ductwork in attics should be insulated.

Causes   Missing insulation is usually an installation problem. Occasionally the insulation has been damaged and inappropriately removed, rather than repaired. Loose insulation may be the result of poor installation or damage from people or animals.

Leaky or disconnected ducts may be the result of poor installation or mechanical damage.

Implications   Implications include a loss of heat and condensation damage inside the ductwork. Warm, moist air in the home is likely to condense inside the attic or the cold ductwork in winter. This can result in water leakage into the house through the ductwork, as well as rusting the ducts themselves. Condensation in poorly insulated ducts is often mistaken for roof leakage.

Condensation may also form in the attic itself. Leaking ducts add to heating and cooling costs, and reduce comfort.

In summer, condensation may occur on the outside of ducts. Again, this can look like leakage.

Strategy   Look for complete and intact insulation on heating and cooling ductwork in attics or other unconditioned spaces. In heating climates, the air/vapor barrier should be on the warm side. The ductwork acts as an air/vapor barrier in most cases. In cooling climates, the air/vapor barrier should be on the warm side. This is the outside of the insulation. The vapor barrier will be covering the insulation as you look at it from the attic in a cooling climate.

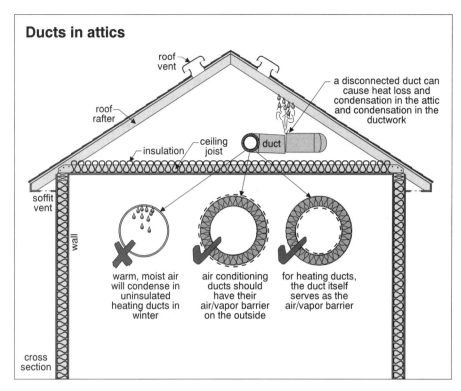

Ducts in attics

roof vent

roof rafter

a disconnected duct can cause heat loss and condensation in the attic and condensation in the ductwork

ceiling joist

insulation

duct

soffit vent

wall

warm, moist air will condense in uninsulated heating ducts in winter

air conditioning ducts should have their air/vapor barrier on the outside

for heating ducts, the duct itself serves as the air/vapor barrier

cross section

Leaky Or Disconnected Ducts   Watch for ducts that are disconnected or leaking.

8.3.13 DUCT AIR/VAPOR BARRIER MISSING OR DAMAGED

The air/vapor barrier on a duct in the heating system is typically the duct itself. The insulation on the outside of the ducts may be exposed. In cooling climates, the air/vapor barrier should be on the outside, or warm side, of the insulation. When inspecting the attic, you should only see the air/vapor barrier covering the insulation. Ductwork is often sold with integral insulation and an external air/vapor barrier.

Air/vapor barriers must be intact to do their job.

Causes   A missing air/vapor barrier is an installation issue.

Damaged air/vapor barriers are typically the result of –

• Careless installation,
• Animal or bird activity in the attic, or
• Careless work by people in the attic.

Implications   Condensation within the insulation or on the outside of the duct is the implication. In a cooling climate, the warm, moist air in the attic coming in contact with the cool sheet metal ductwork will form condensation on the outside of the duct. This can rust the duct and damage interior finishes below. Again, this can be mistaken for a roof leak.

Strategy   In a heating climate, the duct is the air/vapor barrier. In a cooling climate, look for an air/vapor barrier on the outside of the insulation.

Where an external air/vapor barrier is present, check that it is intact. Recommend that damaged areas be repaired.

8.3.14 INSULATION TOO CLOSE TO CHIMNEYS

Masonry chimneys should have non-combustible insulation surrounding them. Metal chimneys or vents should have no insulation around them for a distance of one to two inches, depending on the type of vent or chimney. Generally speaking, B-vents and L-vents for gas and oil appliances require a one-inch clearance. Metal chimneys for wood stoves and fireplaces typically require a two-inch clearance. There are some exceptions to these guidelines. A boxed-in area is usually constructed to keep insulation away.

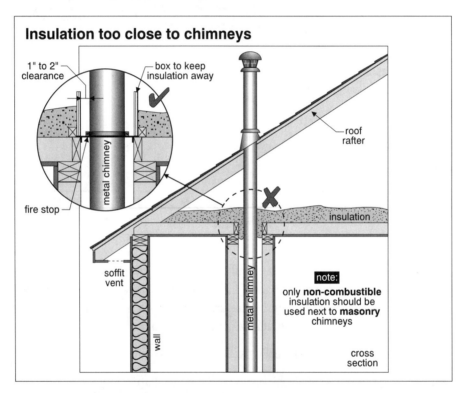

Causes   Insulation too close to chimneys is an installation issue.

Implications   Combustible insulation adjacent to a masonry chimney is a fire hazard. Insulation packed around a metal chimney can lead to overheating and deterioration of the chimney, even if the insulation is non-combustible. This is also a fire hazard.

Strategy   Look at masonry and metal chimneys to verify that the insulation is kept away from metal chimneys and that insulation against masonry chimneys is non-combustible.

Check also that an air barrier is provided around the chimney. In many cases, we can look down around metal chimneys through one or even two floor levels. This is not only an air barrier problem but also a fire-stopping problem. There should be a fire stop at each floor level to prevent the quick spread of fire up around the outside of a chimney. Chimney chases that are open over the entire height of the house can lead to very fast spread of fires.

8.3.15 AIR/VAPOR BARRIER MISSING, INCOMPLETE OR WRONG LOCATION

There should be an air/vapor barrier on the attic floor. Most attics have air/vapor barriers although they are typically less than perfect.

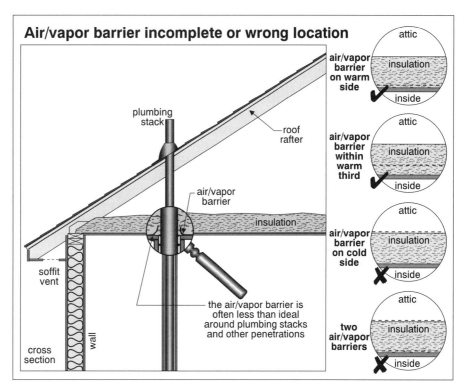

Causes   Air/vapor barriers may be missing because of original construction, or because original insulation batts with kraft paper have been removed. Interior finishes can form an air/vapor barrier in theory, although in practice very few homes rely on painted drywall or plaster and caulking at penetrations to form the air/vapor barrier.

Air/vapor barriers should be on the warm side or within the warm third of the insulation. Location problems are the result of poor installation.

Implications   Increased air leakage and possible rot damage from condensation are the implications of a missing air/vapor barrier. An air/vapor barrier on the cold side of the insulation may cause condensation in the insulation.

Strategy   Lift up the insulation to look for an air/vapor barrier. It is usually kraft paper (older homes) or polyethylene film (1970s and newer homes).

177

Look For
Mold, Mildew
And Rot

Check the air/vapor barrier around roof penetrations such as plumbing stacks. Can air get up into the attic between the air/vapor barrier and the penetration? In most cases the air/vapor barrier is less than perfect. Recognizing this, it's more important to look for evidence of problems (mold, mildew and rot) than to criticize a less than perfect installation. Rot, mold and mildew are more likely if attic ventilation is poor. Very often, several systems in a house can be slightly less than perfect and we won't have any problems; however, if there are significant deficiencies in a number of systems and house humidity levels are high, rot, mold and mildew are likely.

Double vapor barriers or a vapor barrier on the cold side of the insulation are red flags. Look for evidence of condensation damage in and below the insulation. Recommend the cold-side vapor barrier be removed.

8.3.16 AIR LEAKAGE EXCESSIVE

This is another way to say that air/vapor barriers are incomplete; however, it goes further than this. Poor caulking and weather stripping around openings can create air leakage. Common points where air leaks in significant quantities into attic include –

• Access hatches
• Plumbing stacks
• Light fixtures (especially recessed fixtures)
• Chimneys and vents
• Ducts
• Partitions
• Wall/ceiling intersections at the building perimeter
• Exhaust fans and their ducts

S9

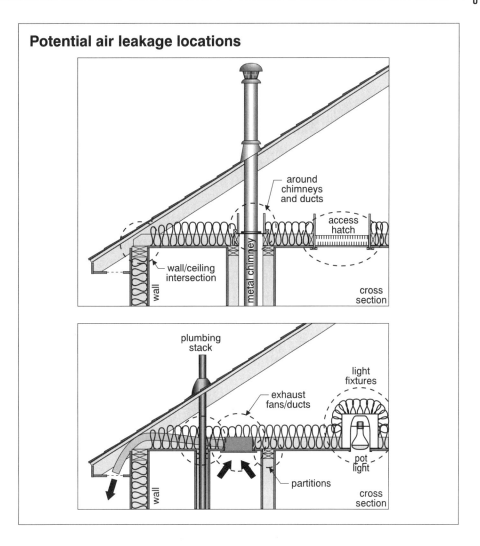

Potential air leakage locations

Causes These are installation issues. Good caulking and weather-stripping of these penetrations is rare.

Implications Again, mold, mildew and rot are the risks, along with increased heat loss.

Strategy It's very difficult to do a comprehensive air leakage inspection in the attic as part of a conventional home inspection. We recommend you spot-check to get an overall sense of the detail here. More importantly, look for evidence of the rot, mold and mildew that indicates damage being caused by the excessive air leakage.

8.3.17 ATTIC VENTING MISSING OR INADEQUATE

Attic venting includes –

• Soffit and roof vents
• Soffit and ridge vents
• Soffit and gable vents
• Opposing gable vents.

1:300 Vent Ratio

The recommended venting area is often 1/300 of the floor area of the attic. This is a free-vent area and the actual size of the vents has to be increased because all vents have louvers, screens or both. Many vents have a net area stamped into them so you can get an idea of how much venting is present.

Causes

Missing or inadequate venting is usually an installation issue. Some houses are designed with open rafters. There are no soffits and no conventional venting. In these cases, unintentional ventilation is usually relied upon, with varying degrees of success.

Implications

The implications of missing or inadequate venting are condensation and subsequent mold, mildew and rot damage.

Low And High Level

We've talked about 50 percent of the venting being at the soffits and the other fifty percent high on the roof. If the soffit venting is less than 50 percent, we run the risk of some of the high level vents being inlet vents, and of rain and snow getting into the attic area. We also move less air if the soffit vents are missing or obstructed, and the net ventilation effect is greatly reduced.

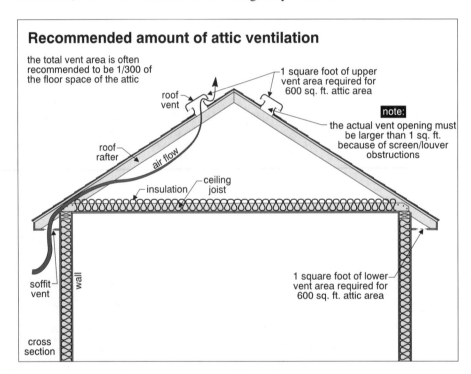

Recommended amount of attic ventilation

the total vent area is often recommended to be 1/300 of the floor space of the attic

1 square foot of upper vent area required for 600 sq. ft. attic area

roof vent

note:

the actual vent opening must be larger than 1 sq. ft. because of screen/louver obstructions

roof rafter

air flow

insulation

ceiling joist

soffit vent

wall

1 square foot of lower vent area required for 600 sq. ft. attic area

cross section

Strategy

Look for soffit vents and high level vents, or gable vents at opposing ends of the attic. Is there adequate total ventilation? Check that roughly 50 percent of the ventilation is at the soffits. When on the roof you may be able to read the net vent area off roof or ridge vents.

8.3.18 VENTING OBSTRUCTED AT SOFFITS OR ROOF VENTS

This is one of the most common problems we find in attics.

Causes   Soffit vents may be obstructed by –

- insulation filling the space between the ceiling and underside of the roof sheathing,
- a retrofit soffit treatment added over existing soffits with no provision for venting, or
- vents with fine louvers or screens becoming clogged.

Roof vents may be obstructed by –

- birds' nests;
- mechanical damage from foot traffic on the roof, for example;
- missing or undersized openings in roof sheathing; or
- failure to remove building paper beneath ridge vents.

If the hole or slots cut in the roof sheathing aren't big enough, this seriously impairs the vent performance.

Implications   The implications of inadequate venting are mold, mildew and rot as a result of condensation, and rain and snow being drawn in through high level roof vents, creating water damage.

Ice Dams   Poor venting may lead to ice damming, since the attic will be warmer than it should be. Good venting maintains the attic temperature close to the outdoor temperature. This minimizes melting of roof snow and helps prevent ice dams.

Strategy   Check that soffit vents are open. Where insulation is deep near the eaves, look for expanded polystyrene, cardboard or wood baffles that maintain an air space between the insulation and the underside of the roof sheathing. As you look around the perimeter of the roof, you can often see daylight filtering in through the soffit vents.

You can check roof vents when outside the house and from inside the attic. It's easiest to see mechanical damage and birds' nests from the outside. It's easiest to see undersized sheathing holes from the inside.

8.3.19 SNOW OR WET SPOTS BELOW ROOF VENTS

We've touched on this already.

Causes   When soffit ventilation is inadequate, or roof vents are only on the windward side of the roof, the roof vents may be sources of air being drawn into the attic rather than escaping. When winds are accompanied by snow or rain, moisture can get into the attic through the roof vents.

Implications

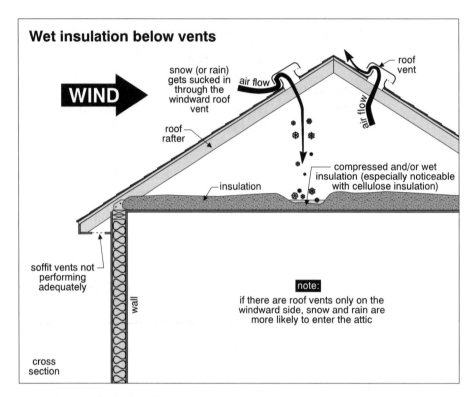

Water damage is the implication of snow and rain getting into the attic.

Strategy   Look for evidence of wet spots, including discoloration and compressed insulation below roof vents.

8.3.20 TURBINE VENTS NOISY OR SEIZED

Turbine vents are wind-driven roof vents designed to enhance roof ventilation. We do not recommend the use of turbine vents. On calm days they do not help ventilation of the roof to any great extent. On windy days they may lead to excessive ventilation and depressurization of the attic. As we've discussed, excessive depressurization of the attic increases the air leakage from the house into the attic. This increases heating costs and can promote condensation.

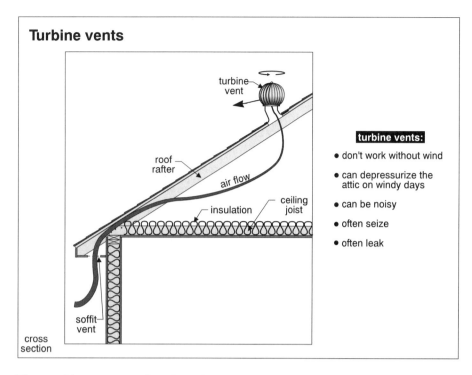

Turbine vents

turbine vent

roof rafter

air flow

insulation

ceiling joist

soffit vent

cross section

turbine vents:
- don't work without wind
- can depressurize the attic on windy days
- can be noisy
- often seize
- often leak

Low Quality   These turbine vents usually rely on low quality mechanicals to provide the movement. Because they are exposed to a hostile environment, and they are substantially inaccessible for servicing by many homeowners, we find their performance is often poor. Noisy turbine vents are a common problem.

Sometimes the turbine vent cannot be made to move. This is not a matter of great concern unless leakage is experienced.

Causes   Noisy turbine vents are caused by poor alignment and performance of adjacent moving parts.

Turbine vents may be seized because –

- they are obstructed,
- they are bent, or
- they are corroded.

Implications   The noise is a nuisance, and usually suggests that the system will seize shortly. There are usually no serious performance implications to a seized turbine vent. In some cases, the vent is vulnerable to rain or snow penetration. We see many turbine vents that are covered with plastic garbage bags because of leakage into the attic through the vents.

Strategy   On a calm day, the vents probably won't be turning, although on a hot summer day there may be enough convective airflow out of the attic to move the vents. If you move the vent with your hand, you can sometimes detect a noise problem. In some cases, lubrication can solve the problem, at least temporarily. If the vent is seized or has been bent, damaged or is corroded, replacement may be necessary. We recommend replacement with conventional roof vents.

8.3.21 POWER VENTS OPERATING IN WINTER

Electric exhaust fans mounted on the roof are useful during the summer to help remove heat from the attic. In the winter, they can depressurize the house and actually increase heat loss. We recommend these not be used during the winter.

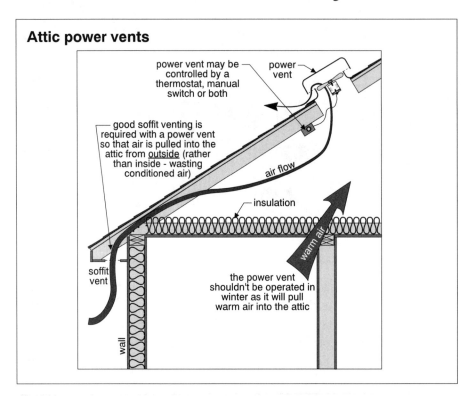

Attic power vents

power vent may be controlled by a thermostat, manual switch or both

power vent

good soffit venting is required with a power vent so that air is pulled into the attic from <u>outside</u> (rather than inside - wasting conditioned air)

air flow

insulation

warm air

soffit vent

the power vent shouldn't be operated in winter as it will pull warm air into the attic

wall

Causes	A power vent operating in the winter is an original arrangement or homeowner education issue.
Implications	Increased heat loss and air leakage into the attic are implications of a power vent operating in the winter.
Strategy	Where you see a power vent in the attic, determine whether it's operating. Trace the wiring back to the controls to see how the system is activated (manual switch or thermostat, for example). You may want to explain to the client the reasons for not operating the unit in the winter.
Good Soffit Venting Needed	When the power vent is operating in the summer, good soffit venting is necessary because we want to pull outdoor air into the attic and exhaust the super-heated attic air. We don't want to pull the cool air out of the house into the attic. If soffit vents are missing or obstructed, the negative pressure in the attic will naturally move more air out of the house. If the house is air conditioned, we are wasting energy by pulling cool air up out of the house and into the attic.

8.3.22 POWER VENTS – POOR WIRING

The electrical wiring to a power vent should follow all of the good practices we talked about in the Electrical Module. We are looking for wiring that is well secured, free of mechanical damage, and appropriately sized and connected.

Causes   Poor wiring is usually an installation issue.

Implications This can be an electrical shock or fire hazard.

Strategy   Follow the wiring to the power vent. Make note of undersized, damaged or poorly supported wire, or poor connections. Look for evidence of corrosion at junction boxes or on the vent itself.

8.3.23 POWER VENTS INOPERATIVE IN SUMMER

In some cases, the power vent fails to respond to its controls.

Causes   The problem can be mechanical or electrical.

Implications The attic is likely to be warmer than it would be with the vent operating. This is not a serious problem.

Strategy   When you see a power vent in an attic, follow the wiring back to the controls and ensure the vent responds to these. If it does not, recommend the problem be investigated and corrected.

8.3.24 RAFTERS AND SHEATHING – ROT, MOLD OR MILDEW

Mold, mildew and rot are common problems inside and outside houses. Mold and mildew are usually surface conditions while rot is caused by fungi capable of decaying wood by breaking down the wood fibers.

All Are Fungi   Mold, mildew and rot are all fungi. There are also staining fungi. Fungi grow from spores or seeds released by mushrooms. These spores are present in the air almost everywhere. When spores settle on surfaces that favor growth, they will establish small filaments called hyphae. These hyphae release enzymes that break down the organic matter to become food.

Favorable Conditions for Fungi Growth

1. Oxygen

There has to be an oxygen supply. These fungi cannot be established under water. Submerged wood does not rot.

2. The Right Temperature

Temperature ranges from 40° to 100°F are ideal. Some fungi can tolerate temperatures up to 115°F. While higher temperatures can kill the fungi, lower temperatures simply make it go dormant, but do not kill it.

3. Moisture

Mold and mildew need moisture contents in wood of roughly 20 percent. Rot fungi need about 28 percent moisture content in wood to get a foothold.

How Wood Gets Wet

Let's review how the moisture content in wood changes. When wood is absolutely dry, it has zero percent moisture content. When all of the fibers in wood are fully saturated, its moisture content is about 30 percent. Wood can be wetted by water on its surface, of course. Wood will draw moisture into the fibers more quickly through the end grain than across the grain. Wood posts and deck boards typically rot first near end grains because that's where the moisture is drawn in.

Wood Draws Moisture From Air

This point is subtle and a little hard to understand. Wood will draw moisture from air independent of the air temperature and independent of its absolute humidity. Wood responds to the relative humidity in the air. This seems hard to believe, but that's the way it works. If we have air at 70°F and 40 percent relative humidity, the wood only pays attention to the 40 percent relative humidity. This bundle of air has much more moisture in it than a bundle of air at 0°F and 80 percent relative humidity. Despite that, the wood will extract moisture from the air at 80 percent relative humidity, even though there is less moisture in the air.

Equilibrium Moisture Content

This alone isn't enough to cause a problem. When the relative humidity is 40 percent, the equilibrium moisture content in wood goes up to less than 8 percent. As the relative humidity rises to 60 percent, the equilibrium moisture content goes up to about 11 percent. When the relative humidity reaches 80 percent, the equilibrium moisture content is only about 16 percent. Even at 90 percent relative humidity, the moisture content will approach a level just over 20 percent.

Where's The Problem?

It's unusual to have relative humidities in excess of 90 percent for extended periods, except in some very hot, humid climates, such as Florida. However, the relative humidity is an issue when we look at condensation as well. We've said that wood will take moisture from liquid water in contact with the wood. Air at 70°F and 50 percent relative humidity will form condensation when the temperature drops to about 50°F. At this temperature, the relative humidity is **100 percent** and the air has reached its dew point.

4. Food

The last criterion for growing fungi is **food**. The wood provides this.

If any one of these conditions is not met, fungi will not grow. Unfortunately, all of these conditions are typically present in homes, with the exception of the moisture content. When we elevate the moisture content of the wood in the home to 20 percent or more, we risk mold, mildew and rot. This is why interstitial condensation in buildings is such a big concern. This is the condensation in walls and roof spaces that we don't see. Condensation on interior surfaces is easy to see and most people won't tolerate mold and mildew growth on walls and ceilings for very long before reacting. It's the concealed mold, mildew and rot that is the danger.

Mildew   Mildew is most often black, although there are also green, red, blue and brown mildews. Although mildews can form on almost any surface and will discolor it, mildew does not typically damage the wood. Mildew can cause allergic reactions in people. It can be readily cleaned off with bleach.

Mold   Molds are typically green, but can also be black and orange. Again, they don't damage wood, accumulating primarily on the surface. Cleaning with bleach is effective for molds as well.

Rot   Rot is more serious than either mold or mildew. The rot actually attacks the wood fibers. Brown rot and white rot are the most common. Brown rot is most common on softwoods. White rot is most common in hardwoods. As a result, **brown rot is the enemy of most framing members**. We indicated that a 28 percent moisture content in wood is necessary to get a foothold. This is almost the saturation point of wood (30 percent). However, once rot is established, it will continue to flourish with moisture contents as low as 20 percent.

Identifying Rot   In the early stages, the wood surface becomes dull and discolored. There's often a musty odor. As the rot advances, the wood becomes spongy and crumbly. It can also be stringy. Brown rot can grow to a considerable size on the surface of wood. The large growths of the fibrous hyphae and fruiting bodies may be apparent. Wood loses its strength.

Brown Rot Can Look White   White fluffy growths on the surface of wood may indicate brown rot.

Dry Rot – Isn't   Many people refer to dry rot. This is confusing, since dry wood can't rot. What you are usually seeing is wood that has been wet and rotted and subsequently dried out. It's probably best to avoid the term dry rot.

Other Rots   Brown and white rots are not the only two types. Soft rots, for example, can attack cedar roofing. However, brown rot is the most common enemy of house structures.

Control Of Mold, Mildew and Rot   Simply put, if we keep the moisture content of wood below 20 percent, the problem goes away. This is by far the most effective solution and why it's so important that we control moisture leaking into walls and attics in homes.

Causes   We've talked extensively about condensation and how it can cause rot. Sometimes there is a fungal growth on the sheathing or rafters, but no evidence of rot. This will generally occur over a large area unless the source of warm, moist air is an exhaust fan or clothes dryer, for example. Localized damage is possible near the source of moisture. Rot may also result from chronic minor roof leakage, common around chimneys because of poor flashing details, for example.

Implications   Mold and mildew are often accompanied by rot. The implications of rot are weakened structural performance and ultimate failure of the roof structure.

Strategy *Rot*   Look for rot on structural members throughout. If you can get to the structural members, probing with a carpenter's awl or screwdriver is helpful in determining the presence and extent of rot. A strong light is essential to a proper attic inspection.

Mold And Mildew

Look for a discoloration on structural members on the attic. Mold and mildew are usually black or green, and, if you can touch them, form a slightly raised surface on the wood that can usually be wiped off.

FRT Plywood Darkening

Mold and mildew are not the same as the premature darkening or charring that occurs with some fire-retardant treated (FRT) plywood. Darkened FRT plywood cannot be wiped clean; mold and mildew can usually be wiped off.

Inspection Limitations

Unless you can move through the entire attic area, there will be some sections you don't see. This means there may be some rot you will miss. Make sure you document these limitations.

8.3.25 RAFTERS AND SHEATHING DELAMINATING OR BUCKLING

One of the implications of high moisture levels, from either condensation or leakage, is delamination of plywood sheathing. This is sometimes accompanied by buckling. Panel-type roof sheathing may lift off the rafters in a regular pattern.

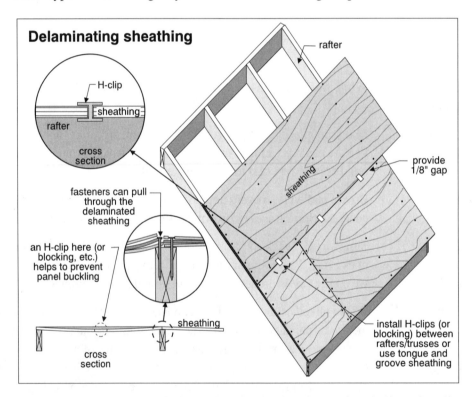

Causes

High moisture levels break down the glue in the plywood and allow the plies to delaminate.

Buckling is usually the result of elevated moisture levels in the sheathing, combined with inadequate room for expansion in adjacent panels. H-clips are commonly used with panel-type roof sheathing to allow for some expansion and contraction of adjacent panels. Where these are not used, buckling is more likely.

Implications   Delaminated plywood is weakened and may not be able to carry its intended loads. Ultimately the plywood may give way under the weight of foot traffic, for example.

Buckling can weaken the roof by separating the sheathing from the rafters or trusses. If fasteners pull through the sheathing, the entire roof covering and sheathing becomes loose and may be blown off during high winds.

Strategy   Check the plywood sheathing, particularly at the panel edges, for evidence of delamination. In many cases, you'll need to make a judgment call as to how severe the delamination is. Where you are unsure, recommend a specialist for further investigation.

Check that the sheathing sits tightly on the top of rafters or trusses over its entire length. Where the sheathing has clearly buckled, make recommendations to reduce moisture levels in the attic and possibly refasten the sheathing.

8.3.26 WHOLE-HOUSE FAN WITH NO INSULATED COVER

Whole-house fans are helpful in the summer to make the home cooler and more comfortable. In the winter, whole-house fans that are not in use can be a source of considerable heat loss and air leakage into the attic.

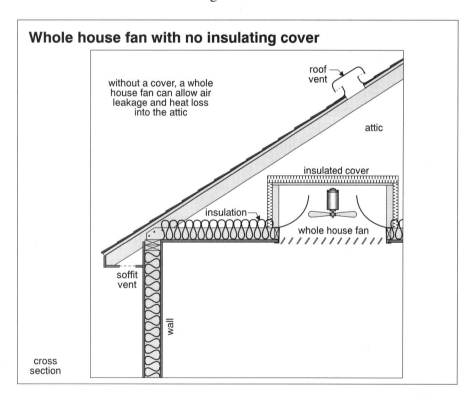

Whole house fan with no insulating cover

without a cover, a whole house fan can allow air leakage and heat loss into the attic

roof vent

attic

insulated cover

insulation

whole house fan

soffit vent

wall

cross section

Causes   The nature of the fan is a loose-fitting hole in the ceiling. Insulation can't be poured over the fan because it wouldn't work in the summer.

Implications   The heating costs will be higher than they should be, and we may have condensation and resultant problems in the attic because of the considerable air leakage.

Strategy   Where there is a whole-house fan, look for an insulated cover that can be placed over the fan in the attic during the winter months. Where none is found, recommend that one be added. A tightly constructed plywood box with caulked joints and insulation on the outside is a common approach.

Summary

You should be getting the impression that there's a lot to do in an attic. It's not our favorite part of the house, but it is important. You may want to do a little research to find out how other local inspectors approach attics. You don't want to unknowingly create an inspection strategy for attics that's dramatically different from others.

Don't Crawl   We do not crawl through attics unless we can see the tops of the ceiling joists or
Through   truss chords that we are walking on, or there is a walkway. We believe the risk of
Insulation   damaging the home and injuring ourselves is high enough to avoid this approach. Not all home inspectors will agree with this.

Insulation
& Interiors
M O D U L E

QUICK QUIZ 9

☑ INSTRUCTIONS

• You should finish Study Session 9 before doing this Quiz.

• Write your answers in the spaces provided.

• Check your answers against ours at the end of this Section.

• If you have trouble with the Quiz, re-read the Study Session and try the Quiz again.

• If you did well, it's time for Study Session 10.

1. What should you wear when inspecting attics? (list three things)

    _____

    _____

    _____

2. List at least three things that may make attics dangerous.

    _____

    _____

    _____

3. What safety issues are there surrounding permanent staircases to attics?

    _____

    _____

    _____

4. List at least two safety concerns related to pull-down stairs.

_____

_____

_____

5. List six other house systems that may have components to inspect while in the attic.

_____

_____

_____

_____

_____

_____

6. List four common access hatch problems and the implications of each.

_____

_____

_____

_____

7. List five attic staircase problems and the implications of each.

_____

_____

_____

_____

8. List three common pull-down stair problems and the implications of each.

_____

_____

_____

9. What are the implications of insulation that is only two inches thick, wet, compressed, interrupted by gaps or voids, or missing?

    _____

    _____

    _____

    _____

10. Under what circumstances should insulation cover recessed lights? What is one inspection strategy to help identify recessed lights while in the attic?

    _____

    _____

    _____

11. Will you ever see walls in attics? Where? Should these be insulated?

    _____

    _____

    _____

12. What two things should you be looking for on ducts in attics?

    _____

    _____

13. How should insulation be handled around masonry chimneys?

    _____

    _____

14. How should insulation be handled around metal vents?

    _____

    _____

15. What are the implications of air/vapor barriers that are missing or incomplete?

_____

_____

_____

16. List three common problems with roof vents and their implications.

_____

_____

_____

17. What is suggested by snow on attic insulation below roof vents?

_____

_____

18. List three possible concerns with turbine vents.

_____

_____

_____

19. Should power vents be operating in the winter? Why or why not?

_____

_____

20. What is suggested by mold, mildew or rot on rafters and sheathing?

_____

_____

21. How is heat loss and air leakage controlled around whole-house fans?

_____

_____

If you had no trouble with this Quiz, you are ready for Study Session 10.

Key Words:

- *Mask*
- *Goggles*
- *Gloves*
- *Access hatch*
- *Staircase*
- *Pull-down stairs*
- *Insulation*
- *Dropped ceilings*
- *Recessed lights*
- *Knee walls*
- *Skylights*
- *Duct insulation*
- *Air/vapor barrier*
- *Chimneys*
- *Soffit vents*
- *Roof vents*
- *Ridge vents*
- *Gable vents*
- *Turbine vents*
- *Power vents*
- *Rot, mold and mildew*
- *Delaminated or buckled sheathing*
- *Whole house fans*

Insulation
& Interiors
MODULE

STUDY SESSION 10

1. You should have completed Study Sessions 1 to 9 before starting this Session.

2. This Session covers the inspection of flat and cathedral roofs.

3. At the end of this Session, you should be able to—
- list four general approaches to insulating flat and cathedral roofs.
- list seven common problems and their implications.
- describe in one sentence the inspection strategy for each of these problems.

4. This Study Session may take you roughly 45 minutes.

5. Quick Quiz 10 is at the end of this session.

Key Words:
- *Attic*
- *Rot*
- *Mold*
- *Mildew*
- *Insulation*
- *Air/vapor barrier*
- *Ventilation*
- *Cross ventilation*

► 9.0 FLAT AND CATHEDRAL ROOF INSPECTIONS

Roofs with no attic area are hard to inspect and are difficult to build correctly. There are several approaches to insulating flat and cathedral roofs. Let's look at some of them.

Treat As An Attic

1. Treat as an attic. One of the most common approaches to insulating flat and cathedral roofs is simply to treat them like an attic. A little insulation is laid on the ceiling and an air space is left between the top of the insulation and the underside of the sheathing. On older homes, minimal attention was paid to ventilation in houses with attics. Accidental ventilation was often enjoyed because of the loose construction techniques. The use of roof planks, for example, leads to more natural ventilation than the use of plywood or waferboard sheathing.

Flat Roofs Don't Breathe

Cathedral roofs have a slight advantage over flat roofs in that the roof covering on a cathedral roof is usually a shedding system. Some air leakage through asphalt shingles, for example, is common. Flat roofs form a tight membrane. There is typically no air leakage through a flat roof membrane, and as a result less accidental ventilation.

Completely Fill Roof Space

2. Some people treat flat and cathedral roofs by filling them with a fairly dense blown-in insulation, such as cellulose fiber. If the cavity can be filled and air leakage stopped in this way, the amount of insulation can be maximized and air leakage into the roof space will be prevented. Sprayed in-place insulations are very effective air barriers and may be the best for this situation; however, conventional sprayed in-place insulations cannot be sprayed into enclosed cavities. This makes these impractical as retrofit insulations.

The elimination of ventilation carries some risk: If air leakage into the roof space takes place, the chances of moisture damage are significant.

Insulation Above Sheathing

3. Some take the approach that is commonly used on commercial buildings. Insulation can be added above the roof sheathing. There may be little or no insulation between rafters, roof joists or trusses. On commercial and industrial buildings, the roof membrane typically goes on top of the insulation, which sets on top of the roof deck.

Insulation Above Roof Membrane

An alternative is an exposed insulation approach, where the insulation is laid on top of the roof membrane and is held in place with ballast. Some people call this an inverted roof membrane assembly (IRMA) or an upside down roof.

Not Common

These approaches are not common in residences. It has the disadvantage of considerable air movement in the joist or rafter spaces below the roof deck. If insulation is not provided at the ends of these cavities, there can be considerable heat loss through the ends of the roof, despite a good amount of insulation on the top of the roof. Convective heat loss established in this situation can be dramatic, especially in a cathedral roof where natural convective loops can be very strong.

*Insulate Below
Roof Structure*

4. Insulation can also be added on the underside of the roof rafters, ceiling joists o
trusses. This lowers ceiling heights, but does allow the space between the framing members to be well ventilated. This does not mean that good ventilation is
always provided.

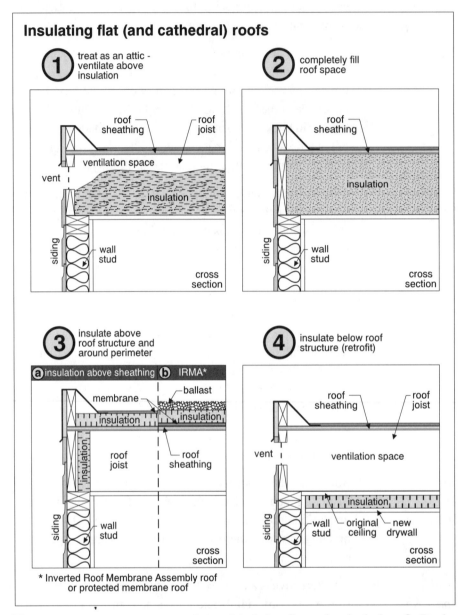

Insulating flat (and cathedral) roofs

① treat as an attic -
ventilate above
insulation

② completely fill
roof space

③ insulate above
roof structure and
around perimeter

④ insulate below roof
structure (retrofit)

* Inverted Roof Membrane Assembly roof
or protected membrane roof

Hard To See

These are some common approaches to insulating flat and cathedral roofs. During
a home inspection you may not know which approach, if any, has been taken. You
should probably accept the fact you are going to have a tough time identifying the
insulation approach, material, total R-value, presence and effectiveness of an
air/vapor barrier, and the effectiveness of ventilation. As we move through the
common problems, we'll discuss some of the inspection strategies, but they are
limited, especially when compared to the inspection strategies for attics.

Watch For Water Damage Rot is the big risk to a home with a flat or cathedral roof. We've discussed the mechanism by which warm, moist air gets carried into roof spaces, cools, condenses and allows mold, mildew and rot to get a foothold. Flat and cathedral roofs are vulnerable to this, particularly because visual inspection is usually impossible. Extensive damage can occur before it's apparent in the home. We'll look at some of the conditions that may make a roof susceptible to damage, and some of the clues that suggest damage may have occurred.

It's A Short Inspection The good news is that it doesn't take nearly as long to inspect flat or cathedral roofs as it does attics. There just isn't very much we can see.

Now let's look at some common problems.

9.1 CONDITIONS

1. Insulation – too little
2. Insulation – wet, compressed or voids
3. Air/vapor barrier – missing or incomplete
4. Air leakage – excessive
5. Venting – missing or inadequate
6. Venting – obstructed
7. Mold, mildew or rot suspected

9.1.1 INSULATION – TOO LITTLE

Since there's no access hatch and usually no way to get a look at the insulation, we won't know whether there is insulation at all, let alone whether the amount is appropriate or not. Incidentally, many authorities specify lower insulation levels for flat or cathedral roofs than they do for attics. This isn't because there is less heat loss; it's because there is usually less depth available to accommodate insulation and ventilation. From an energy efficiency standpoint, we'd like to have the same amount of insulation in a flat or cathedral roof as we do in a conventional attic.

Causes Low insulation in flat or cathedral roofs is usually an installation issue. The older the house, the less insulation there is likely to be.

Implications Low insulation levels result in higher heating costs, but no damage to the structure.

Strategy Unless there are holes or gaps in ceilings, there isn't usually an opportunity to look at insulation in flat or cathedral roof cavities. Removing ceiling-mounted light fixtures or exhaust fans can allow a look around electrical junction boxes for insulation. Most home inspectors do not go this far. If you chose to do this kind of investigation, make sure there is no electrical power to the fixtures you are working around

Melting Snow   One of the crude methods that inspectors in cold climates can use during some parts of the year is an observation of the amount of snow on the roof. The less insulation there is in a roof, the warmer it will be, and the more quickly snow will melt. When snow coverings are light, and there are many similar houses on a street, it is possible to get a sense of which roofs are well insulated, relative to others on the street. Houses with more insulation tend to have more snow accumulate, and the snow lasts longer. You have to be careful using this tool. Consideration should be given to orientation, slope, wind, shading, etc.

Evidence Of
Insulation
Added

In some cases you may be able to tell your client that insulation has been upgrades. Clues would include –

• plugged holes in roof coverings.
• plugged holes in ceilings.
• plugged holes in fascia boards.
• vents added to the roof.
• an extra thickness to the roof.
• a lower ceiling height than expected.

None of these clues are conclusive, nor do they tell you what the insulation material is or its total R-value. The addition of insulation by blowing material in has probably eliminated any venting within the roof space and may increase the risk of rot within the roof structure.

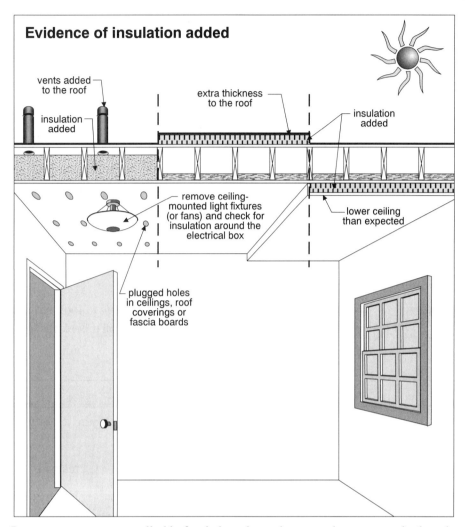

Evidence of insulation added

vents added
to the roof

insulation
added

extra thickness
to the roof

insulation
added

remove ceiling-
mounted light fixtures
(or fans) and check for
insulation around the
electrical box

lower ceiling
than expected

plugged holes
in ceilings, roof
coverings or
fascia boards

Remove Vents
To Check
Insulation

In some cases, vents applied in fascia boards can be popped out to get a look at the insulation. In some cases, you may be able to determine not only the material but also the amount of insulation. Be careful not to do any damage while trying to remove vents.

9.1.2. INSULATION – WET, COMPRESSED OR VOIDS

Again, you won't be able to see much. The evidence is usually indirect.

Causes

Insulation may be wet because of –

• roof leaks, or
• condensation.

Insulation may be compressed because of an original installation issue or moisture in the insulation. Voids may be the result of poor original installation or insulation batts slipping down the roof cavity on steep cathedral roofs.

Implications

The implications are reduced insulating values, and if the insulation is chronically wet, mold, mildew and rot are risks.

Strategy   You may see evidence of wet insulation as staining, sponginess or dampness on the ceiling finishes below. If the evidence is localized, you might suspect a roof leak. If the evidence is widespread, condensation is more likely the culprit. You won't usually see compressed areas or voids, except perhaps as localized areas of melting snow outside when weather conditions allow.

9.1.3. AIR/VAPOR BARRIER – MISSING OR INCOMPLETE

Again, in most cases you won't be able to identify the presence of an air/vapor barrier, let alone problems with it. Remember that many ceiling finishes are effective air/vapor barriers if they are reasonably tight.

Causes   Missing or incomplete air/vapor barriers are an installation issue.

Implications   The implications are increased risk of mold, mildew and rot, owing to air leakage into the roof space.

Strategy   We talked earlier of possibly finding openings to look at the insulation in roof spaces. These same opportunities may allow you to look for an air/vapor barrier. In most cases, however, you simply won't know.

9.1.4. AIR LEAKAGE – EXCESSIVE

Home inspectors do not measure air leakage. Again, it's difficult to quantify this problem.

Causes   This is either an original installation issue, or one created by people working on ceilings, creating openings.

Implications   Again, mold, mildew and rot are the implications.

Strategy   You can get a sense of how tight the ceiling finishes are by looking at wall/ceiling intersections, and looking at openings for lights, fans, etc. In most cases, you will find gaps. Improving the airtightness around any obvious openings always makes sense.

Your comments about air leakage can't be definitive. You can't see what kind of air sealing has been done where partition walls meet ceilings. It's common for plumbing stacks and electrical wires to run up through partition walls into the roof space. Air in partition walls can usually find its way into the roof space through holes in the top plate to accommodate wires, pipes, etc. Air gets into partition wall cavities at wall/ceiling, wall/wall and wall/floor intersections, as well as openings for electrical switches, outlets and so on. There will typically be some air leakage into the roof space.

Two Very Different Strategies

There are two very different strategies builders have used to address air leakage into roof spaces.

1. Vent the roof space. This traditional approach is similar to what we do in attics. The idea is to move air through the roof space to flush out any warm, moist air that gets into the cavity before it condenses and does damage.

2. Some have decided to try to seal the top and sides of the roof cavity. If no air can leak out of the roof space, there won't be an opportunity for house air to leak into the roof space. If you think of the roof space as a balloon that is already filled with air, there won't be much more air that can move in. Keeping the air still helps minimize convective loops and air changes between the roof space and the house. Even if this strategy is used, it makes sense to air seal at the ceiling as well as possible.

In practice, you often won't know whether a builder used either of these philosophies, or simply didn't think much about venting roof spaces.

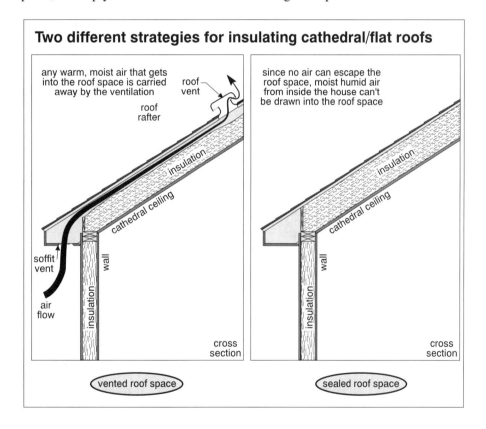

9.1.5. VENTING – MISSING OR INADEQUATE

Just as we look for vents on attics, we should be looking for vents on flat or cathedral roofs. The venting opportunities on a cathedral roof generally include soffit and ridge vents. On flat roofs, the venting is most often at opposing fascias.

Causes   Missing or inadequate vents are an installation issue.

Implication   Warm, moist air condensing in the roof and causing damage is the implication.

Strategy   On the outside of the building it's relatively easy to look for vents on flat or cathedral roofs. The presence of a good venting system is the exception rather than the rule, especially on those homes built before 1975. Obvious venting on flat or cathedral roofs older than this usually indicates upgraded insulation in the roof space.

Channeled Vents Versus Cross Ventilation   In many flat or cathedral roofs, vents are at either end of roof joist or rafter runs. Ventilation can only move in a straight line through the roof space. Where strapping is added on top of and perpendicular to roof joists and rafters, cross ventilation is possible. Open web trusses also allow cross ventilation. Generally speaking, the more directions ventilation can flow, the better things are.

Channeled vents versus cross ventilation

for cathedral ceilings and flat roofs the recommended vent area is 1 square foot for every 150 square feet of roof area

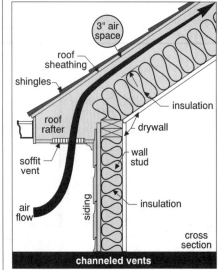
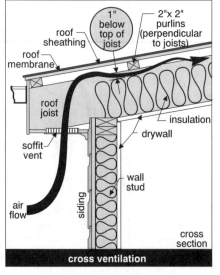

open web trusses also permit cross ventilation

Lots Of Venting

For flat and low-sloped roofs, many recommend ventilation ratios of 1:150. This means there should be one square foot of venting for every 150 square feet of roof space. This ratio refers to net free area of the vents. These are sometimes stamped onto the vents themselves, but in many cases it is difficult to determine.

Two Sides Of Roof Minimum

Vents must be on at least two sides of the roof to promote cross ventilation.

9.1.6. VENTING – OBSTRUCTED

Vents need to be free and open to operate properly.

Causes

New insulation or shifting original insulation can cause vents to become obstructed. Insulation often slides down cathedral roofs. Dust and dirt can also accumulate and block vents. Birds and animals can build nests that obstruct vents.

Implications

Obstructed vents increase the risk of mold, mildew and rot damage in the roof space.

Strategy

Look for vents on the outside of the house. If you can get close to them, make sure they are not obstructed. Try to determine the net free area of the vents. On soffit vents, try to look up through the vents with a flashlight. It's not unusual to find that insulation has either been blown right onto the soffit vents or has slipped down onto them. If insulation is touching the soffit vents, they are not likely to be effective.

9.1.7. MOLD, MILDEW OR ROT SUSPECTED

This is the most important problem that you may find with flat or cathedral roofs.

Causes

The causes of mold, mildew and rot are typically –

• roof leaks, or
• condensation as a result of air leakage into the roof space.

Implications

Damage to structure members is the implication. In severe cases, the roof can loose its ability to carry live loads.

Strategy   Because there is no access into the roof space, this can be a very difficult thing to determine. Some of the clues to look for include –

- Sagging or spongy roof surfaces. With practice, you will get a sense when walking on roofs of unusual deflection or sponginess. This can mean rotted or delaminated sheathing.
- Sagging plaster or drywall ceilings. We recommend you scan the ceiling surface below flat or cathedral roofs with a flashlight beam parallel to the ceiling to reveal sags. Drywall or plaster that is wet will often sag. A repetitive pattern outlining the underside of the ceiling joists, rafters or trusses may be visible.
- Mold or mildew on the ceiling surface. In some cases, the ceiling itself is cool enough to condense air in the house on the lower surface. Mold and mildew may get a foothold here. If this is the case, you may suspect more problems above the ceiling.
- Rusted nail heads on ceiling finishes. Sometimes you can see small rust patterns on plaster or drywall ceilings. This usually indicates high moisture levels in the roof space condensing on, and rusting, nails and screws. Again, this should be a red flag.

Can't Be Conclusive   In most cases, you won't be able to determine whether there is a problem or how extensive it is.

Summary

We recommend you advise clients that flat and cathedral roofs are prone to concealed damage from mold, mildew and rot. Without destructive testing, it's almost impossible to know whether there's a serious problem. We've talked about some of the clues, but we've been surprised at the amount of damage found in some roof spaces with no evidence on the interior or exterior of the building.

Moisture Scanners   Moisture scanning devices may provide information about elevated moisture levels in roof spaces. High moisture levels obviously suggest mold, mildew and rot problems. The danger is that a low moisture reading does not necessarily indicate there is no problem. During a dry summer, for example, the framing members may dry out. However, during the heating season, condensation may cause high moisture levels. Some building scientists say that early winter is the time that moisture levels in framing systems are likely to be highest.

Check In Early Winter   Some building experts maintain that early winter is the best time to check roof spaces for high moisture levels. If you are going to perform or recommend these tests, you should be aware of the likely seasonal variations in moisture content in roof spaces.

Describe Limitations   It's important with flat and cathedral roofs to let your clients know that most of your comments are based on indirect evidence because there is very little you can see.

Insulation
& Interiors
M O D U L E

QUICK QUIZ 10

☑ INSTRUCTIONS

- You should finish Study Session 10 before doing this Quiz.
- Write your answers in the spaces provided.
- Check your answers against ours at the end of this Section.
- If you have trouble with the Quiz, re-read the Study Session and try the Quiz again.
- If you did well, it's time for Study Session 11.

1. Describe in one sentence each, four general approaches to insulating flat and cathedral roofs.

   _____

   _____

   _____

   _____

2. List seven common problems found with flat roof insulation systems.

   1. _____

   2. _____

   3. _____

   4. _____

   5. _____

   6. _____

   7. _____

3. For each of the problems listed above, describe in one sentence, the implications.

1. _____

2. _____

3. _____

4. _____

5. _____

6. _____

7. _____

4. Describe one possible way to get a look at insulation in a flat roof.

_____

_____

5. The amount of snow on a flat roof may tell you something about its insulation. How can you use this tool?

_____

_____

6. List six things that may suggest insulation has been added to a flat or cathedral roof.

_____

_____

_____

_____

_____

_____

7. What things might suggest wet insulation?

_____

_____

8. Describe two strategies used to deal with venting of flat roofs.

_____

_____

9. On a flat roof, where will you most often see the vents?

_____

_____

10. What is an appropriate ratio of vent area to roof area?

_____

11. Rot in flat and cathedral roofs is most often caused by what two things?

_____

_____

12. List four things you might look for as clues that there may be rot in a flat or cathedral roof.

_____

_____

_____

_____

If you didn't have any difficulty with the Quiz, then you are ready for Study Session 11.

Key Words:
- *Attic*
- *Rot*
- *Mold*
- *Mildew*
- *Insulation*
- *Air/vapor barrier*
- *Ventilation*
- *Cross ventilation*

Insulation & Interiors

M O D U L E

STUDY SESSION 11

1. You should have completed Study Sessions 1 to 10 before starting this Session.

2. This Session covers the inspection of above grade walls, basements and crawlspaces, and the inspection of floors over unheated areas.

3. By the end of this Session, you should be able to—
  • list three common problems with wall insulation.
  • list two locations for insulating basement walls.
  • list seven advantages for each approach.
  • list two general approaches to insulating crawlspaces.
  • state which crawlspace insulation approach is preferred, and explain why in two sentences.
  • list nine common problems with basement and crawlspace insulation and ventilation.
  • describe in one sentence the implications of each of the nine common problems.
  • describe in one sentence each the inspection strategy for the nine conditions.
  • describe in one sentence the **cold floor effect**.
  • list seven common areas where insulation may be provided over unheated spaces.

4. This Study Session may take you roughly an hour and a half.

5. Quick Quiz 11 is at the end of this session.

Key Words:

• *Insulating sheathing*

• *Basement*

• *Crawlspace*

• *Interior*

• *Exterior*

• *Cold wall effect*

• *Drying potential*

• *Basement moisture control*

• *Moisture barrier*

• *Cold floor effect*

• *Earth floor*

• *Vented rain screen principle*

► 10.0 INSPECTING WALLS ABOVE GRADE

Above-grade walls are not any easier to inspect than flat and cathedral roofs. We don't get much of a look. During your inspection of the structure, you should have determined the makeup of the exterior walls. While there are other possibilities, most are wood frame or masonry. We'll concentrate on these. Log walls and stacked plank walls have no insulation, typically. Stress skin wall panels are mostly insulation.

Wood Frame Walls

Wood frame walls can be insulated three ways:

• The space between studs can be filled.
• Insulation can be added on the interior face of the studs.
• Insulation can be added on the exterior face of the studs.

Traditionally, insulation was either omitted or provided between the studs. Starting in the 1970s, builders in some areas began to provide insulation between the studs and insulating sheathing on the outside of the studs. Other builders moved from two by four to two by six studs to accommodate more insulation without insulating sheathing.

Insulating Sheathing

Insulating sheathing increased wall insulation levels and stopped thermal bridging through studs. On the other hand, it sometimes added a second vapor barrier on the exterior part of the wall – not a good thing. It made siding attachment more difficult, and in replacing conventional sheathing, usually needed diagonal bracing of walls to prevent racking.

Avoid Short Circuits

In some cases, insulation is added to the exterior of a wood frame wall without adding insulation into the wall stud cavity. This approach may be used, for example, where original siding is left in place, insulation is added over the siding, and then a new siding material is provided. The results may not be good. The wall stud cavity creates a large convective loop that can short circuit the new insulation on the exterior. Air in the cavity picks up heat from the living spaces and convective loops allow the heat to escape out through the top of the wall cavity. In some cases, the addition of exterior insulation to wood frame walls has no measurable heat loss reduction or comfort improvement.

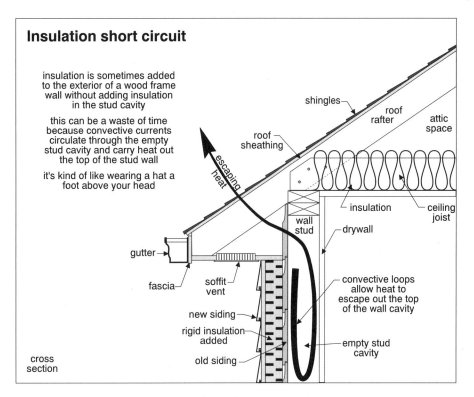

Insulation short circuit

insulation is sometimes added to the exterior of a wood frame wall without adding insulation in the stud cavity

this can be a waste of time because convective currents circulate through the empty stud cavity and carry heat out the top of the stud wall

it's kind of like wearing a hat a foot above your head

escaping heat

shingles

roof rafter

attic space

roof sheathing

insulation

ceiling joist

wall stud

drywall

gutter

soffit vent

convective loops allow heat to escape out the top of the wall cavity

fascia

new siding

rigid insulation added

old siding

empty stud cavity

cross section

Interior Insulation For Renovation

Adding insulation to the interior face of a wood stud wall is not common. Even during a renovation, it is disruptive and requires the relocation of window frames and casings and electrical outlets and switches. It also makes rooms smaller.

Masonry Walls

Masonry walls typically do not have enough space within them to add insulation. As a result, most masonry walls have no insulation. Where insulation is provided, it is typically on the interior or exterior face of the wall. Insulation on solid masonry houses is typically only added during renovations. Modern solid masonry walls may have insulation integrated as part of interior or exterior finishes. *Note: we are talking about solid masonry wall construction here, not wood-frame brick veneer or cavity wall construction.*

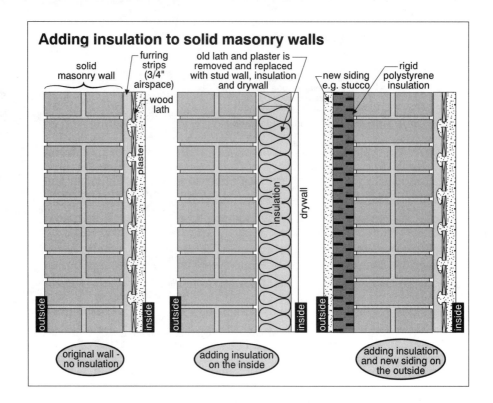

Adding insulation to solid masonry walls

10.1 CONDITIONS

Let's look at some of the common problems related to wall insulation. Again, there isn't much that we can see, and the big enemy is mold, mildew and rot.

1. Insulation – too little
2. Insulation – sagging or voids
3. Air/vapor barrier missing, incomplete or in wrong location
4. Mold, mildew or rot suspected

10.1.1 INSULATION – TOO LITTLE

Recommended levels of wall insulation vary. We talked earlier about typical levels of R-12 to R-20 in new construction. Older homes will rarely have this much and many have no wall insulation.

The R-Value Of An Uninsulated Wall

Walls that have no insulation often have R-values in the neighborhood of 4. This is true of wood frame and masonry walls.

Causes

Inadequate insulation levels are an original installation issue.

Implications

The implications are higher heating costs and, perhaps, lower comfort.

Cold Wall Effect

One of the implications of poor wall insulation levels is the **cold wall effect**. This is a phenomenon that makes you feel cool in a room that's at 72°F (perfectly comfortable), if you are sitting close to an uninsulated wall on a winter day. The cold wall effect is a result of heat radiating directly from your body to a cold surface. Heat flows from warm bodies to cold. Sitting beside a cold wall or a window may make you uncomfortable even though the air in the room is perfectly comfortable. The effect is quite dramatic. You can often notice that the side of your body close to the wall is significantly cooler than the other side.

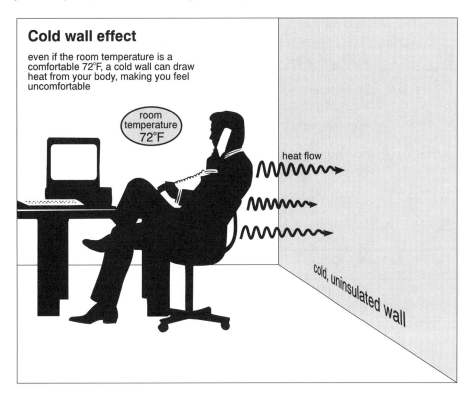

Cold wall effect

even if the room temperature is a comfortable 72°F, a cold wall can draw heat from your body, making you feel uncomfortable

room temperature 72°F

heat flow

cold, uninsulated wall

Strategy

In many cases, you won't be able to find evidence of insulation. We typically remove covers from electrical switches and receptacles on the inside of walls and try to get a look at electrical boxes. You should turn off the electrical power before removing the covers and poking around. You may also find whether or not there is an air/vapor provided as you do this. On masonry walls, electrical boxes are typically shallow and it's unusual to find insulation here. If you don't find any insulation looking from the inside, it is possible that insulation does exist, either within the stud cavities or on the outer surface. You'll have to report that you didn't observe any wall insulation.

Check Bottom Of Exterior Walls

You may be able to see evidence of insulation under the siding on the building exterior. This is sometimes visible on the bottom of the siding, near grade level.

Check Top Plates Of Walls And Attics

If insulation has been blown or poured into wall cavities, there may be evidence in the attic. If the top plates are accessible, check for evidence that these have been drilled or opened up to add insulation.

Foamed-in-place Insulations

Foamed-in-place insulations such as urea formaldehyde foam insulation can often be identified around electrical boxes because the foam typically oozes into the box. Evidence is also often visible at the sill plate and top plate.

Check Wall Thickness

You may be able to determine that insulation has been added to the wall by its thickness, if the wall has clearly been built out relative to the other houses in the neighborhood.

Again, if there is no insulation in the wall, we would not describe it as a deficiency, but would indicate that there is room for improvement.

10.1.2 INSULATION – SAGGING OR VOIDS

In most cases, you won't be able to see what the insulation is doing. However, walls in unfinished spaces, such as knee walls, or insulation on walls in split level homes visible from the attic, can be inspected.

Causes

Insulation may not have been well secured or insulation in some spots may never have been installed.

Implications

Increased heating costs are, of course, the implication.

Strategy

Where you can get into unfinished spaces with insulated walls, look for the presence and continuity of insulation.

10.1.3 AIR/VAPOR BARRIER MISSING, INCOMPLETE OR IN WRONG LOCATION

Again, in most cases you're not going to be able to see this.

Causes

This is an installation issue.

Implications

A missing, incomplete or cold-side air/vapor barrier may lead to increased air leakage and mold, mildew and rot in wall cavities. As we've discussed, good air sealing on a conventional drywall or plaster wall can achieve most of the effectiveness of an air/vapor barrier.

Strategy

If you remove covers from electrical switches and outlets (again, shutting off the power first), you may be able to see evidence of a polyethylene air/vapor barrier. (For those of you in cooling climates, you do not want to see an air/vapor barrier here!) If you are able to look at walls in unfinished areas, such as new walls, you can pull the insulation back slightly and see whether an air/vapor barrier has been provided.

Watch for cold-side vapor barriers. These can cause damage.

Again, in most cases, we would not recommend adding an air/vapor barrier unless we saw evidence of some problems.

10.1.4 MOLD, MILDEW OR ROT SUSPECTED

Causes

Again, this is a serious problem that may result from insulation and air leakage problems in wall cavities.

Implications

The implications are damage to structural members and possible loss of structural integrity over the long term. Damage to interior finishes and indoor air quality problems are also possible.

Strategy

Again, your inspection is going to rely mostly on indirect evidence. Unless there is considerable damage, it's going to be hard to identify this problem without removing interior or exterior finishes.

Damage At Bottom Of Walls

Damage is usually concentrated at the bottom of walls. Condensation tends to run down wall studs and get caught on bottom plates. The end grain of the studs sitting on the bottom plates wicks moisture up and, as a result, the bottoms of studs and sill plates show rot first.

In some cases, there's evidence of staining or rusting nail heads on the inner face of the wall near floor level. Damaged subflooring may be visible. In severe cases, there may also be evidence on the inside of the foundation wall below a wood frame wall, for example.

Vented Rain Screen Principle

You should be able to get a sense of the **drying potential** of the wall. A wall with a vented rain screen is more likely to dry quickly than a face-sealed system. You may be able to see evidence of moisture in walls around electrical boxes. In some cases, you can see staining or rusting if the boxes are metal. You may also see mold and mildew in the box or on the surfaces around them. If you have found a missing air/vapor barrier and lots of air leakage paths into the wall, you should suspect concealed damage. This risk is heightened if the exterior siding is stucco or another face sealed system.

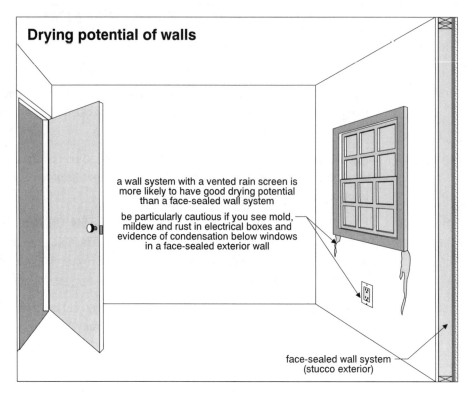

Drying potential of walls

a wall system with a vented rain screen is more likely to have good drying potential than a face-sealed wall system

be particularly cautious if you see mold, mildew and rust in electrical boxes and evidence of condensation below windows in a face-sealed exterior wall

face-sealed wall system (stucco exterior)

Look At Walls And Ceilings
You may also get some clues from the condition of the living spaces. If there is evidence of condensation on walls (especially below windows) or on ceilings, you may be more suspicious of condensation in the wall cavities.

Where To Look
On interior surfaces, look for mold and mildew on upper floor levels, particularly in corners or areas behind furniture or clothing in closets. Areas that are cooler, have lower airflow and are dark are more prone to mold and mildew. If you are in an area with prevailing winds, the windward walls will be colder. Look for mold and mildew on wall and ceiling surfaces on the windward side. The same is not true of damage inside wall cavities. Air leakage from the house into the walls is more likely on the leeward side than the windward side of the house.

If one were going to cut an investigative hole to look for interstitial damage from condensation, you might do it on the leeward side of the home on an upper floor level. Air leakage into walls tends to be greater on the upper floor levels than on lower floors, because of stack effect.

► 11.0 INSPECTING BASEMENTS AND CRAWLSPACES

11.1 BASEMENTS

Is Insulation Necessary?

Many homeowners ask whether basements and crawlspaces need to be insulated. No part of the house **needs** to be insulated. Insulating the basement and crawlspace will help to reduce heating costs and may improve comfort. Some people say that because the earth outside the foundation is not as cold as the air above, this reduces the need for insulation. It's true that the earth is often warmer than the outdoor air. However, the earth is a large heat sink and can draw considerable amounts of heat out of a building. The temperature of the earth may be lower than the outdoor air on spring and fall days, for example. Adding basement or crawlspace insulation is an improvement, rather than a repair, like any other insulation upgrade.

The Basement Floor

In some cases, the builder has provided insulation under the concrete floor slab of the basement or crawlspace. In most cases, you won't know whether this is done and it's not cost effective to add this insulation anyway. Home inspectors are usually silent on the issue of sub-slab insulation.

Insulating Basement Walls – Inside Or Outside?

Basement walls can be insulated on the inside or outside. Both are acceptable. Insulating outside has some advantages:

Exterior Insulation Advantages

• Exterior insulation keeps the foundation warm, which stabilizes the house temperature and eliminates any concern of frost heaving of the foundations.
• If the basement is finished, there is no need to disrupt interior finishes.
• No interior living space is lost.
• If foundation walls have to be exposed anyway to address basement leakage problems, the work can be cost effective.
• Outside walls are more uniform with fewer interruptions. A better insulation job can often be done.
• It may be easier to insulate the rim joist area from the outside of the home. This involves extending the basement wall insulation up above the top of the foundation.
• Exterior water control systems can be added or improved. Drainage layers against foundation walls and new or upgraded perimeter drainage tile systems can be added.

Insulating from the outside has some disadvantages:

Exterior Insulation Disadvantages

- It's usually more expensive than insulating from the inside, unless excavation is necessary for other reasons.
- It may be difficult to excavate around porches, driveways, well-established trees, retaining walls, concrete porches, etc.
- There is a risk of undermining the footings and foundations unless work is done carefully.
- The insulation typically extends above grade level. This part of the insulation needs to be protected from sunlight and mechanical damage. We also want to prevent water getting between the insulation and the foundation wall. This protection can be unattractive and expensive, especially if insulation is carried up above the foundation to cover the ends of the joist spaces. A flashing or cap has to be added at the top of the insulation and its protective coating.
- There is a loss of drying potential for the foundations. Moisture in foundations usually escapes outdoors through the above-grade part of the foundations. If these are insulated and covered, the drying path may be closed off.
- It may be difficult to provide good insulation and detailing around basement windows.
- Where part of the structure is unheated (e.g., attached garage, cold room or attached concrete porch), insulating around the outside is not effective. Insulation for these areas has to be placed on the inside, against the structures. There has to be an overlap of the exterior insulation and the interior insulation.
- Only some insulation materials are suitable for outdoor use below grade. These include—
  - some types of semi-rigid fiberglass and mineral wool
  - some types of polystyrene insulation
  - polyurethane boards
  - polyisocyanurate boards
- The fiberglass, mineral wool and polystyrene boards may have drainage capabilities to help channel surface water away from foundation walls to the perimeter foundation drainage system.

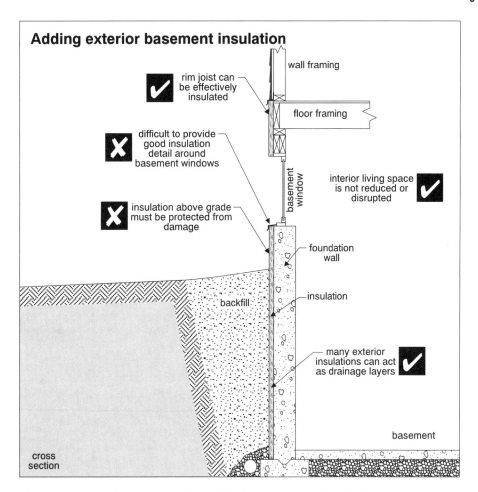

Adding exterior basement insulation

rim joist can be effectively insulated ✔

wall framing

floor framing

difficult to provide good insulation detail around basement windows ✘

basement window

interior living space is not reduced or disrupted ✔

insulation above grade must be protected from damage ✘

foundation wall

insulation

backfill

many exterior insulations can act as drainage layers ✔

basement

cross section

Interior Insulation Advantages

There are some advantages to insulating basement walls on the inside:

• It's typically less expensive.
• For do-it-yourselfers, it is easier work.
• It can be done at any time, irrespective of weather.
• It's less disruptive to the house.
• It's often easier to address basement windows.
• Foundations can still dry to the outside above grade.
• When insulating from the outside, the joist spaces often have to be insulated from the basement anyway.
• Any batt or board insulation material is suitable.
• If homeowners have plans to finish the basement in any case, the cost is very small.
• Porches and garages are not a problem.

There are also some disadvantages to insulating basements from the inside:

Interior Insulation Disadvantages

• If the basement leaks intermittently (and most do), it may be hard to monitor and correct leakage. Damage may also be done to insulation and support structures, such as support walls and interior finishes.
• There are typically many obstructions and wall penetrations to be worked around.

- In some areas with expansive soils, full height basement insulation on the inside may result in structure damage. This is a local issue and not a widespread problem.
- If there are no plans to finish the basement, the costs of the supporting structure for the insulation can be significant.

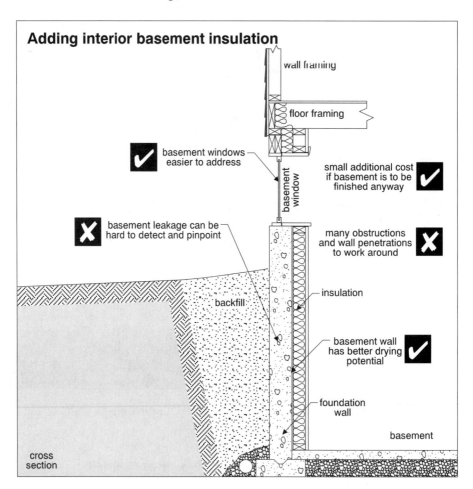

Adding interior basement insulation

wall framing

floor framing

✔ basement windows easier to address

✔ small additional cost if basement is to be finished anyway

basement window

✗ basement leakage can be hard to detect and pinpoint

✗ many obstructions and wall penetrations to work around

insulation

backfill

✔ basement wall has better drying potential

foundation wall

basement

cross section

Insulate Above Unheated Spaces

If there are living spaces above unheated garages or cold rooms, the ceilings of these spaces should be insulated.

Insulation Height

Recommended insulation treatments have changed considerably over the years. Insulation was often installed to just below grade level. Current thinking is that insulation should extend the full height of the basement wall in most areas.

Joist Space Insulation

Emphasis is also placed on insulating the joist spaces at the perimeter of the building. This is an area that historically has been ignored with respect to insulation and air leakage control.

Basement Moisture Control

We've already said that most basements leak at some point, and finishing basement walls on the inside should take this into consideration. There are a number of strategies that have been used. Our current favorite is a moisture barrier applied to the foundation wall that extends from the floor level up to **grade level only**. The moisture barrier shouldn't be carried above grade level for a couple of reasons:

• Moisture penetration through the foundation above grade is not likely.

• Foundation walls can store and release moisture. Stopping the moisture barrier at grade level allows moisture that ends up in the insulation to escape outdoors through the foundation wall. We don't want the insulation trapped between two plastic films – one against the foundation and wall and one on the inside of the insulation acting as the air/vapor barrier.

Extend Moisture Barrier Under Insulation

The moisture barrier can be one of several products, but is often polyethylene film or building paper. The plastic extends down to floor level and about twelve inches out onto the floor, typically. Wood studs against the foundation wall typically support the insulation. The bottom of the studs should not rest directly on the basement floor. Extending the plastic moisture barrier out from the wall ensures that the stud wall will sit on the plastic film, not directly on the concrete. Some people also use sill gaskets under the stud walls. Others use pressure treated sill plates to avoid rot. These techniques are employed because we recognize that most basement floors will get wet at some point.

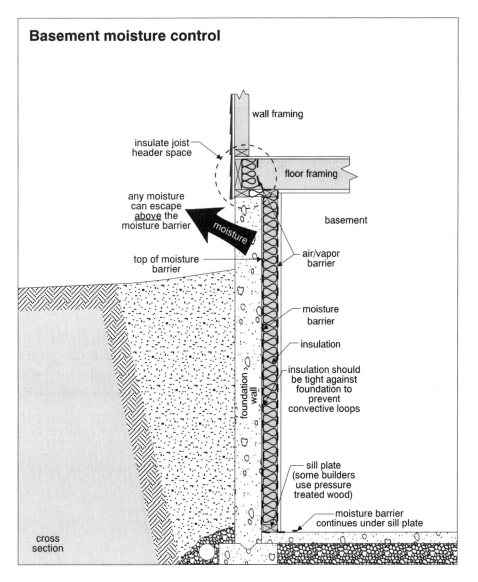

Basement moisture control

wall framing

insulate joist header space

floor framing

basement

any moisture can escape <u>above</u> the moisture barrier

moisture

top of moisture barrier

air/vapor barrier

moisture barrier

insulation

insulation should be tight against foundation to prevent convective loops

foundation wall

sill plate (some builders use pressure treated wood)

moisture barrier continues under sill plate

cross section

Insulation Should Be Tight Against Foundation — The insulation is typically installed between wood studs. The insulation works better if it is fit tightly against the foundation wall. We don't want a space behind the insulation that creates convective loops, dramatically reducing the value of the insulation.

Air/Vapor Barrier In Conventional Spot — A plastic air/vapor barrier is typically installed on the inside face of the studs covering the insulation before a wall finish material is added.

Joist Spaces — The joist header spaces can be insulated by fiberglass batts tucked into the spaces, ideally after the joints in the wood assembly have been caulked to minimize air leakage.

Joists Embedded In Foundation Wall — Joists are sometimes embedded in the top of a foundation wall. In this case, the joist connection points to the foundation wall should be well caulked, but insulation should not be provided in the joist spaces. This may cause rot at joist ends.

Insulation Material — We've been talking about batt insulation, which is most often fiberglass or mineral wool. Basement walls can also be insulated with rigid plastic insulation. Polystyrene is commonly used. Polystyrene insulation is available with prefinished wall systems bonded to the insulation. This can be drywall or other wall finish materials.

Basement Wall Insulation Is A Bonus — Where no basement wall insulation has been provided, we do not describe the house as having a defect. Basement wall insulation is an improvement, not a necessity.

11.2 CRAWLSPACES

There are two approaches to insulating crawlspaces, depending on whether the crawlspace is to be part of the heated area of the home or not. We can insulate the walls, creating a heated crawlspace, or insulate the floor above, creating an unheated crawlspace.

Prefer Heated Crawlspaces — We prefer heated crawlspaces for a few reasons:
• The floor above is more comfortable if the crawlspace is heated.
• It's easier to insulate the walls than attach insulation to the crawlspace ceiling.
• Less material is needed to insulate the crawlspace walls on the underside of the floor, unless the crawlspace walls are very tall.
• There is no risk of freezing pipes or excessive heat loss through ducts in crawlspaces if the space is heated.
• Insulation can be added on the outside or inside of the walls.

Unheated Crawlspaces   If the crawlspace is not to be heated, insulation is usually provided on the underside of the floor above. Insulation is typically tucked up between floor joists. The subfloor and flooring material above are usually relied on as the air/vapor barrier. We do not want an air/vapor barrier on the cool underside of the insulation. The insulation can be supported a number of ways, including chicken wire and housewrap. We frequently find insulation sagging or falling out of place.

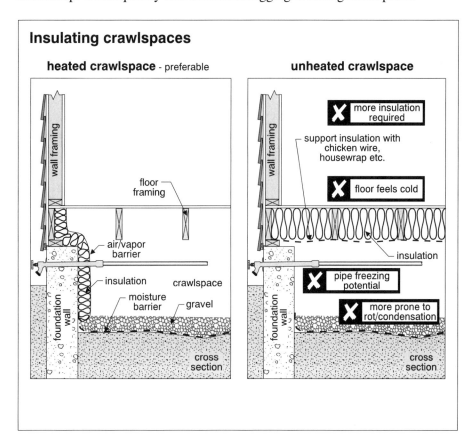

Rot More Likely   Unheated crawlspaces may be more susceptible to rot than heated crawlspaces. Warm, moist air leaking into a cool crawlspace may condense, depositing moisture on wooden structural members.

Freezing Pipes And Cold Ducts   Freezing plumbing and excessive heat loss through ducts are a risk in unheated crawlspaces in cold climates.

Floor Feels Cold   Even when the floor above the crawlspace is very well insulated, the floor will often feel cool.

*To Vent Or
Not To Vent?*

There are two schools of thought on venting crawlspaces. Most people agree that if the crawlspace is heated, it should not be vented to the outdoors, even in the summer. Unheated crawlspaces are typically vented to the outdoors. A vent ratio of one square foot for every 500 square feet of crawlspace floor area is a common recommendation. Recently, people have challenged the wisdom of venting unheated crawlspaces in hot, humid climates. If the crawlspace is cooler than the outdoor air, leakage of warm, high humidity outdoor air into the crawlspace may result in condensation on building components inside the space. Crawlspaces in drier climates may benefit from ventilation to the outdoors. You should become familiar with local conditions and recommended practices in your area before starting to do home inspections.

*Unfinished
Crawlspace
Floors*

We talked earlier about the tremendous amount of moisture that can be added to a home through an earth floor. If the crawlspace has a concrete floor, this is not a big issue. However, where the floor is earth, a moisture barrier should be provided. This is typically polyethylene film sealed at joints and at the perimeter. It's often covered with sand or round gravel to keep the plastic in place. This cuts down tremendously on the amount of moisture in the crawlspace and dramatically reduces the chances of rot in subgrade wooden structural components.

*Getting Into
Crawlspaces*

Home inspectors have long discussed whether or not they have to enter crawlspaces. Two inspectors discussing the issue might come up with three different positions. Our philosophy is that if the crawlspace is safe, we should enter it. If there is less than about 18 inches of clearance between the floor and structural members, it may be almost impossible to move through the crawlspace. Different inspectors have different tolerances for how low a crawlspace they will get into. You have to be comfortable that you are not going to be stuck in the space.

*Common
Access Points*

The access into a crawlspace may be a hatch in the floor of the living space, an exterior hatch in the wall near grade level, or an opening in an adjacent basement wall.

*Removing
Covers*

If the access cover to the crawlspace is secured and we are likely to damage the cover or the area around it in opening it, we do not remove the cover. Some inspectors will not remove any covers that require tools. Others are happy to remove four or six screws on an access cover. Still others use a wrecking bar to get into crawlspaces.

Wet Floors

Where the crawlspace has standing water on the floor, we consider that an unsafe condition. The water may be stagnant. In most cases, we don't know the source of the water. It could be raw sewage and a health hazard. Water may also pose an electrical hazard. It's not unusual to find wires on the floor of a crawlspace. If these wires contact wet earth or water, there is a risk of electrical shock.

Animals

We consider a crawlspace unsafe if we can identify animals in the crawlspace or animal waste. Snakes, scorpions, foxes, raccoons, etc., can all be hazardous to your health. Again, different inspectors have different tolerances for the conditions they will endure in a crawlspace.

Dark And Dirty	Most crawlspaces are unpleasant. They are dark and you will get dirty crawling through them. We don't consider either of these conditions reason to avoid entering a crawlspace. We carry strong lights and disposable coveralls for crawlspace inspections.
Rewards Can Be Great	The information gained from a crawlspace is often significant. Because crawlspaces are unpleasant, most people do not go into them regularly. Problems may develop that in other parts of the home would be found and corrected promptly. Crawlspace problems may go undetected for a long time. We encourage you to access crawlspaces wherever possible.
Access From Basement	In houses with partial basements, the crawlspace is often accessed from the basement, as we mentioned. We encourage homeowners to heat crawlspaces and remove any covers so that the crawlspace communicates directly with the basement. This helps ventilate the crawlspace with warm area and reduces the risk of concealed rot.

11.3 CONDITIONS

Let's look at some of the common problems we find related to insulating basements and crawlspaces.

1. Insulation – too little or incomplete
2. Exterior insulation not suitable for use below grade
3. Exterior insulation not protected at top
4. Insulation missing at rim joists
5. Insulation sagging, loose or voids
6. Exposed combustible insulation
7. Air/vapor barrier missing, incomplete or wrong location
8. No moisture barrier on basement walls
9. No moisture barrier on earth floor

11.3.1 INSULATION – TOO LITTLE OR INCOMPLETE

Let's start by repeating that many houses have no basement or crawlspace wall insulation. This is not a problem. Homeowners can add insulation as a home improvement.

Recommended Levels	You should find out what levels of insulation (if any) are recommended in your area for basement and crawlspace walls. In our area, R-13 is called for in new construction.
Insulate To The Floor	You may also want to research whether insulation is recommended all the way down to floor level on basement walls. In our area, the recommendations have changed over the last several years. Currently, the recommendation is to extend insulation down to within about six inches of floor level.
Causes	Too little or incomplete insulation is an installation issue. Less than ideal insulation levels lead to higher heating costs.

Strategy   In some cases, it's easy to determine whether insulation has been provided. In other cases, it can be tough to tell.

Inside Or Outside?   Insulation can be provided on the inside or outside of foundation walls. It can also be provided on the ceiling, although this is more common in a crawlspace than in a basement.

Look Behind Wall Finishes   Where basement walls are finished, you may be able to determine whether there is insulation by removing covers from electrical receptacles and light switches on perimeter walls. Turn off the electricity first.

Exterior   On the exterior, you may see evidence of the protective coating (e.g., parging, fiber-cement board or pressure-treated plywood) and a flashing at the top of an insulation material. You'll rarely see the insulation itself on the outside.

Exposed Insulation   Sometimes you can see insulation exposed on basement or crawlspace walls. It may be held in place with tack strips, metal hooks, or a polyethylene vapor barrier and wood strapping. Where the batts are installed between wood studs, the insulation is usually covered by wood paneling or drywall.

Insulation At Rim Joists   You may be able to see whether insulation has been provided around the perimeter of the joist cavity at the top of the basement or crawlspace. It's not unusual to find the basement walls insulated, but no insulation at the rim joists.

11.3.2 EXTERIOR INSULATION NOT SUITABLE FOR USE BELOW GRADE

Some types of rigid fiberglass, mineral wool, and polystyrene boards are suitable for use outside below grade. These are the most common materials used. Conventional fiberglass board or semi-rigid cladding is not suitable for use below grade. Some types of expanded polystyrene are also not suitable for below grade use. Cellulose fiber insulation is not suitable for use below grade, nor is most mineral wool insulation.

Causes   The use of inappropriate insulation is an installation issue.

Implications   The insulation's performance will be dramatically reduced by moisture if the insulation is not suitable for below grade use. (Some experts have questioned the ability of some materials designed for this use to dry quickly. If the insulation stays wet for long periods, it won't be an effective insulator.)

Strategy   In most cases, you won't be able to see what type of insulation has been used on the exterior. However, since an inappropriate material suggests an amateurish installation, you may be able to see some of the insulation itself.

11.3.3 EXTERIOR INSULATION NOT PROTECTED AT TOP

The top of an exterior installation board should be mechanically protected with a parging, fiber cement board or pressure treated plywood, for example. The top of the opening should be covered with a cap or flashing unless the siding extends down over it.

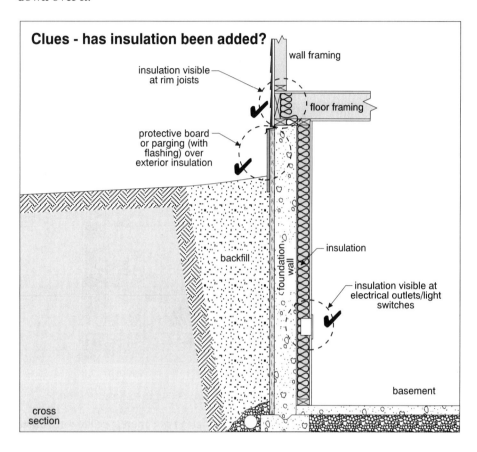

Clues - has insulation been added?

Causes	Inappropriate protection of the insulation above grade is an installation issue. It may also be the result of mechanical damage.
Implications	Insulations materials do not stand up well to mechanical damage from lawn mowers, weed eaters, etc.
Insulation Vulnerable	Many types of insulation are adversely affected by sunlight (polystyrene, for example). Water penetrating the insulation or between the insulation and the foundation can dramatically reduce the insulating value.
Strategy	You should not be able to see foundation insulation as you look at the building exterior. Where insulation can be seen, improvements are probably necessary.

11.3.4 INSULATION MISSING AT RIM JOISTS

We've talked about insulating the perimeter of the building above the foundation wall where the joists rest on top of the foundation.

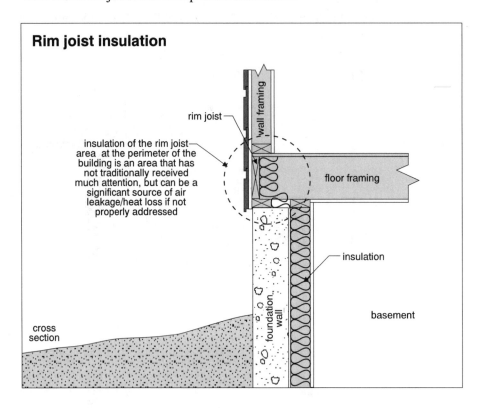

Rim joist insulation

rim joist

insulation of the rim joist area at the perimeter of the building is an area that has not traditionally received much attention, but can be a significant source of air leakage/heat loss if not properly addressed

wall framing

floor framing

insulation

foundation wall

basement

cross section

Causes	Omission of insulation in this area is an installation issue.
Implications	Implications of missing installation are increased heating costs. If air sealing is not well done, moisture damage to structural components is possible. The rim joist area is a vulnerable part of the building. If exterior grade is too high, the rim joist may be exposed to constant dampness from earth. The exterior grade levels should be at least six inches below the top of the foundation wall. Water accumulating in wall cavities as a result of leakage and condensation will flow by gravity to the bottom of the wall cavity and may end up at the rim joist area.
Strategy	Where basement walls are not insulated, you would not expect the rim joist area to be insulated. However, where basement walls have been insulated, a consistent approach would include good air sealing of the rim joist area and insulation. The most common approach is to stuff fiberglass batts between the joists.

Don't Insulate Embedded Joists

We touched earlier on the problem of joists embedded in the concrete foundation itself. Where this is done, the perimeter joist bases should not be insulated. This promotes rot damage to the ends of the joists. Good air sealing of the joist/concrete joint with caulking is recommended.

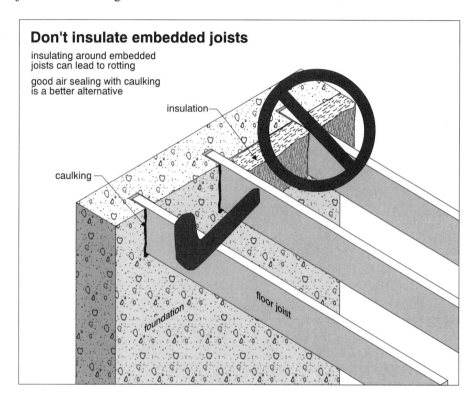

Don't insulate embedded joists

insulating around embedded joists can lead to rotting

good air sealing with caulking is a better alternative

insulation

caulking

foundation

floor joist

11.3.5 INSULATION SAGGING, LOOSE, OR VOIDS

Insulation hung on walls is always subject to the forces of gravity. This is particularly true where no wall finish is provided. Wall finishes help to hold the insulation in place.

Causes

This is typically a result of poor installation or mechanical action pulling insulation out of place.

Implications

Increased heat loss is the implication of insulation that is sagging, loose or has voids. We mentioned earlier that if the insulation is not fit tightly against the wall, convective loops may be formed that dramatically reduce the insulating value.

Strategy

Where insulation is provided on basement or crawlspace walls, make sure that it is intact. Look for continuity of the insulation and a tight fit against the foundation wall.

Crawlspace Or Basement Ceiling Insulation

Where the ceiling of the subgrade area is insulated, it's usually easy to identify sagging or missing pieces of insulation. You'll also want to make sure that an air/vapor barrier has not been provided on the cold side of the insulation.

11.3.6 EXPOSED COMBUSTIBLE INSULATION

Insulation in basements, crawlspace, garages or any other interior spaces accessible to people, electricity and heating sources should be covered with drywall or other interior finishes. Combustible wood paneling is acceptable in most areas.

Polystyrene Insulation

When people think of combustible insulations, they usually think of polystyrene. Polystyrene is difficult to ignite but releases a great deal of heat when it burns and produces a toxic black smoke. Cellulose, polyurethane, and polyisocyanurate boards are also combustible.

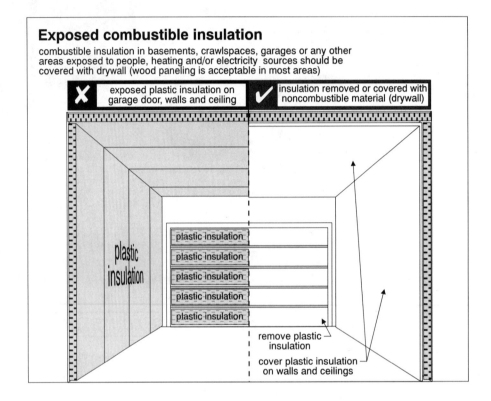

Exposed combustible insulation

combustible insulation in basements, crawlspaces, garages or any other areas exposed to people, heating and/or electricity sources should be covered with drywall (wood paneling is acceptable in most areas)

X exposed plastic insulation on garage door, walls and ceiling

✓ insulation removed or covered with noncombustible material (drywall)

plastic insulation

plastic insulation
plastic insulation
plastic insulation
plastic insulation
plastic insulation

remove plastic insulation

cover plastic insulation on walls and ceilings

11.3.7 AIR/VAPOR BARRIER MISSING, INCOMPLETE OR WRONG LOCATION

Insulation on basement or crawlspace walls should have an air/vapor barrier. It may be visible polyethylene film. If the insulation is covered with interior finishes, you won't know whether there is an effective air/vapor barrier.

Causes   Failure to provide a complete air/vapor barrier on the warm side of the insulation is an installation issue.

Implications   The implications are condensation on the cold side of the insulation. This won't usually damage the foundation wall, but the moisture can reduce the effectiveness of the insulation and damage a wood strapping or stud wall system used to support the wall insulation.

Strategy   If the insulation is not covered, you'll be able to identify the presence of an air/vapor barrier. If finishes are provided, you may be able to see the presence of an air/vapor barrier when removing electrical cover plates (after shutting off the power). However, you will be able to evaluate the effectiveness of the air/vapor barrier.

Watch for double vapor barriers or vapor barriers on the cold side only. Plastic moisture barriers on basement walls should stop at grade level. Insulation on crawlspace ceilings should not be supported by a vapor barrier.

11.3.8 NO MOISTURE BARRIER ON FOUNDATION WALLS

The moisture barrier is applied directly to the foundation wall extending from the floor up to grade level (but not to the top of the foundation wall). If the basement wall has been insulated, you will rarely be able to see this moisture barrier. You may see evidence of it under a stud wall at the floor level.

Causes   The omission of a moisture barrier is a common installation issue.

Implications   Moisture coming through the foundation wall may wet the insulation, reducing its insulating ability. The water may also damage the wood strapping or stud wall assembly that supports the insulation.

Strategy   In most cases, you won't be able to see the moisture barrier. If there are gaps in the insulation assembly, you may be able to identify a moisture barrier. You may also be able to see the floor level below the stud wall, as mentioned earlier.

Condensation On Outer Side Of Air/Vapor Barrier   A common situation in relatively new homes is condensation of the outer side (closest to the foundation) of a polyethylene air/vapor barrier on an insulated basement or crawlspace wall. It is common to find fiberglass batts held in place with polyethylene film and no moisture barrier on the concrete foundation walls behind.

Concrete
Drying

New concrete contains considerable moisture. That moisture may leave the foundation wall on the inner side, often driven by the sun's heat. The moisture is trapped behind the polyethylene air/vapor barrier and condenses. This is very distressing to homeowners and some builders. Water may run down the foundation walls below the insulation.

Normal
Condition

There is no real problem here, and in most cases this situation will correct itself within the first several months of occupancy. The amount of moisture in the concrete, in the soil outside, and in the house air all affect this situation. It's one reason that we recommend basements not be finished for at least one year after a new home has been occupied.

11.3.9 NO MOISTURE BARRIER ON EARTH FLOOR

We have discussed a couple of times how important it is to control moisture migrating out of the soil below a house.

Causes

Failure to provide a moisture barrier is an installation issue.

Implications

Elevated moisture levels in the air in crawlspaces and in the living space are the implications. These elevated moisture levels can cause damage to the structural members in the crawlspace and other parts of the home.

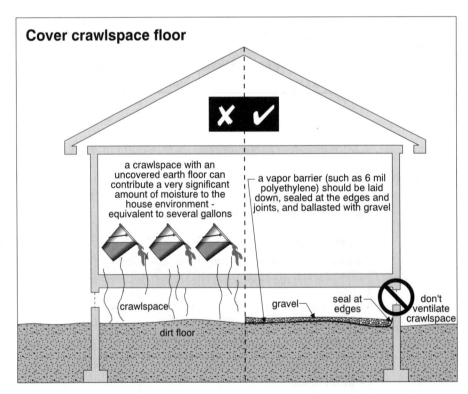

Strategy   Basement and crawlspace floors should be concrete or some other impervious material. Where the floor is earth, recommend that it be covered with at least a polyethylene sheet. The polyethylene sheet should be sealed at the joints and around the perimeter to the foundation walls. A protective gravel or sand coating may be added.

Don't Be Fooled   When looking at a crawlspace or basement floor, make sure you dig down to ensure that there is no moisture barrier, before criticizing the system.

Summary

The insulation of basements and crawlspaces may or may not be an issue in your area. An understanding of generally accepted practices is essential before performing this part of your inspection.

► 12.0 INSPECTING FLOORS OVER UNHEATED AREAS

Floors over unheated areas are typically cold during the heating season, relative to other floors in the home. This is true even though the room itself is well heated and the air is the same temperature as the other rooms in the house.

Cold Floor Effect   We talked earlier about **cold wall effect** and how heat radiates from people's bodies to cold wall surfaces. The same is true of floors, with the additional complication of conduction. Because our feet typically contact the floor, not only does our body radiate heat to a cold floor surface, our feet transmit heat directly to the floor by conduction.

Insulation Doesn't Stop Heat Flow   We've said that insulation slows the flow of heat, but does not stop it. Even with a floor that is well insulated (our area currently calls for R-27), the floor can feel cool on a cold winter day. Underpad and carpeting help to diminish the cold floor effect, but they often do not eliminate it altogether.

Auxiliary Heat Common   It's not unusual in northern climates to find auxiliary electric baseboard heaters, for example, installed in rooms over unheated spaces. Because these auxiliary heaters rarely heat the floor itself, they are often not particularly effective. Most heating systems are designed to sense the air temperature and keep it comfortable. The air temperature is often normal and acceptable, although the floor feels cold.

Convective Loops And Air Leakage   Cold floors are commonly caused by convective loops through gaps in the insulation and by air leakage. If the neutral pressure plane in the house is above the floor level (this is common), outside air may leak into the home through gaps in the floor system. Obviously, this outdoor air will feel cold.

Garages Are Cold Too   Cantilevered floor systems with nothing below them can definitely be cold. Surprisingly, floors over garages are almost as cold. It's tempting to think of the garage as a semi-heated area, but in many northern climates floors above garages can be very cold indeed.

*Possible
Solutions*

Several solutions have been tried to make floors more comfortable over unheated spaces. The heated cavity between the floor and insulation below outlined in Section L of the Insulation Chapter of **The Home Reference Book** is one approach. As the text suggests, the success of this system depends on a number of things, including –

• unobstructed airflow
• good balancing of supply and return air
• good air sealing

The perimeter of this space needs to be well insulated – a detail that was often overlooked.

*Sprayed-in-
place Foams*

A popular approach in many areas is the use of isocyanate or polyurethane foam insulations. These insulations are foamed in place and covered on the underside. The advantage of these systems is that very good air sealing is achieved by the insulation. This dramatically reduces convective heat loss problems and also reduces air leakage.

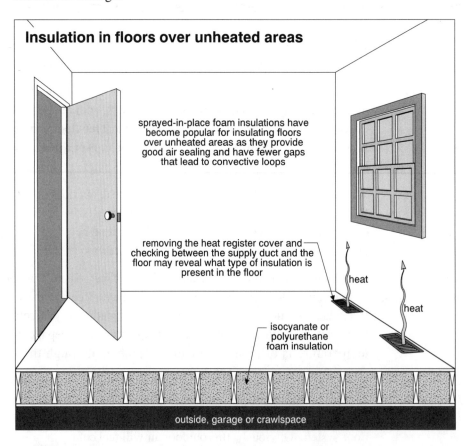

Insulation in floors over unheated areas

sprayed-in-place foam insulations have become popular for insulating floors over unheated areas as they provide good air sealing and have fewer gaps that lead to convective loops

removing the heat register cover and checking between the supply duct and the floor may reveal what type of insulation is present in the floor

heat

heat

isocyanate or polyurethane foam insulation

outside, garage or crawlspace

*Inspection Not
Possible*

In most cases, you won't be able to determine what insulation, if any, has been provided in floors over unheated spaces. You should point out to your client the potential for cool floors and the limitations to your inspection here.

Strategy   Where insulation is exposed, you can identify loose or missing insulation, as well as gaps. If the insulation is exposed, you'll want to make sure there is no air/vapor barrier on the cold side. Remember the air/vapor barrier should be on the warm side of the insulation. In most cases, no air/vapor barrier is necessary, since the subfloor and finished flooring usually provide an adequate air/vapor barrier.

Plugs Are   In some cases, you may see clues that insulation has been added in these areas.
Clues   Plugged holes in the underside of the openings or new finishes both suggest that insulation has been added.

Summary

Floors over unheated areas are difficult to inspect and can lead to unhappy clients. Make sure you identify all such areas. They may include spaces –

• above garages
• above porches
• in cantilevered areas
• over breezeways
• below windows that project out from the building
• over unheated crawlspaces or other subgrade areas
• over open areas below houses with pier foundation and no skirting

In most areas, these will be inaccessible areas where you'll simply be raising a question. Insulation improvements in these areas are usually possible at moderate cost. The important thing is to advise your client of the possibility.

Insulation
& Interiors
M O D U L E

QUICK QUIZ 11

☑ INSTRUCTIONS

• You should finish Study Session 11 before doing this Quiz.

• Write your answers in the spaces provided.

• Check your answers against ours at the end of this Section.

• If you have trouble with the Quiz, re-read the Study Session and try the Quiz again.

• If you did well, it's time for Study Session 12.

1. Basement wall insulation is necessary. True or false?

   _____

2. Describe in two sentences each, two strategies for insulating basement walls.

   a) _____

   _____

   b) _____

   _____

3. Describe in one sentence each, seven advantages to each approach.

   a) _____

   _____

   _____

   _____

   _____

   _____

   _____

   b) _____

   _____

   _____

   _____

   _____

   _____

   _____

4. Describe in one sentence each, two general approaches to insulating crawlspaces.

   _____

   _____

5. Indicate which approach to insulating crawlspaces is preferred.

   _____

6. Give four reasons why this approach may be better.

   _____

   _____

   _____

   _____

7. All crawlspaces should be vented. True or false? Explain.

   _____

   _____

8. List three safety issues involved in inspecting crawlspaces.

_____

_____

_____

9. List nine common problems with insulation and ventilation in basement and crawlspace areas.

1. _____

2. _____

3. _____

4. _____

5. _____

6. _____

7. _____

8. _____

9. _____

10. List the implications of each of these problems.

1. _____

2. _____

3. _____

4. _____

5. _____

6. _____

7. _____

8. _____

9. _____

11. Why does condensation sometimes show up in new homes on the outer side of a polyethylene vapor barrier on the inside of a basement wall?

_____

_____

12. Describe the cold floor effect.

_____

13. Describe two possible strategies for keeping floors over unheated spaces warm.

_____

_____

14. List seven areas where floors over unheated spaces may feel cool.

_____

_____

_____

_____

_____

_____

_____

If you didn't have any difficulty with the Quiz, then you are ready for Study Session 12.

Key Words:
- **Insulating sheathing**
- **Basement**
- **Crawlspace**
- **Interior**
- **Exterior**
- **Cold wall effect**
- **Drying potential**
- **Basement moisture control**
- **Moisture barrier**
- **Cold floor effect**
- **Earth floor**
- **Vented rain screen priciple**

Insulation & Interiors

MODULE

STUDY SESSION 12

1. You should have completed Study Sessions 1 to 11 before starting this Session.

2. This Session covers ventilation systems, including heat recovery ventilators.

3. By the end of this Session, you should be able to—
   • list five signs of high humidity in homes.
   • list two ways humidity levels can be controlled.
   • list eight ways to reduce moisture sources in homes.
   • list ten problems with exhaust fans, and their implications.
   • describe the inspection strategy for each problem listed above.
   • list 25 common HRV problems and their implications.

4. This Study Session may take you roughly one and a half hours.

5. Quick Quiz 12 is at the end of this session.

Key Words:

- *Condensation*
- *Staining*
- *Mold*
- *Mildew*
- *Odors*
- *Backdraft*
- *Reduce moisture sources*
- *Ventilation*
- *Termination point*
- *Duct insulation*
- *Weather hood*
- *Exhaust fan*
- *HRV*
- *Cabinet*
- *Exhaust grilles*
- *Flow measuring stations*
- *Flow collars*
- *Balancing dampers*
- *Ventilation fan switch*
- *Filters*
- *Heat exchanger*
- *Condensate drain*
- *Trap*

► 13.0 INSPECTING VENTILATION SYSTEMS

13.1 EVALUATING AND CONTROLLING HOUSE HUMIDITY

Most home inspectors do not use tools or instruments to determine house humidity. We look for evidence that humidity levels are high. These include –

• condensation on windows
• staining or streaking on window sills and areas below, indicating condensation problems during cold weather
• mold and mildew on cool, dark or moist house surfaces, including bathrooms and closets
• stale odors and stuffy air in the house
• backdraft of combustion appliances, including fireplaces

Controlling
House Humidity

House humidity is typically controlled one of two ways:

1. Reduce moisture sources.
2. Ventilate the home.

Reducing
Moisture

The following steps can be taken to reduce moisture levels in the home:

• Don't store firewood in the home.
• Correct foundation leakage problems.
• Disconnect or remove humidifiers.
• Cover any earth floor in a basement or crawlspace.
• Cover sump pits.
• Don't hang laundry to dry inside the home.
• Vent clothes dryers to the outside.
• Use exhaust fans when showering or cooking.
• Limit the use of misters and steam generators (common for people with respiratory problems).

Reducing moisture

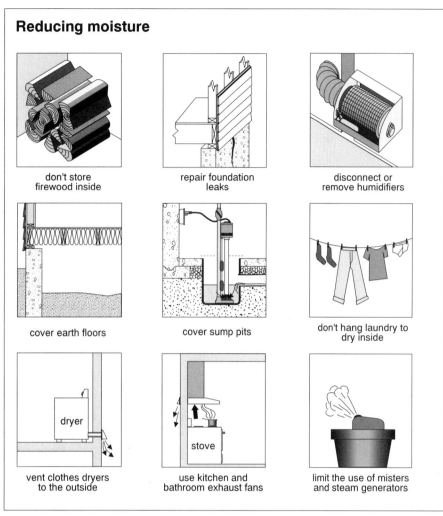

don't store
firewood inside

repair foundation
leaks

disconnect or
remove humidifiers

cover earth floors

cover sump pits

don't hang laundry to
dry inside

vent clothes dryers
to the outside

use kitchen and
bathroom exhaust fans

limit the use of misters
and steam generators

Since some moisture will always be generated, most homes require some ventilation system to rid the house of excess moisture. We've discussed exhaust-only, supply-only and balanced ventilation systems. The most common ventilation systems include –

• exhaust fans
• energy recovery ventilators or heat recovery ventilators, often combined with exhaust fans

We'll focus on these two systems. We've talked about how exhaust and ventilation systems should work. Let's look at some of the common problems.

13.2 EXHAUST FAN CONDITIONS

Common problems with exhaust fans include the following:

1. Inoperative
2. Noisy
3. Inadequate air movement
4. Fan cover missing
5. Wiring unsafe
6. Ducts leaky, damaged, disconnected or missing
7. Ducts not insulated in unconditioned space
8. Termination point not found
9. Poor termination point location
10. Weather hood missing, damaged or loose
11. Inadequate backflow prevention (flap)

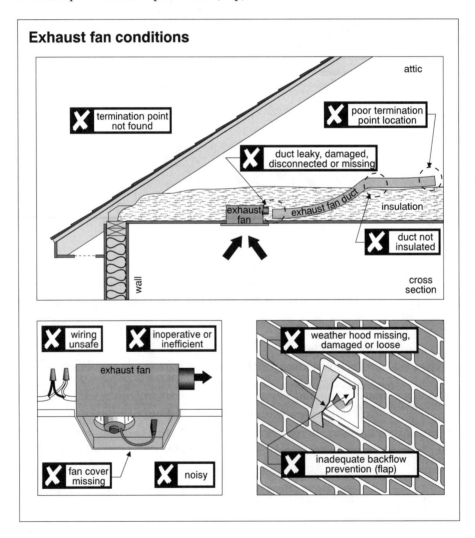

13.2.1 INOPERATIVE

Causes   Exhaust fans may be inoperative because of an electric problem or because the fan or motor is seized.

Implications If the fan is inoperative, it won't help reduce house humidity.

Strategy   Fans may be operated by switches, timers or dehumidistats. Locate the controls and operate the fan.

13.2.2 NOISY

Many inexpensive exhaust fans are quite noisy. As a result, they are often disconnected by homeowners who find the noise irritating.

Causes   With inexpensive fans, the noise may be the nature of the fan. When other fans become noisy, it may be a bearing or alignment problem.

Implications Noisy fans may be on the verge of failure or may indicate inexpensive fans.

If the fans fails on its own or is shut off because the homeowner objects to the noise, the ventilation system is defeated.

Strategy   As you test the fan operation, listen for excessive noise.

13.2.3 INADEQUATE AIR MOVEMENT

Causes   Inadequate air movement may be the result of –

• obstructed, crimped or crushed ductwork
• obstructed exhaust hood
• dirty fan blades
• obstructed intake grille on the fan cover

Implications Exhaust fans that don't move much air won't perform their ventilating function.

Strategy
Check At
Exhaust Point

There are two ways to check that the exhaust fan is moving air adequately. The best, but most difficult, way is to check at the exhaust point. This is often difficult because the exhaust point may be on the roof, at a soffit or high on a wall on the outside of the building. Most home inspectors don't turn exhaust fans on and go out and check for good airflow at the exhaust point.

Check At
Intake

The second, and less desirable, way to check the fan is to make sure that air is being pulled in at the fan itself. Some people use their hand or a piece of tissue to check for airflow. If you use a piece of tissue, don't let it be pulled into the fan. The test at the inlet point is not as positive because good flow may be indicated here but the duct may be disconnected at some point. It's possible that the warm moist air is being blown into an attic or other unconditioned space, which can lead to problems.

13.2.4 FAN COVER MISSING OR DIRTY

Fan covers are often inexpensive plastic grilles.

Causes   Fan covers may be missing because they

- were never installed originally
- have fallen off and been lost
- are broken
- were removed because they clogged

Implications   The fan and ductwork are likely to get dirty faster with the cover missing. The fan blades and wiring are exposed and may be touched inadvertently.

Strategy   Check that the exhaust fans have covers in place.

13.2.5 WIRING UNSAFE

This is usually an original installation issue.

Implications   Electric shock or fire hazard are the implications of unsafe wiring.

Strategy   We've talked about how to look for wiring problems in the Electrical Module. Poor connections, inappropriate wire types and poorly supported wires are common problems related to exhaust fans.

13.2.6 DUCT LEAKY, DAMAGED, DISCONNECTED OR MISSING

Duct runs from exhaust fans to the building exterior should be as short and straight as possible. Long duct runs reduce airflow.

Causes   Ducts may be leaky, damaged, disconnected or missing as a result of poor original installation, mechanical damage or renovation activity in the house.

Implications   Reduced airflow and warm, moist air being dumped into unconditioned spaces are the implications.

Strategy   Where possible, follow the duct from the fan to the termination point on the building exterior. Watch for sags in ducts running through attics and other unconditioned spaces. Condensation and lint can build up in the low points and clog the duct. This problem is common with clothes dryers, as well as exhaust fans.

13.2.7 DUCT NOT INSULATED IN UNCONDITIONED SPACE

Any duct passing through cold spaces (or warm spaces in cooling climates) should be insulated. A minimum insulation of R-3 is typically recommended.

Causes   Missing insulation is usually an installation issue.

Implications   Condensation inside the duct in the heating season and on the outside of the duct in the cooling season is the implication.

Strategy   Follow the ductwork where possible and ensure that it is insulated where it runs through unconditioned spaces. A common practice with exhaust fans at ceiling levels is to run the uninsulated ductwork across the ceiling, below the attic insulation. This may effectively insulate the ductwork, although if often creates voids in batt insulation.

13.2.8 TERMINATION POINT NOT FOUND

Sometimes, you can't identify the point where the exhaust fan vents outside. While it is possible that the vent is concealed, it's also possible that it does not extend outside. This is obviously a problem, as we've discussed before.

Causes   This is an installation issue.

Implications   Warm, moist air may be dumped into the building.

Strategy   Try to follow the ducts from exhaust fans to try to find their termination points. In some cases, the ductwork is simply carried out through the eaves and turned down against the soffit vents. While this may perform adequately, the chances of the duct getting moved and/or the soffit vent being obstructed are high. The other disadvantage to this arrangement is that there is typically no flap at the end of the duct to prevent backflow. Cool drafts coming back through the fan into the house are common with this configuration.

13.2.9 POOR TERMINATION LOCATION

The exhaust point should not be located near a fresh air inlet, near a spot where it may be obstructed or where it is exposed to mechanical damage (along a driveway, for example).

Causes   This is an installation issue.

Implications   The exhaust from the fan may be drawn back into the building if it is too close to a fresh air inlet. It will not vent properly if obstructed by snow, for example.

Strategy   Look for exhaust vents that are at least four inches above grade. They should be typically three feet away from fresh air inlets. Exhaust hoods should not be in enclosed spaces, including under decks and porches. The moisture may accumulate here and cause damage.

13.2.10 WEATHER HOOD MISSING, DAMAGED OR LOOSE

Where an exhaust fan duct terminates outside, it is usually protected by a weather hood. This stops rain and snow driven by wind from getting into the fan. A flap is usually provided to prevent backdraft. In some areas, screens are also used to prevent animals, birds and large insects from getting into ducts. Screening is more commonly used on inlets than exhaust ducts, because inlets can't have backdraft flaps.

Causes   Weather hoods may be missing, damaged or loose because of poor installation practices or mechanical damage. Weather hoods close to grade level are particularly vulnerable to mechanical damage.

Implications   If the weather hood is missing, damaged or loose, the possibility of rain and snow getting into the building at this point is significant.

Strategy   Check that the weather hood is present and tightly installed.

13.2.11 INADEQUATE BACKFLOW PREVENTION

We've talked about this flap already. This is located at the end of the duct as part of the weather hood assembly. It allows exhaust air to go out through the hood but stops backflow. The flap may be stuck open or closed, broken or missing.

Causes   This may be an installation issue, although it's more often a maintenance problem.

Implications   If the flap remains open, cold air can flow back into the house through the ductwork and exhaust fan. If this flap is stuck in the closed position, we won't be able to exhaust the warm, moist air.

Strategy   When outside, if the exhaust hood is accessible, operate the flap with your finger. Lift the flap out and let go of it to see if it drops back into place and closes appropriately. If the flap is difficult to open or does not close properly, repair or replacement is in order.

13.3 HEAT RECOVERY VENTILATOR CONDITIONS

Heat recovery ventilators or **energy recovery ventilators** or **air exchangers**, as they are often called, are more complex than exhaust fans, although they share many of the same components. We've talked about these devices and our listing of the problems will help you review the components and functions of HRVs. Let's start by listing the common problems we find in the field:

1. Inoperative
2. Noisy
3. Inadequate air movement
4. Cabinet cover missing or dirty
5. Wiring unsafe
6. Ducts leaky, damaged, disconnected or missing
7. Cold-side ducts not insulated
8. Termination or inlet point not found

9. Poor termination or inlet point location
10. Weather hood missing, damaged or loose
11. Inadequate backflow prevention on exhaust or screening on inlet
12. Warm-side fresh air duct not properly connected to furnace duct
13. Exhaust grilles missing, poor location or obstructed in house
14. Exhaust grille missing grease filter in kitchen
15. Flow collars missing on warm-side ducts
16. Dampers missing on warm-side ducts
17. Vapor barrier missing, damaged or incomplete on cold-side ducts
18. HRV not interlocked with furnace fan
19. Ventilation fan switch not found or not labeled
20. Filters dirty or missing
21. Heat exchanger core dirty or missing
22. Cabinet damaged, rusted, poorly supported
23. Condensate drain missing, leaking or clogged
24. No trap in condensate drain
25. Poor discharge point for condensate drain

13.3.1 TO 13.3.6 (SAME AS 13.2.1 TO 13.2.6)

Items 13.3.1 to 13.3.6 are similar to the discussions on exhaust fans in 13.2.1 to 13.2.6. We won't repeat the discussions here.

13.3.7 COLD-SIDE DUCTS NOT INSULATED

A fresh air inlet duct running from the outdoor wall to the HRV should be insulated. Similarly, the exhaust air outlet from the HRV to the outside wall should be insulated. Insulation values of approximately R-3 are adequate unless the duct runs are long.

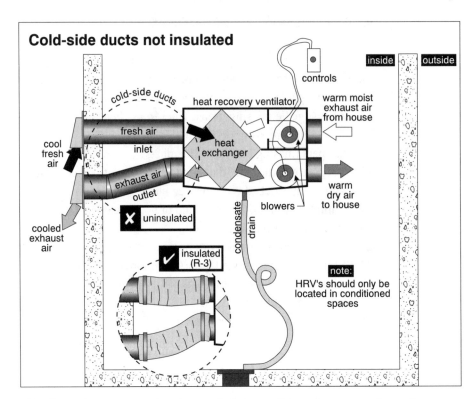

HRVs In Conditioned Spaces

This discussion assumes that the HRV is in a conditioned space, such as a basement. HRVs should not be located in unconditioned spaces. This is rare but is a possibility.

The balance of the comments are the same as for 13.2.7.

13.3.8 TERMINATION OR INLET POINT NOT FOUND

This is the same discussion as we had on exhaust fans in 13.2.8, except that we are looking for both an exhaust air outlet and a fresh air intake.

13.3.9 POOR TERMINATION OR INLET POINT LOCATION

Again, this discussion is similar to what we had for exhaust fans in 13.2.9. The outlet hood discussion is exactly the same.

Air Inlets

Fresh air inlets should be at least eighteen inches above grade level and away from areas where pollutants may enter the house. For example, intakes should not be adjacent to clothes dryer vents, driveways or carports, garbage storage areas, etc. Inlet and exhaust points are typically separated by six feet or more. Various manufacturers and local jurisdictions have varying recommendations here.

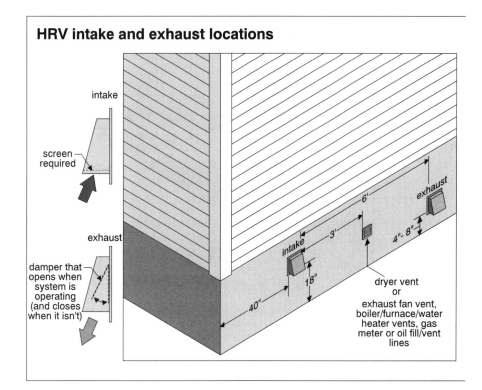

HRV intake and exhaust locations

intake

screen
required

exhaust

damper that
opens when
system is
operating
(and closes
when it isn't)

exhaust

6'

intake

3'

4"- 8"

18"

40"

dryer vent
or
exhaust fan vent,
boiler/furnace/water
heater vents, gas
meter or oil fill/vent
lines

13.3.10 WEATHER HOOD MISSING, DAMAGED OR LOOSE

Both the inlet and outlet should have a weather hood. The comments in 13.2.10 apply.

13.3.11 INADEQUATE BACKFLOW PREVENTION ON EXHAUST OR SCREENING ON INLET

The exhaust vent should have a flap. The inlet should have a screen with openings roughly a quarter-inch square.

13.3.12 WARM-SIDE FRESH AIR DUCT NOT PROPERLY CONNECTED TO FURNACE DUCT

The fresh air duct coming from the HRV very often ties into the return plenum for a forced-air furnace. This connection may be direct or indirect, depending on the recommendations of the manufacturer and the local jurisdiction. As a general rule, the fresh air inlet should be ten feet from the furnace, minimum. Where the exhaust is taken out of the return air plenum, the fresh air inlet should be downstream of the exhaust and, typically, at least three feet away.

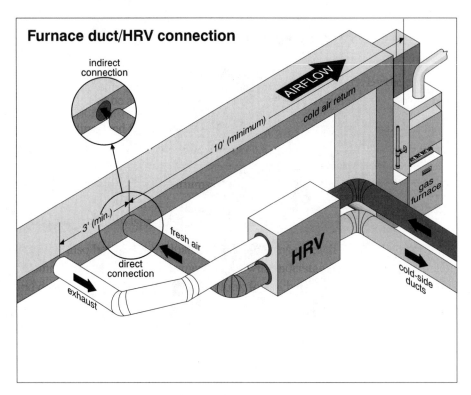

Furnace duct/HRV connection

indirect
connection

AIRFLOW

cold air return

10' (minimum)

3' (min.)

fresh air

direct
connection

exhaust

HRV

gas
furnace

cold-side
ducts

Causes An improper connection is an installation issue.

Implications If the cool, fresh air reduces the temperature in the return plenum below 55° or 60°F, the furnace may see air that is too cold. This may lead to premature furnace heat exchanger failure and may void the manufacturer's warranty on the furnace. The connection should be arranged so that if the HRV shuts off, either intentionally or accidentally, the heating system balance will not be disturbed. Similarly, if the furnace fan shuts down, we don't want the cool, fresh air to move through the house heating system. This could lead to poor comfort in the house and condensation on the duct system.

Strategy Follow the fresh air duct from the HRV to see where it terminates. It may feed individual rooms in the house, although more often it will discharge into the furnace return plenum. Check that it is roughly ten feet from the furnace and downstream of any HRV exhaust duct connections. You won't know whether it's supposed to be a tight connection or an indirect connection unless you have the manufacturer's literature.

13.3.13 EXHAUST GRILLES MISSING, POOR LOCATION OR OBSTRUCTED IN HOUSE

High Level   On many HRV systems, the exhaust grilles are located in the living spaces. These grilles are connected to the exhaust ducts on the warm side that lead to the heat recovery ventilator. The exhaust grilles should typically be high on walls or at ceiling level. They may be at moisture sources such as kitchens or bathrooms. Exhaust grilles should not be range hoods. The high grease content in here makes this an undesirable arrangement. Any exhaust grille in the kitchen, even if remote from the range hood, should typically have a grease filter.

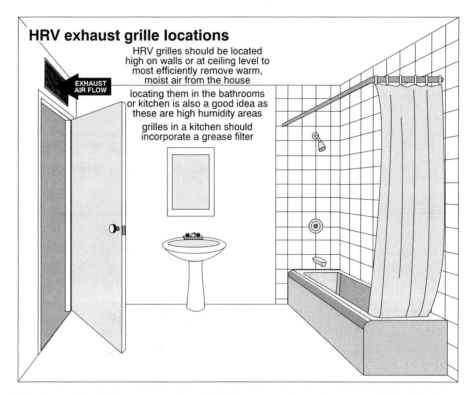

HRV exhaust grille locations

EXHAUST AIR FLOW

HRV grilles should be located high on walls or at ceiling level to most efficiently remove warm, moist air from the house

locating them in the bathrooms or kitchen is also a good idea as these are high humidity areas

grilles in a kitchen should incorporate a grease filter

Non-Combustible Ducts   The ducts should be non-combustible if connected to a furnace or if leading from the kitchen.

Causes   Poorly located exhaust grilles are an installation issue. Missing or obstructed grilles may be an installation or maintenance issue.

Implications   Exhaust grilles poorly located will not remove an appropriate amount of warm, moist air from the house. If the grilles are missing or obstructed, the ducts won't move an appropriate amount of air.

Strategy   Check that exhaust grilles are present, unobstructed and located high on walls or at ceiling levels. If there is an exhaust grille in the kitchen, make sure it is several feet from a cooktop. There may be a separate range hood for the cooktop.

13.3.14 EXHAUST GRILLES MISSING GREASE FILTER IN KITCHEN

Many manufacturers call for a grease filter when there is an exhaust grille in the kitchen.

Cause

The filter may be missing because it was not installed originally or because it was dirty and has been removed for cleaning or replacement.

Implications

If the filter is missing or dirty, grease may accumulate in the duct and become a fire hazard.

Strategy

The grille typically lifts open so that you can remove and clean or replace the filter. Check for a filter at the grille in the kitchen.

13.3.15 FLOW COLLARS MISSING ON WARM-SIDE DUCTS

Flow collars (flow measuring stations) are not required in many jurisdictions. However, it is impossible to balance an HRV without flow collars. These collars should be located on the warm-side ducts close to the HRV. There should be one on the exhaust and one on the fresh air duct. They should be at least twelve inches from balancing dampers.

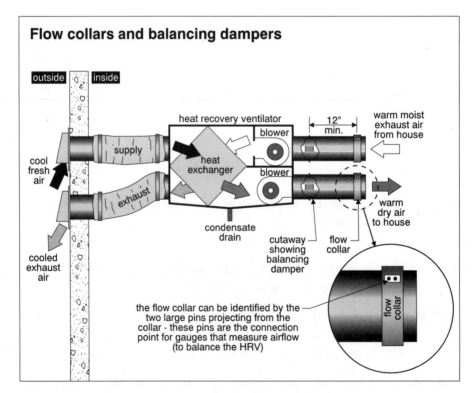

Flow collars and balancing dampers

Cause

If these are not present, this is an original installation issue.

Implications

It's impossible to balance the HRV without these flow measurement stations.

Strategy

Check that these collars are in place. They are usually easy to identify by the two large pins projecting from the collar. These pins are used for connections to gauges.

13.3.16 DAMPERS MISSING ON WARM-SIDE DUCTS

We indicated that we needed to have flow collars so that we could measure the flow on the exhaust and return side. We also need a way to adjust the airflow in each duct so that we can achieve a balance. Dampers should be provided on each duct, on the warm side close to the HRV. These dampers should be at least twelve inches away from the flow collars.

Causes   Missing dampers are an installation issue.

Implications   Without dampers you can't balance the system.

Strategy   Check that dampers are in place.

13.3.17 VAPOR BARRIER MISSING, DAMAGED OR INCOMPLETE ON COLD-SIDE DUCTS

The cold-side ducts should not only be insulated, as we discussed earlier, they should also have a vapor barrier. The vapor barrier should be on the warm side, which in this case is the outside. Pre-manufactured flexible ducts with insulation and an external vapor barrier are often used.

Causes   Missing, damaged or incomplete vapor barriers may be an installation issue or a mechanical damage problem.

Implications   Condensation and reduced insulating value are the short-term implications. Rusted ductwork is the longer-term implication.

Strategy   Check for a tight vapor barrier on the cold-side ducts.

13.3.18 HRV NOT INTERLOCKED WITH FURNACE FAN

HRVs that take exhaust air or provide fresh air into the return plenum should be interlocked with the furnace fan. The HRV should operate on low speed with the furnace fan operating. If the furnace fan is multiple speed, it is often operated on low speed at the same time. When the house Ventilation Fan switch is activated, the HRV shifts to high speed. The furnace fan may also be interlocked to operate at high speed. We don't want the HRV fan operating without the furnace fan operating. Most HRVs are set up to operate continuously at low speed. This means that the furnace fan should operate continuously.

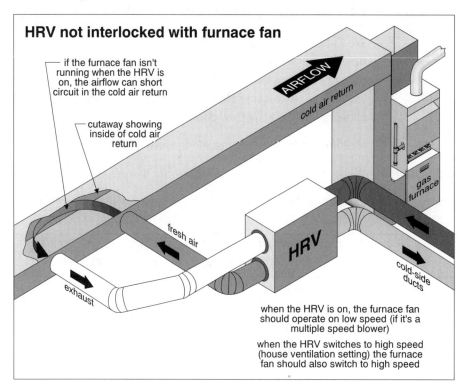

HRV not interlocked with furnace fan

if the furnace fan isn't running when the HRV is on, the airflow can short circuit in the cold air return

cutaway showing inside of cold air return

AIRFLOW

cold air return

gas furnace

fresh air

HRV

exhaust

cold-side ducts

when the HRV is on, the furnace fan should operate on low speed (if it's a multiple speed blower)

when the HRV switches to high speed (house ventilation setting) the furnace fan should also switch to high speed

Cause	The failure to interlock these systems was common on older installations. It is an original system setup issue.
Implications	The HRV may not remove enough warm, moist air or add enough fresh air if the furnace fan is not operating. If both the exhaust and fresh air ducts from the HRV are connected to the return plenum, the HRV may simply short circuit when the furnace fan is off.
Strategy	Check that whenever the HRV is operating, the furnace fan is operating. If you shut off the HRV, the furnace fan may shut off. As long as both operate continuously at low speed, this requirement is satisfied. Again, you'll find many older systems that do not operate this way. We recommend rearrangement to make this happen.

13.3.19 VENTILATION FAN SWITCH NOT FOUND OR NOT LABELED

There should be a switch that allows the HRV to be turned on at high speed to achieve increased ventilation capacity to meet short-term needs. This is a requirement according to many jurisdictions. You should learn whether this is called for in your area.

Cause	The omission of a Ventilation Fan switch is an installation issue.
Implications	The HRV cannot be turned up to high speed as necessary to improve ventilation.
Strategy	If this arrangement is called for in your area, make sure that such a switch is present. It is usually centrally located in the house, often near the furnace thermostat.

An Alternative Arrangement   In some cases, a bathroom exhaust fan serves as a primary exhaust fan and this may be controlled by a centrally located Ventilation Fan switch. Operating this switch will, in some cases, activate the heat recovery ventilator and, in other cases, activate the bathroom exhaust fan. Watch for both possible arrangements.

13.3.20 FILTERS DIRTY OR MISSING

Filters are typically located in the HRV cabinet. Open the cabinet cover and look for filters. You will typically see a slotted tray that accommodates the filter. If they are present, remove them and check their cleanliness. Remember the flow of air will be in two different directions through the heat exchanger.

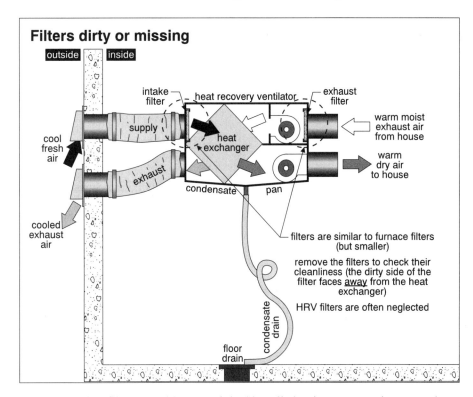

Causes   Dirty or missing filters are either an original installation issue or a maintenance issue.

Implications   If the filter is missing, the heat exchanger is likely to get dirty. If the filters are dirty, airflow may be restricted.

Strategy   Check that the filters are in place and clean.

13.3.21 HEAT EXCHANGER CORE DIRTY OR MISSING

On some heat exchangers, it is very easy to remove the heat exchanger core and check that it is clean. The cores on many types of heat exchangers can be cleaned easily in a dishwasher or laundry tub.

You may not be comfortable removing the heat exchanger unless you are familiar with the particular HRV.

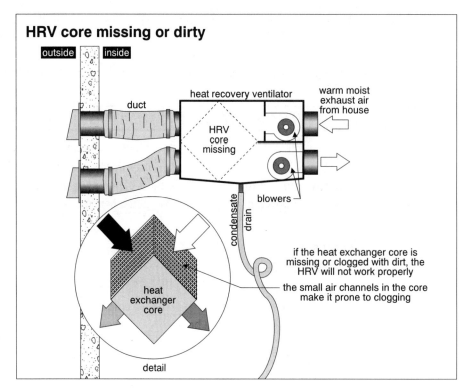

HRV core missing or dirty

outside | inside

heat recovery ventilator

warm moist exhaust air from house

duct

HRV core missing

blowers

condensate drain

if the heat exchanger core is missing or clogged with dirt, the HRV will not work properly

the small air channels in the core make it prone to clogging

heat exchanger core

detail

Causes   A dirty or missing heat exchanger is a maintenance issue.

Implications   If the heat exchanger is missing, the HRV will not work properly. If the heat exchanger is dirty, airflow will be restricted and ventilation will be poor.

Strategy   Check that there is a heat exchanger in place. If possible, remove it and check its cleanliness.

13.3.22 CABINET DAMAGED, RUSTED OR POORLY SUPPORTED

The HRV cabinet houses the fans and fan motor, the filters and the heat exchanger. Cabinets are typically hung from floor joists, often on vibration damping systems. The cabinet may be connected to ductwork with vibration damping collars to avoid transmitting noise and vibration through the house through the ductwork system.

Causes   Damaged, rusted or poorly supported cabinets may be the result of poor installation, poor maintenance or physical abuse.

Implications   If the cabinet is rusted or damaged, the components may also be damaged. If the cabinet is poorly supported, the condensate system may not work and duct connections may be strained. Noise may also be transmitted to the structure.

Strategy   Check that the system is level and free of mechanical damage and rust. Check carefully for rust below the condensate tray.

13.3.23 CONDENSATE DRAIN MISSING, LEAKING OR CLOGGED

Most HRVs require a condensate handling system, although rotary wheels and capillary blower-types do not.

Causes   A missing drain system or poor termination point is an installation issue. Leaking or clogged condensate drains are maintenance issues.

Implications Condensate may collect in the HRV and rust the unit. Condensate may also leak onto the floor below, and may damage finishes or storage.

Strategy   Check for a condensate drain line. It is usually a half-inch plastic tube. Look for a leaking, clogged or crimped line.

13.3.24 NO TRAP IN CONDENSATE DRAIN

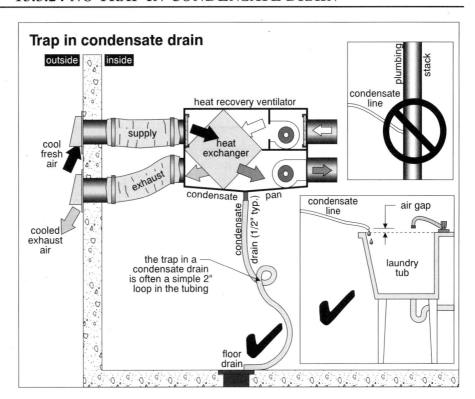

Cause   This is an installation issue.

Implications The trap in the condensate line prevents air being drawn into the house through the condensate line and the heat recovery ventilator.

Strategy   Make sure that a condensate drain trap has been provided. A two-inch loop in the drain line is typical. The line is often simply looped to create the trap.

13.3.25 POOR DISCHARGE POINT FOR CONDENSATE DRAIN

Condensate drains should not discharge outside in cold climates because of the possibility of freeze up. Condensate drains should not discharge into plumbing stacks or directly into waste plumbing piping. This can lead to sewer odors being drawn into the house through the HRV. Condensate drains should terminate with a one-inch air gap above a floor drain or other fixture. If gravity drainage from a condensate pipe is not possible, a condensate pump may be necessary.

Cause   A poor condensate discharge point is an installation problem.

Implications   Sewer gases may enter the home, as discussed.

Strategy   Follow the condensate drain and make sure it terminates appropriately with a one-inch air gap.

Summary

This ends our discussion of insulation and ventilation systems. You should have the message that insulation is not critical, but that concealed condensation can cause considerable damage. You should also understand that ventilation of indoor spaces is different from ventilation of attics.

HRVs – Will They Stay? Will They Be More Popular?   The 1980s and 1990s have seen much evolution in indoor ventilation systems. House air quality has become a big issue and we suspect we'll see more innovations in this area.

Homeowner Knowledge Is A Problem   Current ventilation systems are changing quickly. Many homeowners struggle to understand how their systems should work. Many field installations are not operating or not operating properly. Time will tell what sorts of changes are yet to come.

Rating For HRVs   Energy efficiency and other ratings are available for some HRVs in the "Certified Home Ventilating Products Directory" available from

1. Home Ventilating Institute (HVI)
    30 West University Drive
    Arlington Heights IL

2. Heating, Refrigerating and Air Conditioning Institute of Canada (HRAI Canada)
    5045 Orbitor Dr
    Building 11, Suite 300
    Mississauga ON L4W 4Y4

Insulation
& Interiors
MODULE

QUICK QUIZ 12

☑ INSTRUCTIONS

•You should finish Study Session 12 before doing this Quiz.

•Write your answers in the spaces provided.

• Check your answers against ours at the end of this Section.

• If you have trouble with the Quiz, re-read the Study Session and try the Quiz again.

• If you have no difficulty with the Quiz, you're ready to look at Inspection Tools and the Inspection Procedures.

1. List five signs of high humidity in houses.

_____

_____

_____

_____

_____

2. List two ways that humidity levels in houses can be controlled.

_____

_____

3. List eight ways to reduce moisture sources in houses.

_____

_____

_____

_____

_____

_____

_____

_____

4. List ten problems with exhaust fans.

1. _____

2. _____

3. _____

4. _____

5. _____

6. _____

7. _____

8. _____

9. _____

10. _____

5. In one sentence each, describe the implications of each of these problems with exhaust fans.

   1. _____

   2. _____

   3. _____

   4. _____

   5. _____

   6. _____

   7. _____

   8. _____

   9. _____

   10. _____

6. Describe in one sentence each, an inspection strategy for each of the problems with exhaust fans.

   1. _____

   2. _____

   3. _____

   4. _____

   5. _____

   6. _____

   7. _____

   8. _____

   9. _____

   10. _____

7. List as many common problems with heat recovery ventilators as you can.

1. _____
2. _____
3. _____
4. _____
5. _____
6. _____
7. _____
8. _____
9. _____
10. _____
11. _____
12. _____
13. _____
14. _____
15. _____
16. _____
17. _____
18. _____
19. _____
20. _____
21. _____
22. _____
23. _____
24. _____
25. _____

8. HRVs are typically located in conditioned spaces.
True ☐   False ☐

9. Air inlets should be at least four inches above grade level.
True ☐   False ☐

10. Are flow collars typically on the warm or cold side of the HRV?

_____

11. Where would you find the balancing dampers on an HRV?

_____

12. An HRV should never be interlocked with a furnace fan. True or false?
True ☐   False ☐

13. The Ventilation Fan switch should be centrally located in a home.
True ☐   False ☐

14. A trap in a condensate line is typically a two-inch loop in the drainline itself.
True ☐   False ☐

If you had no difficulty with the Quiz, you're ready to look at Inspection Tools and the Inspection Procedures.

Key Words:
- *Condensation*
- *Staining*
- *Mold*
- *Mildew*
- *Odors*
- *Backdraft*
- *Reduce moisture sources*
- *Ventilation*
- *Termination point*
- *Duct insulation*
- *Weather hood*
- *Exhaust fan*
- *HRV*
- *Cabinet*
- *Exhaust grilles*
- *Flow measuring stations*
- *Flow collars*
- *Balancing dampers*
- *Ventilation fan switch*
- *Filters*
- *Heat exchanger*
- *Condensate drain*
- *Trap*

► 14.0 INSPECTION TOOLS

Ladder

To get to roofs and inspect vents and to get to overhead attic hatches and exhaust fans.

Flashlight

For inspecting attics and crawlspaces.

Tape Measure

To determine depth of insulation in attics and thickness of walls.

Drop Cloth

To protect the home from insulation falling through the access hatch.

Selection of Screwdrivers

To remove access hatch covers for attics and crawlspaces, and to remove covers from HRVS.

Gloves

To wear in attics.

Masks with HEPA (High Efficiency Particulate Arresting) or P-100 type filers
To wear in attics.

Goggles
To wear in attics.

Carpenter's Awl, Screwdriver or Knife

To probe wood for rot.

Coveralls

To protect your clothing while getting into crawlspaces.

Electrical Tester

To ensure power is off before probing for insulation around electrical boxes.

Optional Items

Moisture Meters, Moisture Scanners and Humidity Meters (beyond Standards) To measure moisture content in building materials and house air.

► 15.0 INSPECTION PROCEDURES

Visual Inspection

All Parts Of Home

The inspection of the insulation and ventilation system in the house includes looking at all parts of the home: the exterior, the living space, and attics, basements and crawlspaces.

Exterior

On the exterior you'll look for roof vents and termination points for exhaust fans and clothes dryers. You'll also look for inlet and exhaust vents for heat recovery ventilators. You'll also be looking for evidence of insulation on the exterior of the foundations, and evidence of insulation on exterior walls and roofs. You may see evidence of insulation having been blown into flat roofs, for example.

Living Space Within the living space you'll be looking for evidence of wall insulation and vapor barriers around the perimeter of the home at openings for electrical receptacles and light switches, and any poorly finished details around doors, windows and ceilings. You'll look at, and operate, exhaust fans and heat recovery ventilators. You'll be looking at inlets and exhaust points for ventilation systems and any exposed ductwork. You'll be looking for a Ventilation Fan switch, used to control the ventilation system.

Attics, You'll be looking in attics at insulation, air/vapor barriers and ventilation. In
Basements And basements and crawlspaces you'll be looking at insulation, air/vapor barriers and
Crawlspaces possibly ventilation systems. Heat recovery ventilators are often located in basements.

Sequence Of Inspections

Exterior 1. Most inspectors perform their exterior inspection first. This includes getting onto the roof in most cases. It is during this part of the inspection that you will note soffit, roof, ridge and gable vents, for example. You'll also be looking at the other exterior clues we've touched on.

Basement And 2. The basement inspection and living space inspection are often conducted next,
Living Spaces though not necessarily in that order.

Attic And 3. Attic and crawlspace inspections can be done at any time, although many
Crawlspace inspectors leave them until the end because they are likely to get dirty during this part of the inspection. They may also be hot and uncomfortable spaces. The less dirt tracked through the house, and the less time you spend in the house when you are dirty and sweaty, the more pleasant it is for everyone.

It's Okay To There is no magic to this sequence, and no matter what order you approach things
Go Back in, you should be prepared to double back to check something in another part of the house based on a discovery in one area.

Inspection Your insulation and ventilation inspection is necessarily not complete. There is a
Incomplete great deal you cannot see. In many cases you won't be able to determine whether there is insulation in wall and flat roof areas, for example. You may also not be aware of insulation on the exterior of foundation walls for basements and crawlspaces. Whatever information you can uncover about insulation and ventilation you should treat as a bonus.

Rot Is The We've talked about how heat loss and comfort are the implications of inadequate
Enemy insulation. Neither of these are problems that seriously affect the function of the house in most cases. The serious problem that can be caused by inappropriate insulation, air/vapor barriers and ventilation is rot, which may affect concealed roof and wall components. Your insulation inspection should focus on this potentially serious problem.

What The
Inspection Is
Not
- The inspection of an insulation and ventilation system does not include testing of air leakage using a blower door.
- It does not include heat loss testing using infrared photography.
- It does not include a balance test on a heat recovery ventilator, nor does it include tests of the vapor pressure within the house.
- A home inspection is not a design review. No heat gain or heat loss calculations are typically done.
- The inspection is not destructive nor is it invasive. We don't tear down walls or rip apart roofs to see what is going on inside.
- We do not do indoor air quality tests and do not take samples.
- A home inspection is not an environmental audit.

Should Be
Able To Identify
Insulations
A home inspector should be able to recognize the common insulation materials that one may find on an inspection. In some cases this is easy; in others it may be difficult. As always, home inspectors should never guess. If you find an insulation material you can't positively identify, get some help.

Fans And
Lights On
Some inspectors leave upper story lights and exhaust fans on when going to the attic. It makes it easier to find things.

Operating test

Test Fans And
HRVS
While the insulation and ventilation inspection is primarily visual, operating tests are performed on exhaust fans and HRVs, if present. You'll want to test the controls to ensure these mechanical devices respond appropriately. You'll also want to check that air is moving appropriately. While we don't measure airflows, we do use our hands or a tissue to get a sense of appropriate air movement. In the text, we indicated the best test for air movement through an exhaust fan system is to check the outlet point. In many cases this isn't practical, and most home inspectors settle for testing the inlet.

Test HRV
With Furnace
When HRVs are interconnected with forced air heating and cooling system, the testing of these two systems is often coordinated. These tests are often limited by a lack of complete knowledge of how the systems were designed and set up. As we discussed in the text, there are many variables. HRVs can be controlled by manual switches, dehumidistats, humidistats, timers, etc. They may be one-speed or multi-speed. They may or may not be interlocked with the furnace blower. In many cases, you'll be recommending a specialist to service and balance the HRV.

Testing HRV

Here is one test procedure that can be used.

1. Turn all switches off and set humidistat OFF or to 100%.
2. Turn the Ventilation Fan switch to HIGH speed.
3. Check the airflow at all inlets and outlets, both indoors and out.
4. Listen for noise and vibration.
5. Check for air leaks in ducts.
6. Vary speed of system – set to lower speed
7. Check that all fans respond.
8. Use other controls to turn system on and off. Operate with dehumidistat, for example.
9. Return system to original position.

Document Inaccessible Areas

One of the most common mistakes inspectors make on the insulation and ventilation inspection is failure to notify a client that they could not fully inspect the home.

Common limitations include the inability to–

• move through the entire attic because of the depth of insulation.
• fully inspect the crawlspace because of access or safety concerns.
• inspect crawlspaces at all because you cannot find an access hatch, or the access hatch is tightly sealed.
• get into flat and cathedral roof spaces.
• get into secondary attic spaces (this includes knee wall areas and attics over house extensions).

Make sure you let your clients know what you could not get a look at. For restricted access spaces such as attics and crawlspaces, report how you did your inspection. Did you crawl all the way through? Did you get partway through? Did you simply look from the access hatch?

Emphasize Safety

We've talked about the dangers of inspecting attics:
• Insulation can irritate skin, lungs and eyes.
• You may step on exposed electrical connections.
• You may fall through the ceiling.
• You may come in contact with animal droppings.
• You may come in contact with animals and stinging insects.
• Pull-down staircases may fall onto you or collapse under your weight.

Any way you look at it, attics are not a lot of fun.

Don't Bluff

If HRVs are not common in your area and you come across one, don't bluff your way through the inspection. Explain to the client that these are not common and that you do not have experience inspecting and testing them. Recommend a specialist evaluate the system. Most require regular servicing, in any case. The majority we see in the field are not set up or maintained properly.

More Could
Be Done

As always, a home inspection is not technically exhaustive. You should recognize there are energy consultants who can determine more about a house insulation and ventilation system than you will be able to do as part of as standard home inspection. Help your client understand the limitations to this and every other part of your inspection. There's no need to apologize for being a general practitioner. The danger is in overselling yourself and misleading the client. You cannot be the expert in all fields.

Let's take a look at an inspection checklist that will help guide you through your insulation inspection.

► 16.0 INSPECTION CHECKLIST

N = North   E = East   S= South   W = West
B = Basement   1 = First Story   2= Second story   3=Third story   CS = Crawlspace

ATTICS AND ROOF SPACES				
LOCATION	ACCESS HATCH	LOCATION	INSULATION CONTINUED	
	Missing		Duct insulation loose	
	Inaccessible		Duct air/vapor barrier missing	
	Not insulated		Duct air/vapor barrier damaged	
	Not weather-stripped		Combustible insulation too close to masonry chimney	
			Insulation too close to metal chimney	
	ATTIC STAIRCASE		Air/vapor barrier missing	
			Air/vapor barrier incomplete	
	Inadequate insulation		Air/vapor barrier in wrong location	
	Inadequate weather stripping		Air leakage excessive	
	Excessive rise on steps			
	Inadequate run and tread width		ROOF VENTING	
	Treads loose or broken			
	Treads sloped or not uniform		Missing	
	Handrails or guardrails missing or unsafe		Inadequate	
			Obstructed	
	Lighting missing or ineffective		Snow or wet spots below roof vents	
	Headroom inadequate			
			TURBINE VENTS	
	PULL-DOWN STAIRS			
			Noisy	
	Dangerous to lower or raise		Seized	
	Unsafe to climb			
	Not insulated		POWER VENTS	
	Not weather stripped		Operating in winter	
			Poor wiring	
	INSULATION		Inoperative in summer	
	Too little		RAFTERS AND SHEATHING	
	Wet			
	Compressed		Rot, mold or mildew	
	Gaps or voids		Sheathing delaminating or buckling	
	Missing at dropped ceilings			
	Covering recessed lights		WHOLE HOUSE FAN	
	Inadequate in knee wall areas			
	Inadequate at skylights and light wells		No Insulated cover	
	Duct insulation missing			

WALLS			
LOCATION	INSULATION		
	Too little		
	Sagging or voids		
	Air/vapor barrier missing, incomplete or in wrong location		
	Mold, mildew or rot		

BASEMENTS AND CRAWLSPACES			
LOCATION	**INSULATION**	LOCATION	Exterior insulation not protected at top
			Exposed combustible insulation
	Too little		Air/vapor barrier missing, incomplete
	Incomplete		or in wrong location
	Missing at rim joists		No moisture barrier on basement walls
	Sagging, loose or voids		No moisture barrier on earth floor
	Exterior insulation not suitable for		
	use below grade		

FLOORS OVER UNHEATED AREAS			
LOCATION	**INSULATION**		
	Too little (usually difficult to tell)		

EXHAUST FANS AND HEAT RECOVERY VENTILATORS			
LOCATION	**EXHAUST FANS**	LOCATION	**WEATHER HOOD (CONTINUED)**
	Inoperative		Inadequate screening on inlet
	Noisy		Warm-side fresh air duct not properly
	Inadequate air movement		connected to furnace duct
	Cover missing		Exhaust grilles missing
	Wiring unsafe		Poor location
	Ducts leaky		Obstructed
	Damaged		Exhaust grille in kitchen missing grease filter
	Disconnected		Grease filter dirty
	Missing		Flow measuring stations missing
	Not insulated in unconditioned space		Balancing dampers missing
	Termination point not found		Duct vapor barrier missing, damaged or incomplete
	Poor termination location		HRV not interlocked with furnace fan
	Weather hood missing or loose		Ventilation Fan switch not found
	Inadequate backflow prevention (flap)		Ventilation Fan switch not labeled
	Clothes dryer vented outside?		
			FILTERS
	HEAT RECOVERY VENTILATOR		
			Dirty
	Inoperative		Missing
	Noisy		
	Inadequate air movement		**HEAT EXCHANGER CORE**
	Cabinet cover missing		
	Dirty, rusted		Dirty
	Wiring unsafe		Missing
	DUCTS		**CABINET**
	Leaky		Damaged
	Damaged		Rusted
	Missing		Poorly supported
	Cold-side ducts not insulated		
	Termination or inlet points not found		**CONDENSATE DRAIN**
	Poor termination or inlet location		
			Missing
	WEATHER HOOD		Leaking
			Clogged
	Missing		No trap
	Damaged		Poor discharge point
	Loose		
	Inadequate backflow prevention on exhaust (flap)		

Insulation & Interiors

MODULE

FIELD EXERCISE 1

☑ INSTRUCTIONS

In this Field Exercise we'll combine some research and some inspection work. You should allow yourself about four hours for this Field Exercise. You should assemble the inspection tools and refer to the Inspection Checklist and Inspection Procedure. In this case we're going to start with the research, and then move on to the inspection section.

Part A – Research

For this part of the Exercise you'll want to visit building supply houses. You may also want to talk to local authorities (building inspectors), insulation contractors and energy consultants. Approaches to insulation, air/vapor barriers and ventilation vary dramatically, depending on climate and local building practices. It is important for you to know the generally accepted practices in your area. Develop a set of questions and talk to several people about them. Your questions should include the following:

1. What insulation levels are currently recommended for –
   - attics?
   - flat roofs?
   - cathedral roofs?
   - above-grade walls?
   - below-grade walls?
   - crawlspaces?
   - floors over unheated spaces?
2. What insulation materials are commonly found in this area? Where is each used? (You should obtain samples of each.)
3. What are the most common roof venting approaches?
4. Are basements typically insulated inside or outside the foundation walls?
5. Are crawlspaces typically heated or unheated?
6. Are crawlspaces typically vented or non-vented?

7. What materials are used on unfinished crawlspace floors to prevent moisture rising into the house?
8. What are the most common insulation, air/vapor barrier and ventilation problems found in this area?
9. Are house ventilation systems common? Are they typically exhaust-only, supply-only, or balanced ventilation systems? Are HRVs commonly used? What types are commonly used?

Note: Try to separate authoritative fact from opinion. Don't rely on a single source for all of your information. Look for reinforcement from multiple sources, or authoritative backup.

Part B – Inspection

1. Insulation

For this part of the Field Exercise you should look for five houses that have attics, five houses with unfinished basements, and five houses with crawlspaces. Use the Inspection Procedure and Checklist to check the insulation in roof spaces, walls (to the extent possible) and in basements or crawlspaces or both.

• How much were you able to see?
• Were you able to move through the attics?
• How much time did it take? (You'll want to get a sense for how long to allow for this part of your inspection.)
• What was the temperature in the attic?
• Were you able to identify the air/vapor barrier?
• Did you measure the insulation and multiply it by its approximate R-value per inch to get an overall insulation value in the attic?
• Were you able to convince yourself that soffits and roof vents were open and effective?
• Were there baffles for the soffit vents?
• Were any vents obstructed by birds' nests or other obstructions?
• Did you find any gaps or discontinuities in the air/vapor barrier?
• How was insulation handled around chimneys?
• Were there any recessed lights?
• Did you use the trick of leaving exhaust fans and lights on, to help identify these in the attic?
• How much time would it add to your attic inspection to look at the other house systems that have attic components (structure, plumbing, electrical, heating, air conditioning, and roofing)?

2. Fans and heat recovery ventilators

Try to find three homes with HRVs. If you can't find any, these may simply not be used in your area.

Before operating the HRV, inspect it using the Inspection Checklist. Checking the HRV can be done using the steps set out in the Inspection Procedure. Make sure you reset each HRV to the position you found it in.

All Done?   When you're finished with this Field Exercise then you're ready for the Final Test.

► ANSWERS TO QUICK QUIZZES

Answers to Quick Quiz 1

1. 1. Insulation and vapor retarders in unfinished spaces
2. Ventilation of attics and foundation areas
3. Kitchen, bathroom and laundry ventilation systems

2. 1. Insulation and vapor retarders in unfinished spaces
2. Absence of same in unfinished space at conditioned surfaces

3. 1. Concealed insulation and vapor retarders
2. Venting equipment which is integral with household appliances

4. To control heat loss

5. Restrict vapor diffusion

6. 1. Attic
2. Flat roof space
3. Knee wall areas
4. Crawlspace areas
5. Wall cavities

7. 1. To eliminate moisture
2. To remove indoor air pollutants and draw an adequate supply of fresh air into the house

8. 1. In cold climates, it keeps the house warm
2. In hot climates, it keeps the house cool
3. It's more critical in cold climates

9. Moisture may get into the building structure, causing damage.

10. The average temperature for a day is subtracted from 65°F, giving a number of degree days for that day (i.e. 65°F - 45°F = 20 degree days, if the average temperature is 45°F).

11. 1. Air/vapor barriers
2. Ventilation

12. 1. Flushes moisture from unconditioned spaces in winter and heat from these spaces in summer
2. Provides fresh air for building occupants

Answers to Quick Quiz 2

1.   Amount of thermal energy in a body – BTU

2.   Level of thermal energy in a body – Fahrenheit (°F)

3.   Heat that causes temperature to change

4.   Energy used to change from liquid to vapor state without changing temperature

5.   Energy used to change from solid to liquid state without changing temperature

6.   1. Conduction, radiation or convection
     2. Evaporation applies to people only!

7.   1. Rate of heat transfer over a period of time (usually hours)
     2. A unique property for every material

8.   A measure of heat transfer through materials that are not homogeneous or have large air voids

9.   1. Steel
     2. Lead
     3. Copper
     4. Concrete
     5. Plaster
     6. Stone
     7. Glass
     8. Clay

10. 1. Cork
     2. Sawdust
     3. Some plastics

11. Air is a good insulator on its own, so materials with a lot of air voids make good insulators.

12. The inverse of thermal conductivity, a convenient number to describe the resistance of a material to heat transfer.

13. It changes the air in the house, so we don't end up with stale, polluted air and an unhealthy environment.

14. We lose a lot of heat, and energy costs are higher.

15. When wind blows across insulation, it disturbs the air pockets in it, reducing the insulation value.

Answers to Quick Quiz 3

1. 1. Washing faces, hands, brushing teeth
    2. Showers and baths
    3. Cooking and washing dishes
    4. Washing clothes
    5. Washing floors, walls, furniture
    6. Watcring plants
    7. Breathing and perspiring
    8. Damp soil in subgrade spaces
    9. Firewood drying
    10. Pets

2. The actual amount of moisture in the air

3. The amount of moisture in the air relative to the amount it could hold if saturated, expressed as a percent

4. The maximum amount of moisture air could hold at a given temperature without condensing

5. 1. Bulk moisture
    2. Capillary action
    3. Air-transported moisture
    4. Vapor diffusion

6. As warm air rises in a building, it expands, creating a higher pressure near the top of the house. This higher pressure air tries to escape through any opening it can find. The cooler, lower pressure air near the bottom of the house tends to allow outdoor air in through any openings it finds.

7. The point in the house where the positive and negative pressures balance each other out exactly

8. The point at which condensation will start to occur

9. 1. Air leakage is much more important than
    2. Vapor diffusion

10. The ability of a material or assembly to dry out after it has gotten wet

11. More insulation and tighter building cavities allowed less air leakage. The longer the air stays in the building, the more moisture is likely to be in it. Since the walls and roofs are cooler (more insulation), the moist air passing through will cool and condense at a higher rate than before.

12. 1. Produce less moisture
   2. Keep the condensing surfaces warmer
   3. Stop air leakage
   4. Flush the air out faster
   5. Exhaust warm, moist air directly outside

13. The direction of moisture movement is from the outside to the inside. We worry about air leakage into rather than out of the house. Air conditioning creates a drier, cooler interior, thus increasing the potential for moisture problems.

Answers to Quick Quiz 4

1. Air leakage

2. 1. Kraft paper
   2. Polyethylene films (more modern)

3. We were not able to completely stop air leakage into cool roof spaces, so we need to get rid of that warm, moist air.

4. They have further reduced air leakage.

5. Housewrap fits tighter to the wall, with fewer joints and restricts air movement while allowing vapor diffusion.

6. Modern windows fit tighter, and do not allow moisture to escape. Since they are double or triple paned units, the inner pane is warmer, which makes it more difficult to detect moisture buildup inside the home.

7. 1. They throw out more moisture than air
   2. Lower air pressure inside, which increases dry air movement through walls and roofs

8. 1. Wastes energy, heated air thrown outside
   2. Can cause backdraft of combustion appliances

9. We don't have to worry about backdraft, so we can use our fans and HRVs to control moisture.

10. To transfer some of the heat from the air exhausted from the house into the fresh air coming into the house

11. 1. Two duct systems and fans to move air
   2. Heat exchanger
   3. Defrost mechanism
   4. Balancing dampers and flow collars
   5. A drain
   6. Controls for fan speed and amount of air movement

12. 1. Venting of roof and wall spaces to flush warm moist air out of the building components

2. Exhausting stale air and supplying fresh air to the living space

13. Outside or exterior to the house – the warm, humid air leaking into the cool air conditioned interior

14. Outer part of the wall, or not at all

Answers to Quick Quiz 5

1. Slow the rate of heat transfer

2. 1. High resistance to heat flow
2. Inexpensive
3. Durable
4. Completely fills cavities
5. Air barrier
6. Vapor barrier
7. Moisture and rot resistant
8. Non-combustible
9. Chemically inert

3. Points where insulation is not continuous, allowing heat to leak out. Wood wall studs form thermal bridges, for example.

4. If there are gaps in the insulation, convective loops can allow air to move freely through the wall cavity, lowering the R-value.

5. On a windy day, a lot of air can blow through the insulation, reducing its R-value.

6. 1. Loose fill
2. Batts or blankets
3. Rigid boards
4. Foamed-in-place

7. 1. Fiberglass
2. Cellulose fiber
3. Mineral wool
4. Vermiculite/perlite
5. Expanded polystyrene (EPS)
6. Extruded polystyrene
7. Closed cell phenolic plastic
8. Polyisocyanurate
9. Polyurethane
10. Isocyanate/polyisocyanate

8.
1. Loose, fill, batt, board
2. Loose fill
3. Loose fill, batt, board
4. Loose fill
5. Board
6. Board
7. Board
8. Board
9. Foamed-in-place
10. Foamed-in-place

9.

Advantages:	Disadvantages:
Fiberglass	
1. Inexpensive 2. Versatile 3. Non-combustible (almost) 4. Commonly used	1. Skin/lung irritant 2. R-value may vary depending on installation 3. Can get wet and reduce R-value 4. Compressible, reduce R-value
Mineral Wool	
1. Non-combustible 2. Water resistant 3. Rot resistant 4. Inexpensive	1. Compressible 2. Not an air/vapor barrier
Cellulose	
1. Treated to resist fire 2. Recycled material 3. Less susceptible to wind washing and convective air movement than fiberglass	1. Absorbs water, reduced R-value 2. Combustible if not treated
Vermiculite or Perlite	
1. Non-combustible 2. Resists sunlight	1. Heavy and expensive 2. Absorbs moisture if not treated 3. Not common on new construction
Polystyrene (Expanded)	
1. Used in pre-fabricated wall panels 2. Almost an air/vapor barrier 3. Good R-value per inch	1. Damaged by sunlight, chemicals 2. Combustible

Advantages:	Disadvantages:
Polystyrene (Extruded)	
1. Better than expanded polystyrene, higher R-value 2. Can be used below grade 3. Water resistant	1. Damaged by sunlight, chemicals 2. Combustible
Phenolic Board	
1. High R-value per inch 2. Air barrier if joints taped	1. Combustible 2. Expensive 3. Corrosive acids form if wetted 4. Not a vapor barrier
Polyurethane/ Isocyanurate Boards	
1. High R-value 2. Air/vapor barrier	1. Expensive 2. Combustible 3. Not resistant to moisture
Polyurethane – Foamed in Place	
1. Expands to reach tough-to-insulate spaces 2. Air barrier	1. Deteriorates in sunlight 2. Combustible 3. R-value deteriorates over time 4. Can't use in closed cavities
Isocyanate or Polyisocyanate or Polyicynene	
1. Blow into tough to insulate spaces 2. Air/vapor barrier 3. Resists sunlight	1. Combustible 2. Can't use in closed cavities

10.　1. Installed as a resin with an acid foaming agent, and air as a propellant. Used primarily in existing buildings.
　　2. Controversial because the formaldehyde gas released during installation and curing was said to have long term health implications, although it was never proven.

Answers to Quick Quiz 6

1. To stop air movement through building walls and roof

2. 1. Minimize heat flow to exterior/interior
   2. Minimize moisture flow to building components

3. To protect the building from moisture damage. Designed to protect from moisture due to vapor diffusion, not air movement.

4. Vapor retarder, vapor diffusion retarder (VDR)

5. 1. Stops air movement
   2. Durable
   3. Strong, and either rigid or well enough supported to stay in place
   4. Continuous
   5. Inexpensive
   6. Resistant to moisture, rot, chemicals

6. 1. Polyethylene
   2. Housewraps
   3. Foam insulation boards
   4. Drywall, plaster, wood paneling
   5. Sheathing
   6. Asphalt impregnated fiberboard
   7. Building paper
   8. Sill gaskets
   9. Gaskets for electrical boxes and plastic enclosures around electrical boxes
   10. Backer rods
   11. Caulking and weatherstripping
   12. Polyurethane foams
   13. Duct tape and duct mastic

7. 1. Retards vapor diffusion
   2. Durable
   3. Moisture and rot resistant
   4. Chemically inert
   5. Inexpensive

8. Air barrier

9. Warm side

10. Warm side

11. Imperial – the number of grains of water that will move through one square foot of a material per hour under a 1-inch (Mercury) pressure differential.

12. 1. Polyethylene film
    2. Kraft paper
    3. Aluminum foil

4. Oil-based paints and vapor retardant paints
5. Insulations
6. Vinyl wallpaper
7. Plywood and OSB sheathings

13. To keep the moisture in the ground, instead of diffusing into the air and adding to the moisture levels in the crawlspace and house

Answers to Quick Quiz 7

1. 1. Allow warm, moist air out of the attic
   2. Reduce attic temperatures in summer
   3. Helps prevent ice dams in winter by keeping attic cold

2. 1. Soffit
   2. Ridge
   3. Roof
   4. Gable

3. At least 50%

4. 50% or less

5. If there are gable vents at opposing ends

6. 1. Don't work on calm days
   2. Often noisy or seized
   3. Can depressurize the attic on windy days
   4. Often covered up to prevent water leakage into attic

7. Prevent insulation from covering the vents

8. 1/300 of the floor space of the attic

9. 1/150 of the floor space of the attic

10. 1. Required at the ridge and bottom of upper section
   2. None is required for steep section

11. 1. Yes, if the ceiling is not well sealed
   2. Can cause negative pressure in attic that promotes warm, moist air to enter at a faster rate

12. They tend to depressurize the attic in winter, again promoting entry of warm, moist air

Answers to Quick Quiz 8

1. 1. Eliminate moisture, odors and other indoor air pollutants
    2. Bring fresh air into the home

2. Newer homes are more energy efficient and tighter. We get less air changes than we used to.

3. 1. Exhaust
    2. Supply
    3. Balanced

4. Controlling both the exhaust air and fresh air supply

5. Improve the energy efficiency in a balanced ventilation system by controlling the pressure, and transferring heat from exhaust air to the fresh air supply

6. 1. Cabinet
    2. Heat exchanger
    3. Inlet and exhaust fans
    4. A duct system
    5. Flow measuring stations
    6. Controls
    7. Air filters
    8. A condensate system
    9. A defrost system

7. The warm, stale air is brought in one side of the heat exchanger while the cool, fresh air is brought in the other side. As the warm air passes through, it gives up some of its heat indirectly to the cool air. The warm air cools down, and the cool air warms up.

8. 1. The duct upstream of the HRV which brings cool fresh air in from outside, and the duct downstream of the HRV exhausting the warm, stale air outside should be insulated.

    2. This is because we don't want to use the ducts as a path for heat to escape out of the house.

9. These allow for balancing the fresh air supply with the warm air exhaust. There is one on the fresh air supply, and one on the warm air exhaust, both on the warm side ductwork.

10. Balancing dampers are located on the warm side ducts, and balance the system so we don't over-pressurize or depressurize the house.

11. 1. Through dedicated ducts to various rooms of the house
    2. Into the cold air return plenum of the furnace

12. 1. Kitchens and bathrooms
    2. Any other room in the house
    3. The return (cold) air plenum of the furnace

13. 1. Six feet away from exhaust
    2. 18 inches above grade
    3. 40 inches away from corners of buildings
    4. Three feet from gas meters
    5. Well away from driveways and garages
    6. Three feet from dryer vents, furnace, boiler or water heater vents, and oil fill and vent lines

14. 1. Away from fresh air intakes, attics, garages, crawlspaces
    2. Four to eight inches above grade, and protected from the elements with a hood
    3. Screens
    4. A damper that opens easily when the system is running

15. 1. Manual operation
    2. Automatic operation, with timers or dehumidistats
    3. Continuous operation

16. 1. Incoming air preheated with an electric duct heater
    2. Exhaust air recirculated through the fresh air inlet
    3. Exhaust fan stops and fresh air intake is blocked
    4. Fresh air fan stops and exhaust fan continues to move warm air

17. The HRV and furnace fan operate continuously at low speed. Sometimes there is a switch to operate the HRV. Switching the HRV fan to high speed may also switch the furnace fan to high speed.

18. 1. It is needed because the warm air is often cooled to the saturation point, and we need to handle that condensation
    2. It is not needed on some ERVs, which allow some moisture transfer as well as heat transfer

19. It is more important in a cold climate because the temperature differential between inside and outside is greater in the winter than summer.

Answers to Quick Quiz 9

1. 1. Mask
    2. Goggles
    3. Long sleeves with tight cuffs

2. 1. Fall through the ceiling
    2. Electric shock
    3. Irritate lungs, eyes, skin

3. 1. Should be uniform with adequate tread width and rise
    2. Headroom should be adequate
    3. Handrails and a guardrail should be at the top
    4. Adequate lighting

4. 1. Mechanical components loose or broken when pulling down
    2. Treads or stringers may be loose or broken

5. 1. Roofing
    2. Structure
    3. Electrical
    4. Heating
    5. Air conditioning or heat pump systems
    6. Plumbing

6.
Problem:	Implications:
1. Not insulated	Results in heat loss
2. Not weather stripped	Results in heat loss and air leakage
3. Missing	Results in heat loss and air leakage
4. Inaccessible	Limits inspection

7.
Problem:	Implications:
1. Inadequate insulation/ weatherstripping	Heat loss, air leakage
2. Stair rise, run and tread problems	Safety concerns
3. Handrail and guardrail problems	Safety concerns
4. Lighting problems	Safety concerns
5. Headroom problems	Safety concerns

8.

Problem:	Implications:
1. Stairs come down too fast	Safety/dangerous
2. Not solid and stable	Safety/dangerous
3. Not insulated	Heat loss, air leakage

9.

Problem:	Implications:
1. Two inches thick	Heat loss
2. Wet	Won't work well, damage to finishes
3. Gaps and voids	Localized heat loss, ice damming
4. Compressed	Reduced R-value
5. Missing	Heat loss, ice damming

10. 1. Only if lights are proper type, or installed in drywall boxes
2. Look for double shell, or "IC" stamp

11. 1. Walls indicate skylights or light wells
2. Yes, should be insulated

12. 1. Insulation
2. Air/vapor barrier

13. Masonry chimneys should only have noncombustible insulation surrounding them.

14. Vent is contained in a boxed-in area to keep insulation away

15. 1. Air leakage
2. Rot damage
3. Insulation gets wet from condensation

16. Problems:
1. Birds' nests
2. Mechanical damage
3. Undersized openings

Implications:
1. Inadequate venting
2. Mold/mildew/rot from condensation

17. If soffit venting is missing, rain and snow may actually be drawn into the roof vents.

18. 1. No ventilation on calm days
2. Too much ventilation on windy days
3. Noisy or seized

19. No, they can depressurize the attic, increasing heat loss

20. An exhaust fan dumping warm, moist air directly into the attic

21. With an insulated cover

Answers to Quick Quiz 10

1. 1. Treat as an attic – with a little insulation and an air space
2. Completely fill roof space – in theory it stops airflow
3. Insulate above sheathing – roof membrane applied over insulation
4. Insulate below roof structure – lowers ceiling heights, but provides good ventilation

2. Problem:
1. Too little insulation
2. Wet, compressed or voids
3. Missing or incomplete air/vapor barrier
4. Excessive air leakage
5. Missing or inadequate venting
6. Venting obstructed
7. Mold/mildew/rot suspected

3. Implications:
1. Heat loss, no damage to structure
2. Reduced R-values, mold/mildew/rot
3. Mold/mildew/rot
4. Mold/mildew/rot
5. Warm, moist air condensing and causing mold, mildew and rot
6. Mold/mildew/rot
7. Damage to structural members

4. 1. Remove ceiling light fixture or exhaust fan covers (with power off) OR
2. Pop fascia vents off and look through holes

5. Melting snow may indicate lots of heat loss and little insulation. Compare to similar homes nearby.

6. 1. Plugged holes in roof covering
2. Plugged holes in ceilings
3. Plugged holes in fascia boards
4. Vents added to roof
5. Extra roof thickness
6. Low ceilings

7. Staining, sponginess or dampness on ceilings

8. 1. Vent the roof space
   2. Seal the roof cavity

9. Opposing fascias

10. One square foot of venting for every 150 square feet of roof area

11. 1. Roof leaks
    2. Condensation

12. 1. Sagging or spongy roof surface
    2. Sagging ceilings
    3. Mold or mildew on ceilings
    4. Rusted nail heads on ceilings

Answers to Quick Quiz 11

1. False

2. a. Outside – keeps walls warm, no interior space lost, no interior living space lost, etc., but only certain types of insulation are suitable
   b. Inside – inexpensive, can be done at any time, any type of insulation is suitable, etc., but can be difficult to detect leaks

3. a. 1. Keeps foundation warm, stabilizing the structure
      2. No need to disrupt interior finishes
      3. No interior space lost
      4. If exterior walls exposed, can be cost-effective
      5. Outside walls are more uniform with fewer interruptions
      6. May be easier to insulate the rim joist area
      7. Exterior water control systems can be added or improved

   b. 1. Less expensive, a do-it-yourself project
      2. Can be done at any time, regardless of weather
      3. Less disruptive to the house
      4. Easier to address basement windows
      5. Any batt or board insulation is acceptable
      6. If homeowners plan to finish basement, cost is small
      7. Porches and garages not a problem

4. 1. Insulate walls, creating a heated crawlspace
   2. Insulate floor above, creating a cold crawlspace

5. Insulate walls

6.   1. Floor above is more comfortable if crawlspace is heated
          2. Easier to insulate walls than attach insulation to ceiling
          3. Less material is often needed to insulate walls than ceiling
          4. No risk of freezing pipes or excessive heat loss through ducts

7.   True if unheated – vent to outdoors to remove warm, moist air in cold climates

8.   1. Wet floors
          2. Animals
          3. Dark and dirty

9.   1. Insulation – too little or incomplete
          2. Exterior insulation not suitable for below grade use
          3. Exterior insulation not protected at top
          4. Insulation missing at rim joists
          5. Insulation sagging, loose or voids
          6. Exposed combustible insulation
          7. Air/vapor barrier missing, incomplete or wrong location
          8. No moisture barrier on basement walls
          9. No moisture barrier on earth floor

10.   1. Higher heating costs
          2. Lower R-value due to moisture
          3. Mechanical damage from lawn mowers, weed eaters, etc.
          4. Increased heating costs
          5. Increased heat loss
          6. Risk of fire
          7. Condensation on the cold side of insulation and damage
          8. Wet insulation, reducing R-value
          9. Elevated moisture levels in the crawlspace, possible water damage

11.   Moisture from the curing concrete condenses on the outer side of the air/vapor barrier.

12.   Heat is transmitted by conduction directly from our feet to the floor

13.   1. Heated cavity between floor and insulation below
          2. Spray-in-place foams that fill entire cavity

14.   1. Above garages
          2. Above porches
          3. In cantilevered areas
          4. Over breezeways
          5. Below windows projecting out from the building
          6. Over unheated crawlspaces
          7. Over open areas below houses with pier foundation and no skirting

Answers to Quick Quiz 12

1. 1. Condensation on windows
    2. Staining or streaking on window sills and areas below
    3. Mold or mildew on cool, dark or moist house surfaces
    4. Stale odors and stuffy air in the house
    5. Backdraft of combustion appliances

2. 1. Reduce moisture sources
    2. Ventilate the home

3. 1. Don't store firewood in the home
    2. Correct foundation leakage problems
    3. Disconnect or remove humidifiers
    4. Cover any earth floor in basement or crawlspace
    5. Cover sump pits
    6. Don't hang laundry to dry inside the home
    7. Vent clothes dryers to the outside
    8. Use exhaust fans when showering or cooking
    9. Limit the use of misters and steam generators

4. 1. Inoperative
    2. Noisy
    3. Inadequate air movement
    4. Fan cover missing
    5. Wiring unsafe
    6. Ducts leaky, damaged, disconnected, missing
    7. Ducts not insulated in unconditioned space
    8. Termination point not found, poor location
    9. Weather hood missing, damaged, loose
    10. Inadequate backflow prevention

5. 1. Won't help reduce humidity
    2. Inexpensive fan – won't perform its function
    3. Fan and ductwork are likely to get dirty faster
    4. Electric shock, fire hazard
    5. Electric shock hazard
    6. Reduced airflow, moist air being dumped into unconditioned spaces
    7. Condensation inside and outside duct in winter/summer
    8. Exhaust from fan drawn back into building; warm moist air dumped into building
    9. Rain or snow entry into building
    10. Cold air back into building, exhaust can't get outside

6.
1. Operate the fan
2. Listen for noise
3. Check at exhaust or intake point, use a tissue
4. Check that covers are in place
5. Check for wiring problems
6. Look for sags in ductwork
7. Check for insulation
8. Try to find termination points
9. Watch for location of exhaust vents
10. Check for weather hood present and tightly installed
11. Operate flap with your finger

7.
1. Inoperative
2. Noisy
3. Inadequate air movement
4. Cabinet cover missing or dirty
5. Wiring unsafe
6. Ducts leaky, damaged, etc.
7. Cold ducts not insulated
8. Inlet/outlet points not found
9. Inlet/outlet point locations poor
10. Missing flap or screen on inlet/outlet points
11. Weather hood missing, etc.
12. Duct not properly connected to furnace
13. Grille problems
14. No grease filter on kitchen grille
15. Flow collars missing
16. Dampers missing
17. Duct vapor barrier missing
18. HRV not interlocked with furnace fan
19. Fan switch not found, etc.
20. Filters dirty, missing
21. Core dirty, missing
22. Cabinet problems
23. Condensate drain problems
24. Condensate trap missing
25. Condensate trap discharge in poor location

8. True

9. False

10. Warm side

11. Warm side

12. False

13. True

14. True

2 INTERIOR

Insulation
& Interiors
MODULE

► TABLE OF CONTENTS

FIELD EXERCISE 1

► 1.0 OBJECTIVES

In this Section, we are going to look at the interior of the home. You will learn how to identify common materials and systems used on interiors. By the end of this Section, you will be able to identify common problems with:

• Walls, ceilings and floors
• Trim, counters and cabinets
• Stairs
• Windows, skylights and solariums
• Doors
• Basement and crawlspace leakage

You will also be able to evaluate the implications of each and make appropriate recommendations for improvements. You will understand the causes of problems and develop an inspection strategy to look for each.

Not Code Compliance

As with any home inspection system, this is not a code compliance inspection. There are lots of code issues that apply and you should be familiar with your local codes.

Not The Last Word

As always, our depth is sufficient to allow the general practitioner home inspector to perform a visual inspection. We are not shooting for service technician level and there is always more material that you can study. There are courses available in many areas and we would encourage you to expand your knowledge and keep up to date with local practices and requirements in your area.

Clients Come With You

While we always encourage clients to follow us on inspections, we find that they are most interested in being with us as we move through the interior. This is the area where they will have spent most of their time looking at the house and most of their questions will be related to this area.

Remodeling Questions

You will often be asked remodeling and renovation questions. You'll have to decide whether you are going to address these during the course of a home inspection or not. There is a fine line between being helpful and going beyond your scope. If you are going to offer renovation or remodeling advice, you might consider doing that as a separate service for a different fee.

Form And Function Are Different

As you inspect the interiors, you'll have to be careful to separate functional issues from cosmetic issues. Home inspectors deal with only the function side. Equally important, your client should appreciate the scope of your work.

Insulation & Interiors

MODULE

STUDY SESSION 1

1. This Session covers the Scope and Introduction to the Interior inspection.

2. At the end of this Study Session you should be able to:
- list six components included in a Standard Interior inspection
- describe how many doors and windows should be operated in a home as part of an inspection
- describe how to handle water penetration and condensation as part of an inspection
- list five items not required for the interior inspection according to the Standards
- list three limitations to interior inspections

3. This Session may take roughly 45 minutes.

4. Quick Quiz 1 is at the end of this Session.

Key Words:

* *Walls*

* *Ceilings*

* *Floors*

* *Water penetration*

* *Condensation*

* *Doors*

* *Windows*

* *Steps*

* *Railings*

* *Counters*

* *Cabinets*

* *Garage walls*

* *Room-by-room*

* *Finishes*

* *Function*

* *Cosmetics*

► **2.0 SCOPE AND INTRODUCTION**

2.1 SCOPE

THE ASHI® STANDARDS OF PRACTICE

The following are excerpted from the ASHI® Standards of Practice, effective January 1, 2000.

2. PURPOSE AND SCOPE

2.1 The purpose of these Standards of Practice is to establish a minimum and uniform standard for private, fee-paid home inspectors who are members of the American Society of Home Inspectors. Home Inspections performed to these Standards of Practice are intended to provide the client with information regarding the condition of the systems and components of the home as inspected at the time of the Home Inspection.

2.2 The inspector shall:

A. *inspect*:

1. readily accessible systems and components of homes listed in these Standards of Practice.
2. *installed systems* and *components* of homes listed in these Standards of Practice.

B. report:

1. on those systems and components inspected which, in the professional opinion of the inspector, are *significantly deficient* or are near the end of their service lives.
2. a reason why, if not self-evident, the *system* or *component* is *significantly deficient* or near the end of its service life.
3. the inspector's recommendations to correct or monitor the reported deficiency.
4. on any *systems* and *components* designated for inspection in these Standards of Practice which were present at the time of the *Home Inspection* but were not inspected and a reason they were not inspected.

2.3 These Standards of Practice are not intended to limit inspectors from:

A. including other inspection services, *systems* or *components* in addition to those required by these Standards of Practice.

B. specifying repairs, provided the *inspector* is appropriately qualified and willing to do so.

C. excluding *systems* and *components* from the inspection if requested by the client.

10. INTERIOR

10.1 The inspector shall:

A. *inspect*:

1. the walls, ceilings, and floors.
2. the steps, stairways, and railings.
3. the countertops and a *representative number* of *installed* cabinets.
4. a *representative number* of doors and windows.
5. garage doors and garage door operators.

10.2 The inspector is NOT required to:

A. *inspect*:

1. the paint, wallpaper and other finish treatments.
2. the carpeting.
3. the window treatments.
4. the central vacuum systems.
5. the *household appliances*.
6. *recreational facilities*.

13. GENERAL LIMITATIONS AND EXCLUSIONS

13.1 General limitations:

A. Inspections performed in accordance with these Standards of Practice

1. are not *technically exhaustive*.
2. will not identify concealed conditions or latent defects.

B. These Standards of Practice are applicable to buildings with four or fewer dwelling units and their garages or carports.

13.2 General exclusions:

A. The *inspector* is not required to perform any action or make any determination unless specifically stated in these Standards of Practice, except as may be required by lawful authority.

B. *Inspectors* are NOT required to determine:

1. the condition of *systems* or *components* which are not *readily accessible*.
2. the remaining life of any *system* or *component*.
3. the strength, adequacy, effectiveness, or efficiency of any *system* or *component*.
4. the causes of any condition or deficiency.
5. the methods, materials, or costs of corrections.
6. future conditions including, but not limited to, failure of *systems* and *components*.
7. the suitability of the property for any specialized use.

8. compliance with regulatory requirements (codes, regulations, laws, ordinances, etc.).
9. the market value of the property or its marketability.
10. the advisability of the purchase of the property.
11. the presence of potentially hazardous plants or animals including, but not limited to wood destroying organisms or diseases harmful to humans.
12. the presence of any environmental hazards including, but not limited to toxins, carcinogens, noise, and contaminants in soil, water and air.
13. the effectiveness of any *system installed* or methods utilized to control or remove suspected hazardous substances.
14. the operating costs of *systems* or *components*.
15. the acoustical properties of any *system* or *component*.

C. *Inspectors* are NOT required to offer:

1. or perform any act or service contrary to law.
2. or perform *engineering* services.
3. or perform work in any trade or any professional service other than *home inspection*.
4. warranties or guarantees of any kind.

D. *Inspectors* are NOT required to operate:

1. any *system* or *component* which is *shut* down or otherwise inoperable.
2. any *system* or *component* which does not respond to *normal operating controls*.
3. shut-off valves.

E. *Inspectors* are NOT required to enter:

1. any area which will, in the opinion of the *inspector*, likely be dangerous to the *inspector* or other persons or damage the property or its *systems* or *components*.
2. The *under-floor crawl* spaces or attics which are not *readily* accessible.

F. *Inspectors* are NOT required to *inspect*:

1. underground items including, but not limited to underground storage tanks or other underground indications of their presence, whether abandoned or active.
2. *systems* or *components* which are not *installed*.
3. *decorative* items
4. *systems* or *components* located in areas that are not entered in accordance with these Standards of Practice.
5. detached structures other than garages and carports.
6. common elements or common areas in multi-unit housing, such as condominium properties or cooperative housing.

G. Inspectors are NOT required to:
1. perform any procedure or operation which will, in the opinion of the *inspector*, likely be dangerous to the *inspector* or other persons or damage the property or its *systems* or *components*.
2. move suspended ceiling tiles, personal property, furniture, equipment, plants, soil, snow, ice, or debris.
3. *dismantle* any *system* or *component*, except as explicitly required by these Standards of Practice.

GLOSSARY OF ITALICIZED TERMS

Alarm Systems
Warning devices, installed or free-standing, including but not limited to; carbon monoxide detectors, flue gas and other spillage detectors, security equipment, ejector pumps and smoke alarms

Architectural Service
Any practice involving the art and science of building design for construction of any structure or grouping of structures and the use of space within and surrounding the structures or the design for construction, including but not specifically limited to, schematic design, design development, preparation of construction contract documents, and administration of the construction contract

Automatic Safety Controls
Devices designed and installed to protect *systems* and *components* from unsafe conditions

Component
A part of a *system*

Decorative
Ornamental; not required for the operation of the essential *systems* and *components* of a home

Describe
To *report* a system or *component* by its type or other observed, significant characteristics to distinguish it from other *systems* or *components*

Dismantle
To take apart or remove any component, device or piece of equipment that would not be taken apart or removed by a homeowner in the course of normal and routine homeowner maintenance

Engineering Service
Any professional service or creative work requiring engineering education, training, and experience and the application of special knowledge of the mathematical, physical and engineering sciences to such professional service or creative work as consultation, investigation, evaluation, planning, design and supervision of construction for the purpose of assuring compliance with the

specifications and design, in conjunction with structures, buildings, machines, equipment, works or processes

Further Evaluation
Examination and analysis by a qualified professional, tradesman or service technician beyond that provided by the *home inspection*

Home Inspection
The process by which an *inspector* visually examines the *readily accessible systems* and *components* of a home and which *describes* those *systems* and *components* in accordance with these Standards of Practice

Household Appliances
Kitchen, laundry, and similar appliances, whether *installed* or free-standing

Inspect
To examine *readily* accessible systems and *components* of a building in accordance with these Standards of Practice, using *normal operating controls* and opening *readily openable access panels*

Inspector
A person hired to examine any *system* or *component* of a building in accordance with these Standards of Practice

Installed
Attached such that removal requires tools

Normal Operating Controls
Devices such as thermostats, switches or valves intended to be operated by the homeowner

Readily Accessible
Available for visual inspection without requiring moving of personal property, dismantling, destructive measures, or any action which will likely involve risk to persons or property

Readily Openable Access Panel
A panel provided for homeowner inspection and maintenance that is within normal reach, can be removed by one person, and is not sealed in place

Recreational Facilities
Spas, saunas, steam baths, swimming pools, exercise, entertainment, athletic, playground or other similar equipment and associated accessories

Report
To communicate in writing

Representative Number
One *component* per room for multiple similar interior *components* such as windows and electric outlets; one *component* on each side of the building for multiple similar exterior *components*

Roof Drainage Systems
Components used to carry water off a roof and away from a building

Significantly Deficient
Unsafe or not functioning

Shut Down
A state in which a *system* or *component* cannot be operated by *normal operating controls*

Solid Fuel Burning Appliances
A hearth and fire chamber or similar prepared place in which a fire may be built and which is built in conjunction with a chimney; or a listed assembly of a fire chamber, its chimney and related factory-made parts designed for unit assembly without requiring field construction

Structural Component
A *component* that supports non-variable forces or weights (dead loads) and variable forces or weights (live loads)

System
A combination of interacting or interdependent *components*, assembled to carry out one or more functions

Technically Exhaustive
An investigation that involves dismantling, the extensive use of advanced techniques, measurements, instruments, testing, calculations, or other means

Under-floor Crawl Space
The area within the confines of the foundation and between the ground and the underside of the floor

Unsafe
A condition in a *readily accessible, installed system* or component which is judged to be a significant risk of personal injury during normal, day-to-day use. The risk may be due to damage, deterioration, improper installation or a change in accepted residential construction standards

Wiring Methods
Identification of electrical conductors or wires by their general type, such as "non-metallic sheathed cable" ("Romex), "armored cable" ("bx") or "knob and tube", etc.

► NOTES ON THE STANDARDS

Inspect   The Standards are clear on the meaning of **inspect**. When we inspect we have to look at and test the components listed in the Standards. We look at them if they are **readily accessible** or if we can get at them through **readily openable access panels**. These are panels designed for the homeowner to remove. They are within normal reach, can be removed by one person, and are not sealed in place.

Testing   We test components and systems by using their **normal operating controls**, but not the safety controls. We turn thermostats up or down, open and close doors and windows, turn light switches and water faucets on and off, flush toilets, etc. We do not test heating systems on high limit switches, test pressure relief valves on water heaters and boilers, overload electrical circuits to trip breakers, etc.

Systems   We do not start up systems that are shut down. If the furnace pilot is off, we don't
Shut Down   light it. If the electricity, water or gas is shut off in the home, we don't turn it on. If the disconnect for the air conditioner is off, we don't turn it on.

Accessible   We have to inspect house components that are **readily accessible**. That means we don't have to move furniture, lift carpets or ceiling tiles, dismantle components, damage things or do something dangerous. The exception is covers that would normally be **removed by homeowners during routine maintenance**. The furnace fan cover is a good example because homeowners remove this to change the furnace filter. Many inspectors use tools as the threshold. If tools are required to open or dismantle the component, it is not considered **readily accessible**.

Installed   We only have to inspect things that are **installed** in homes. This means we don't have to inspect window air conditioners or portable heaters, for example.

Deficiencies   We have to report on systems that are **significantly deficient**. This means they are unsafe or not performing their intended function. Although the Standards are not explicit, we are not required to identify every minor defect in a home. Failing to report a sticking door latch or cracked pane of glass would not be a meaningful breach of the Standards. Some common sense is needed here, determining the effect the issue will have on the safety, usability and durability of the home.

End Of Life   We are required to report on any system or component that in our professional opinion is **near the end of its service life**. This is tricky since we don't know whether inspectors will be held accountable for failed components on the basis that they should have known the component was near the end of its life. With the wisdom of hindsight, it may be hard to argue that the component could not have been expected to fail, when in fact, it did. Time will tell. The situation is also tricky because it includes not only **systems** but individual **components** as well. For many systems there are broadly accepted life expectancy ranges, but these aren't available for some individual components. A reasonable criteria may also be the apparent condition of the component.

Remaining Life   We are not required to determine the **remaining life** of systems or components. This is related to, but different than, the **end of service life** issue. If the item is new or in the middle part of its life, we don't have to predict service life, even though the same broadly accepted life expectancy ranges would apply. It's only when the item is near the end, in your opinion, that you have to report it.

Reporting Implications   We have to tell people in writing the **implications** of conditions or problems unless they are self-evident. A cracked heat exchanger on a furnace has a very different implication for a homeowner than a cracked windowpane, for example. It's not enough to tell a client that they have aluminum wiring. We have to tell them of the potential fire risk.

Tell Client What To Do   We have to tell the client in the report what to do about any conditions we found. We might recommend they repair, replace, service or clean the component. We might advise them to have a specialist further investigate the condition. It's all right to tell the client to monitor a situation, but we can't tell them that their roof shingles are curled and leave it at that. We have to tell them what to do about the aluminum wire to reduce the fire risk.

What We Left Out   We have to report anything that we would usually inspect but didn't. We also have to include in our report why we didn't inspect it. The reasons may be that the component was inaccessible, unsafe to inspect or was shut down. It may also be that the occupant or the client asked us not to inspect it.

Walls, Ceilings And Floors   The Interior Standards are a little tricky. We don't have to inspect interior finishes, but we do have to inspect walls, ceilings and floors. Our interpretation is that we have to inspect these things from the standpoint of performance and function, but not appearance. We have to ensure that walls, ceilings and floors can support themselves and their intended loads, remaining intact under normal conditions.

Problems With Other Systems   As we look at walls, ceilings and floors, we're also looking for problems in other house systems. For example, we're checking these to see if the structure has moved. We're looking for evidence of leakage from roofs, walls, windows, supply and waste plumbing systems, and heating and air conditioning systems.

Testing Plumbing, Heating And Electrical   As we move through the house, we test electrical receptacles, lights and switches in each room. We also test the operation of plumbing fixtures, including sinks, toilets, bathtubs, showers and laundry tubs. We look for an operative heating source in each room and, of course, operating doors and windows in each room.

Paint, Wallpaper, Etc.   We do not have to inspect or comment on the condition of finishes on walls, floors or ceilings. This includes the condition of paint or stain, wallpaper, paneling, ceramic tiles, marble, concrete, etc.

A Gray Area

Some things are borderline. If a ceramic tile floor is broken to the point it has become a trip hazard, most inspectors report it. What about missing grout between tiles? What about hairline cracks in tiles? These are clearly cosmetic issues; however, many homeowners see these as obvious and serious flaws. Decide how far you're going to go in pointing out defects such as these. Whatever you decide, let your clients know what your inspection includes. We have decided to include comments on cracked tiles and missing grout on ceramics, for example. We also point out carpet trip hazards.

Steps,
Stairways,
And Railings

Steps, stairways and railings are safety-related parts of the inspection. Point out significant departuresfrom good practice, without quoting building codes. We use a common sense approach to determine whether stairways are safe.

Counters And
Cabinets

We pay little attention to counters and cabinets. There are very few serious functional issues, although porous countertops could trap food particles and become a health hazard. Cabinets may have doors that don't open and close freely, drawers that don't slide easily, or doors or drawers that are loose or damaged. Drawers without effective stops may fall out and injure someone if pulled too far. Hardware can be missing, loose, damaged or corroded. Perhaps the most serious problem is poorly secured wall-hung cabinet units that may fall off the wall, injuring someone.

Talk To Your
Clients

Many clients focus on cabinetry and countertops, especially in kitchens, as part of their evaluation of a home. As we're inspecting the kitchen and testing plumbing fixtures, we often suggest that clients have a look at the counters and cabinets to satisfy themselves. There are many subjective issues, as well as the "nuts and bolts" issues we've touched on.

Perspective

It's very easy to become absorbed in your inspection and forget that your client is trying to decide whether to buy this home. A cigarette burn in a countertop, a sticking drawer, or a loose cabinet door are usually not things that would change someone's mind about buying a home. While it's nice to have an all-inclusive list of minor house defects, we remind our clients that this is not the purpose of a home inspection. We focus on the major and costly issues that are more likely to affect their purchasing decision.

Doors And
Windows

We have to inspect doors and windows, and operate at least one door and one window in each room. All exterior doors have to be operated. Doors and windows are costly and deserve some attention. In the Exterior Module, we talk about doors and windows. We also indicate these have to be inspected from the interior. Opening doors and windows is usually done as part of the interior inspection.

Hardware

Although it's not explicit in the Standards, inspecting doors and windows includes an inspection of the hardware.

Big Dollars   It is far more expensive to replace all the windows in a house than to replace a furnace, roof or water heater. Doors and windows are subjective, and affect the architectural appeal as much any other component of the home. Let your client know you're going to focus on the function of these components, rather than the subjective issues. Many people are tempted to justify replacing windows based on improved energy efficiency. In our experience, the payback using this criterion alone is very long.

Garage Walls   We have to inspect the wall that separates the house from the garage. In some areas, this has to be a fireproof wall, and in others, it only has to be gas-tight. You should know what applies in your area. You should also know what constitutes a fireproof wall in your area. There are requirements for connecting doors between houses and garages in most jurisdictions; these requirements vary by location. In our area for example, concrete block walls have to be plastered or drywalled on the garage side.

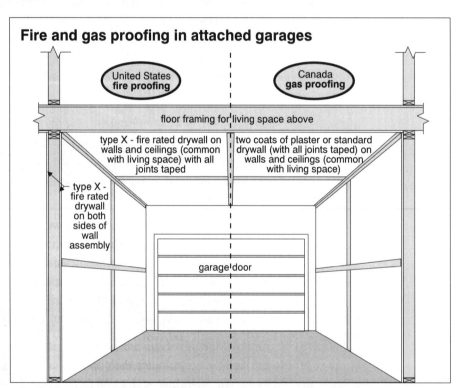

Fire and gas proofing in attached garages

United States fire proofing

Canada gas proofing

floor framing for living space above

type X - fire rated drywall on walls and ceilings (common with living space) with all joints taped

two coats of plaster or standard drywall (with all joints taped) on walls and ceilings (common with living space)

type X - fire rated drywall on both sides of wall assembly

garage door

Ceilings, Too   If garages have living space above or communicate directly with the house attic, garage ceilings are usually treated the same as garage walls, with respect to fire or gas proofing.

Party Walls   Walls separating one home from another in attached housing units usually require some kind of fire separation. Again, this varies and although it's going to hard for you to inspect, you should know what the local requirements are for such walls. For example, virtually all jurisdictions call for these walls to be continuous through attics. When you get into the attic, you should not be able to look into the neighbor's attic.

Window   We are not required to inspect draperies, blinds and other window treatments.
Treatments However, we do encourage you to move these so you can fully inspect windows. Many home inspectors have missed obvious problems because of curtains hanging over and around windows. Moisture damage around windows very often shows up at the lower corners of windows and below; curtains often conceal these corners.

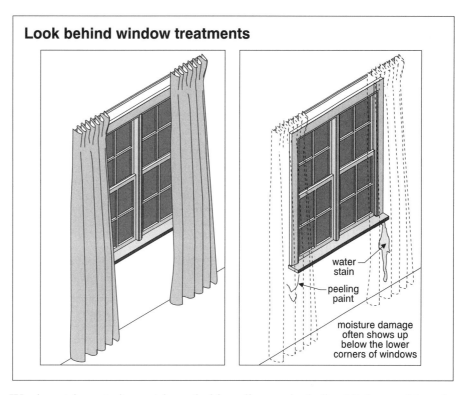

Look behind window treatments

water stain

peeling paint

moisture damage often shows up below the lower corners of windows

Household   We do not have to inspect household appliances, including kitchen and laundry
Appliances   appliances, window air conditioners, etc. We don't have to look at central vacuum systems.

Recreational We do not have to inspect spas, saunas, steam baths, pools, tennis courts, or exercise
Facilities   or entertainment equipment.

2.2 INTRODUCTION

Goal

The goal of the Interior inspection is to evaluate the **function** of the interior components, rather than the **appearance**. We are not concerned with decorating issues during the home inspection. Instead, we focus on the ability of each component to perform its intended task.

Clients May Joke

Many clients make humorous or disparaging remarks about decorating features in a house. They run the risk of offending the sellers if the comments are overheard. Home inspectors risk offending their clients, the sellers and real estate professionals by joining in. There's nothing to be gained and a great deal to lose by making comments about decorating.

Nervous Reaction

There are some people who make jokes to cover their nervousness. Buying a home is more than unsettling, it's often terrifying. People deal with the situation differently. You have to be aware of this and not fall into any traps. This is most likely during the inspection of the interior because it's the most familiar part of the home to your client. Most homeowners don't spend a lot of time on the roof, the exterior, in the basement or in the mechanical rooms of the home. They spend most of their time in the living spaces. Clients will typically become more comfortable and feel more in their own element as you go through these parts of the home. It's also the part of the inspection where home inspectors are most likely to step out of their role.

Clients Most Comfortable Here

Limitations

While the interior is easy to inspect, because most of us have spent a great deal of time in the inside of homes, there are limitations. Interior finishes may conceal the actual condition of walls, ceilings and floors, for example. Carpet, wallpaper and furniture obstruct your view. Drapes, clothes or storage in closets and pictures on walls may conceal valuable information. Even cutting boards on counters can hide damaged countertops.

Intentional Concealment?

We often discover things which were not apparent until we moved something or probed below a surface. We spend no time wondering whether there was an effort made to conceal things. This is a waste of time and energy, and if suspicions are shared with the client, this can color the entire transaction.

You Don't Know

While you may have suspicions, you will never know whether something was intentionally concealed. Even if you are convinced that it was, that doesn't change anything from the standpoint of your inspection. You should look with the same level of diligence whether or not you suspect people have concealed things. You should establish a routine and stick to it. If a real estate professional watches you move a china cabinet or pull up broadloom during one inspection, they will reasonably expect you to do it on every inspection. The Standards spell out what you are expected to do. Stay within the Standards or depart from them (attracting a little more risk), but be consistent in whatever approach you take.

Consistent Approach

Exclusions

Smoke And Carbon Monoxide Alarms

As you move through the interior of the home, you will see and very often be asked about smoke alarms and carbon monoxide alarms. The Standards do not call for you to test these devices. We do not test them. Smoke alarms may be connected to a central station alarm. Testing these without notifying the alarm company may result in the fire department showing up at the door. Further, smoke detectors are sometimes removed by the sellers when leaving the house. Smoke detectors cost less than ten dollars and many inspectors make a general recommendation to install and maintain these. The Standards do not ask us to report on the absense of smoke detectors. There is a discussion of smoke detectors in the Electrical Module.

Carbon Monoxide Sensors

At the time of writing (year 2000), carbon monoxide sensors are somewhat controversial. It's hard to argue against the concept. Anything that you do that may make a house safer must be a good thing. However, the sensitivity and reliability of carbon monoxide sensors are issues. Some changes are taking place and they will get better over time.

Security Systems

We do not test home security systems. These are proprietary systems and there are many different arrangements. In many cases, a central station has to be contacted to test alarm systems. This is something that the homeowner should become familiar with.

Sellers Get Nervous

Sellers don't like people testing their alarm system. They're worried about their own security. The more an outsider knows about a person's alarm system, the easier it would be to defeat the system.

General Inspection Strategy

Throughout most of this program, we've talked about a system-by-system approach. On the interior of the home, a room-by-room approach is more practical. Most home inspectors look at all or most components of each room during a single visit to the room. It would add a lot of time to an inspection to make separate visits to the room to check walls, ceilings and floors; doors and windows; electrical points; plumbing fixtures and piping; and the heating system.

Double Tour Approach

Many inspectors use a double tour approach. During their first tour through the house, they will look at most of the components. Many will leave the heating or air conditioning inspection to a second tour. There are advantages to this approach:

• You can turn the heating or air conditioning system on and check the delivery to each room during a dedicated tour.
• If you turn the system on and include this in your overall room by room inspection, you are likely to overheat or overcool the house before you are finished. You are also less likely to forget to turn the heat or air conditioning back to its original setting.

• The second walk through the house allows you to double check things as necessary. For example, many inspectors move up through the house as they do their room by room inspection. Operating a shower stall in a second floor bathroom should always be followed by another look at the ceiling below the shower. This can be done without wasted steps or time on a second pass that is substantially dedicated to the heating or air conditioning system.

• You can make sure that you have not overlooked a room or closet on your first tour.

Develop Your Own Approach

You will undoubtedly develop your own approach. There is no right or wrong, as long as you cover everything. Be as efficient as you can without cutting corners.

Let The House Tell Its Story

Some inspectors are very good at taking a passive approach to the home, rather than an aggressive, detailed, logical, point by point approach. These inspectors can walk into a room and look around with an open mind. Their eye and mind will be drawn to anomalies and problems without looking for them. This is a skill that can be learned. It may make sense to practice this skill during a second pass through the interior. It is risky to rely on this approach alone, at least until your skill level is such that it can be relied upon not to overlook specific details. Many good inspectors use both techniques.

Finding What's Missing

One of the most difficult things in inspection is to recognize that components are missing. There are no triggers to alert you, as there are with things like stains on ceilings and broken windows. Checklists help you notice what's missing.

Finding Renovations And Additions

The house story is often told through baseboards, casings, window and door styles, and floor, wall and ceiling finishes. Are these consistent throughout the house? If they are different, does this suggest part of the house has been remodeled? Are room sizes what you would expect in a house such as this? If not, have walls been removed? Were they structural? Would there have been electrical, plumbing and heating services buried in the wall that was removed? If so, how do those services get to their destinations now? Where window, door and trim treatments are different, are you looking at an addition or a porch that has been converted to living space?

What It Means

Noticing these details may alert you to inaccessible crawlspaces, secondary attics, different wall and roof construction details, different types of heating systems, a lack of central air conditioning in the addition, and so on. These are the details

Checklists Not So Helpful

that checklists cannot help you with. The history is often there to be seen, but you have to look for it. Clues as subtle as joints in baseboards, seams in flooring and ceiling patches can tell you a great deal, if you listen.

Focus On Problems

The Interior chapter of **The Home Reference Book** does a good job of describing many of the house interior systems. We'll add to those descriptions in some cases, but for the most part, this section will deal with the common problems that we find, their implications and how you can identify them. Let's get to it.

Insulation
& Interiors
M O D U L E

QUICK QUIZ 1

☑ INSTRUCTIONS

- You should finish Study Session 1 before doing this Quiz.
- Write your answers in the spaces provided.
- Check your answers against ours at the end of this Section.
- If you have trouble with the Quiz, re-read the Study Session and try the Quiz again.
- If you did well, it's time for Study Session 2.

1. List five components required in an Interior inspection.

_____

_____

_____

_____

_____

_____

2. How many exterior doors should be operated as part of a Standard inspection?

_____

_____

3. How many interior doors should be operated?

_____

_____

4. How many windows should be operated to meet the Standards?

_____

5. What are you required to do with central vacuum systems?

_____

6. List six items not required for inspection on the Interior section, according to the Standards.

_____

_____

_____

_____

_____

7. List three limitations to an Interior inspection, as described in the Introduction.

_____

_____

_____

8. Many inspectors make _____ (insert number) passes through the interior part of the home during an inspection.

9. What might different style baseboards, windows and construction methods in one part of a house suggest?

_____

10. Does an Interior inspection include carpet and wallpaper?
Yes ☐   No ☐

11. Does an Interior inspection include drapes and curtains?
Yes ☐   No ☐

12. Does an Interior inspection include door hardware?
Yes ☐   No ☐

If you didn't have any difficulty with the Quiz, you are ready for Study Session 2.

Key Words:

• *Walls*

• *Ceilings*

• *Floors*

• *Water penetration*

• *Condensation*

• *Doors*

• *Windows*

• *Steps*

• *Railings*

• *Counters*

• *Cabinets*

• *Garage walls*

• *Room-by-room*

• *Finishes*

• *Function*

• *Cosmetics*

Insulation
& Interiors
MODULE

STUDY SESSION 2

1. You should have finished Study Session 1 and Quick Quiz 1 before starting this Session.

2. This Session covers the floors part of the interior inspection.

3. By the end of this Session, you should be able to:
- list five common floor materials
- list five general flooring problems and their implications
- list seven concrete floor problems and their implications
- list six wood flooring problems and their implications
- list four carpet problems and their implications
- list five resilient flooring problems and their implications
- list five ceramic tile flooring problems and their implications

4. Before you start this Session, you should read Section 1.0 of the Interior Chapter of **The Home Reference Book**.

5. This Session may take roughly one hour.

6. Quick Quiz 2 is at the end of this Session.

Key Words:
- *Concrete*
- *Wood*
- *Carpet*
- *Resilient*
- *Ceramic*
- *Trip hazard*
- *Absorbent material*
- *Squeaks*

► 3.0 FLOORS

Function

Floors are designed as walking surfaces and supports for furnishings. Floors can also be part of the architectural appeal of the home.

Level, Smooth And Durable

From a traffic standpoint, floors should be level so they don't trip us, smooth so they're easy to navigate, and durable. Floors typically take more abuse than walls and walls take more abuse than ceilings. Ceramic tiles designed for walls are not suitable for use on floors, for example.

Materials

Common flooring materials are well described in **The Home Reference Book** so we won't go into a lot of detail. Let's just list some of the common flooring materials:

• Concrete
• Wood
• Hardwood or softwood in strip, plank or parquet styles
• Carpet – wool and synthetic
• Resilient – both tile and sheet goods
• Ceramic and quarry tile
• Stone and marble

Control Joints In Concrete Floors

Concrete floors can be rough, unfinished floors in basements or crawlspaces, or can be immediately below the finished flooring in slab-on-grade construction. Concrete floors may have control joints to help ensure that any cracks develop where we want them to. Control joints are typically cuts in the surface that are roughly a quarter of the depth of the slab. These cuts provide stress concentration points. If the slab cracks due to shrinkage or minor settlement, the cracks will probably occur at the control joints. We can ensure that moisture penetration won't occur by protecting these joints during construction.

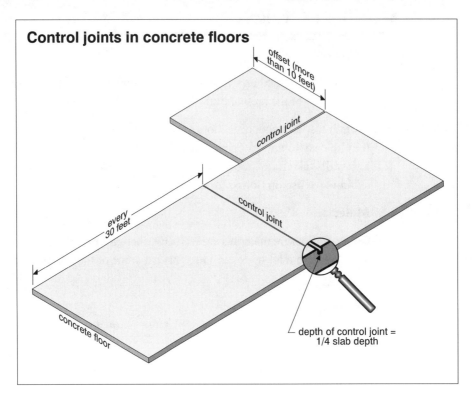

Control joints in concrete floors

offset (more than 10 feet)

control joint

control joint

every 30 feet

concrete floor

depth of control joint = 1/4 slab depth

Every 30 Feet   Control joints are typically about every 30 feet in the slab and at offsets or changes in direction in the slab, especially if the offset is more than 10 feet.

Reinforced Slabs   If the concrete slabs have steel reinforcement, control joints are often omitted.

General Strategy

Home inspectors should always know whether they are walking across a wood or concrete floor. This is something that you can learn to determine with experience by both the sound and the feel of the floor as you move across it. In some cases, you may have to bounce on the floor slightly to confirm your suspicions. Try this everywhere you go for the next few days. You'll find that it's relatively easy to do. It's slightly more difficult on ceramic tile floors because they are often wood flooring with a layer of concrete over them.

New Floors Over Old   In some cases, new flooring material is laid over old. This is not unusual and may only be a matter of mild interest. However, in some cases new flooring has been laid over old because of dramatic settlement or sagging in the old flooring. We have found homes where new flooring systems, including tapered joists, have been installed over old, badly sloping floors. This is not a problem in itself, but it should alert you to considerable movement of the structure. The important question is, "Has the cause of the movement been corrected?"

*Rot Around
Plumbing
Fixtures*

Rot is one of the biggest enemies of wood flooring. It shouldn't be a surprise that rot is most likely to attack wood flooring around and below sources of water. While all plumbing fixtures can contribute, toilets are the most common problem area for wood floors. We find more rotted wood subflooring, joists and beams around toilets than any other plumbing fixture.

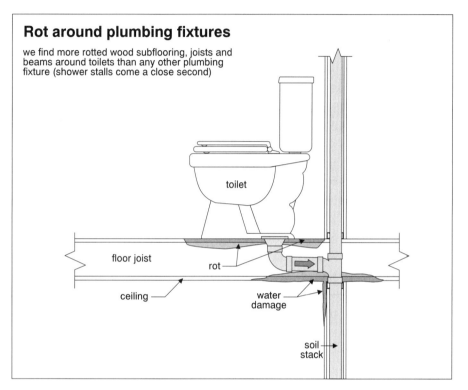

Rot around plumbing fixtures

we find more rotted wood subflooring, joists and beams around toilets than any other plumbing fixture (shower stalls come a close second)

toilet

floor joist

rot

ceiling

water damage

soil stack

*Hardwood
Below
Carpets?*

We are very often asked whether there is hardwood flooring below wall to wall carpeting. You may be able to determine this by lifting heat registers or lifting corners of carpets. (Note: lifting carpets goes beyond the Standards.) In many cases, the carpet is laid over subfloor. If there is hardwood below, you should caution your clients that it may not be in a suitable condition to remove the carpet and expose the hardwood. The carpet may have been installed because the hardwood was damaged. In older homes, hardwood can be sanded to create a new appearance. There is a limit to how often tongue-and-groove floors can be sanded. Modern ⅜ inch thick hardwood flooring can only be sanded once without risk of exposing the tongues. If your clients have their hearts set on hardwood flooring, they can install new hardwood flooring for only slightly more cost than high quality carpet.

Let's look at some things that can go wrong with floors.

3.1 CONDITIONS

These are common flooring problems in houses:

1. Water damage
2. Trip hazard
3. Mechanical damage
4. Loose or missing pieces
5. Absorbent materials in wet areas
6. Concrete
   - cracked
   - settled
   - heaved
   - water penetration
   - efflorescence
   - slopes away from drain
   - hollow below
7. Wood
   - rot
   - warped
   - buckled
   - stained
   - squeaks
   - exposed tongues
8. Carpet
   - rot
   - stained
   - odors
   - buckled
9. Resilient
   - split
   - lifted seams
   - open seams
10. Ceramic, stone and marble
    - cracked
    - broken
    - loose
    - grout missing
    - worn
    - stained

Let's look at each problem more closely.

3.1.1 WATER DAMAGE

Causes   Water damage to floors can be the result of –

• roof or flashing leaks
• plumbing leaks
• heating leaks
• air conditioning leaks
• wall, window or skylight leaks
• door leaks (especially sliding glass doors)
• spills from humidifiers, dehumidifiers, watering plants, aquariums, bathtubs,
  showers, sinks, basins, etc.
• ice damming
• condensation
• melting snow from boots and winter clothing
• careless floor washing

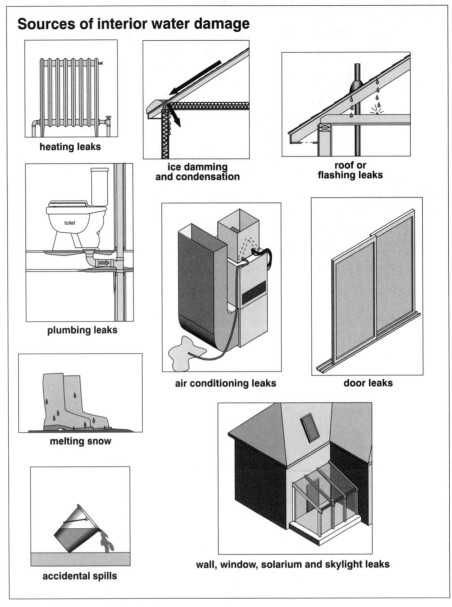

Sources of interior water damage

heating leaks

ice damming and condensation

roof or flashing leaks

plumbing leaks

air conditioning leaks

door leaks

melting snow

accidental spills

wall, window, solarium and skylight leaks

Implications   The implications of water damage may be –

- cosmetic only if the source of water has been contained
- rot, staining or other damage to the floor finish
- rot or other damage to structural components

Strategy   Where you see staining, buckling, warping, rot, efflorescence or wet spots, you should be looking at several things:

- Is the damage localized or widespread?
- Is there concealed damage? This may be difficult to determine, but you should allow for the possibility.
- What is the source of the water? Again, this may be difficult to tell, especially if the problem is not active.
- Is the leak active? If the floor is wet or damp, this is an easy answer. If the floor is dry, the problem may be inactive or intermittent. If it's a roof leak, for example, the floor may only get wet during and after a heavy rain. It may only be after a heavy rain accompanied by wind from a particular direction.

Don't Speculate   Your role as a home inspector has been satisfied once you've identified the problem and let the client know the possible implications. You don't have to troubleshoot the problem. You should give the client some direction. Your recommendations may include more than one of these actions:

- Repair the floor,
- Correct the leak,
- Investigate further, or
- Monitor the situation

3.1.2 TRIP HAZARD

Causes   Uneven floors create trip hazards.

Implications   The implications of trip hazards are personal injury.

Strategy   Look for unevenness in floor systems. Sometimes you'll discover these quite accidentally.

3.1.3 MECHANICAL DAMAGE

Causes   Damage to floors may include mechanical damage due to heavy objects being dragged across floors, impact damage or burns. Again, we aren't worried about cosmetics but are interested in performance issues.

Implications   Mechanical damage may create unevenness that results in trip hazards. It may result in a loss of continuity in a flooring system. Cuts in resilient flooring in kitchens or bathrooms, for example, can allow water into subflooring. Look for evidence of mechanical damage in exposed flooring. Remember that in a furnished house, you are not going to see the entire floor. Let your clients know that things may look very different when they take possession of the house when it is vacant. Most houses look considerably worse with no furniture.

3.1.4 LOOSE OR MISSING PIECES

Pieces of flooring may have come loose and been lost. This is particularly true of parquet flooring and ceramic tile, for example.

Causes   This is usually a result of failed adhesives, mechanical damage or poor installation.

Implications   The implications include trip hazards and moisture penetration to subflooring. There are also cosmetic issues.

Strategy   As you walk across floors, look and feel for loose pieces, particularly with wood parquet flooring and ceramic tile, stone or marble flooring. Many inspectors tap on parquet floors and ceramic tile floors to help identify pieces that are coming loose.

3.1.5 ABSORBENT MATERIALS IN WET AREAS

This is somewhat subjective, but is a common sense issue. The classic example is carpeting in bathrooms. Floors in rooms that are likely to get wet should have non-absorbent, moisture resistant flooring materials.

Cause   This is an installation choice.

Implications   Premature deterioration of the flooring is one implication. Rot damage to subflooring is another. Odors and other indoor air quality issues may create health concerns.

Strategy   We look closely for evidence of problems with flooring or subflooring where we find absorbent materials in kitchens and bathrooms. Wood flooring is marginally acceptable, although, again, the potential for moisture damage is considerable.

Alert Your Clients   You may not want to recommend removal of a flooring system, but you should alert your clients to the disadvantage of the situation.

3.1.6 CONCRETE FLOORS

- Cracked
- Settled
- Heaved
- Water penetration
- Efflorescence
- Slope away from drain
- Hollow below

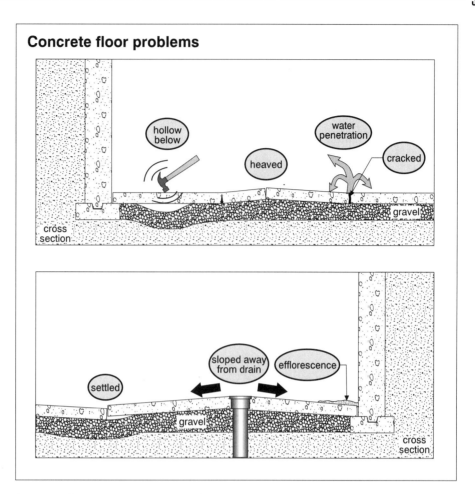

Concrete floor problems

Concrete floors can be the main floor in slab-on-grade construction or may be a rough, unfinished floor in a basement or crawlspace area. There will be concrete floors below most raised or finished flooring in basement areas as well.

Causes	Most concrete floors crack as a result of shrinkage during curing.
Cracking	Floors may settle as a result of inadequate support from the substrate below (building
Settling	on disturbed soil, for example) or excessive loads (from a column without a footing, for example).
Heaving And	Floors may heave as a result of frost below the floor or hydrostatic pressure from a
Water	high water table. Water or efflorescence coming up through floors indicates
Penetration	hydrostatic pressure below. This means the area below the slab is saturated with water and is under some pressure.
Slope To	A floor that slopes away from a floor drain is an original installation issue.
Drains	
Hollow Areas	Floors that are hollow below may be the result of poor original construction or
– Erosion?	sub-slab erosion. The erosion may be the result of surface water or underground streams.

Implications

Cracks

The implications of shrinkage cracks are usually not significant. Water and/or efflorescence may appear at the cracks if there is hydrostatic pressure below.

Settled Or Heaved

Settled or heaved slabs may indicate structural problems or local problems of little significance. The extent of movement is the best clue as to the severity. Settled and heaved slabs may also be trip hazards.

Water And Efflorescence

Water and efflorescence may result in damage to finishes and to the structure. Again, it's a question of the extent and amount of water.

Slope Away From Drains

Floors that slope away from drains suffer more damage to finishes and structures when floors get wet. Floor drains are typically only found on below grade concrete floors. This is primarily a basement issue.

Hollows May Be Serious

Floors that are hollow below may be trivial if the hollow is localized and less than $\frac{1}{2}$ inch, for example. Hollows below concrete floors may also indicate serious erosion and structural problems. It's important to know your local soil conditions and any possibilities of sink holes or unusual features.

Strategy

Cracks, Settling And Heaving

Where floors show typical random shrinkage cracking, no action is typically necessary. Where cracks are accompanied by settlement or heaving, the location and direction of the cracks may be important. Does the pattern suggest a sinking foundation or heaving column, for example? The extent of movement and the age of the building are valuable clues. In most cases, you won't be able to be conclusive about whether the movement is ongoing based on a one time visit. However, 1/2 inch movement in a one year old house is far more likely to be significant than in a 100 year old house. It's often hard to know whether you should be reassuring or alarmist about settled or heaved concrete floors. Common sense tells you to be neither. Document your findings. Explain the limitations and possible implications. Recommend monitoring if the problem is mild and further investigation if the movement is extensive.

Water And Efflorescence

Use a similar approach for evidence of water and/or efflorescence at cracks. Remedial actions may include a sump and pump. It helps tremendously to know local conditions. Are you in an area with a high water table? Is this a seasonal problem associated with melting snow and spring runoff? Is the problem specific to the house because of poor control of roof and surface rainwater?

Hollows

Hollow spaces below floors should be treated much like settled or heaved floors. You can't usually be conclusive about the size and severity of voids below floors. You won't know whether there is progressive erosion or movement of the soil below or whether it's simply an original construction condition that won't get any worse. Again, document your findings, explain the possible implications and recommend monitoring if the condition is localized (one or two square feet). Recommend further investigation if the hollow spaces are more extensive.

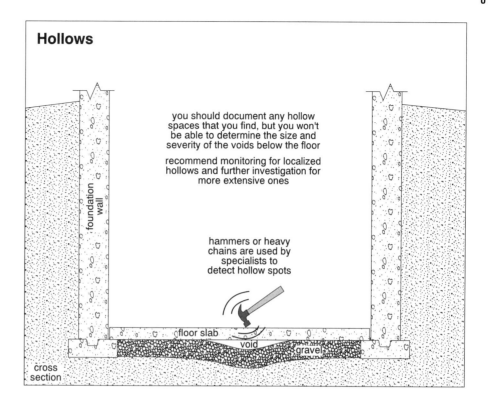

Hollows

you should document any hollow spaces that you find, but you won't be able to determine the size and severity of the voids below the floor

recommend monitoring for localized hollows and further investigation for more extensive ones

hammers or heavy chains are used by specialists to detect hollow spots

foundation wall

floor slab

void   gravel

cross section

3.1.7 WOOD

- Rot
- Warped
- Buckled
- Stained
- Squeaks
- Exposed tongues

Causes   Rotted, warped, buckled or stained floors are the result of water damage. We talked about several possible sources of water damage. Wood is vulnerable to rot attack when the moisture content is above 20 percent.

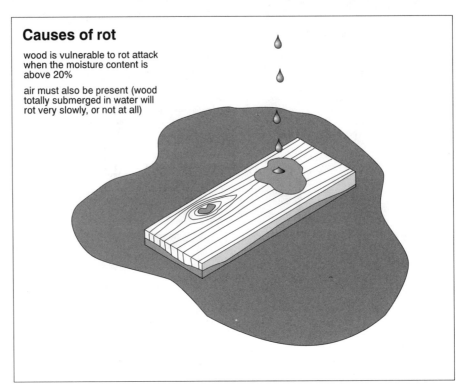

Causes of rot

wood is vulnerable to rot attack when the moisture content is above 20%

air must also be present (wood totally submerged in water will rot very slowly, or not at all)

Squeaks	Squeaks are typically caused by finished flooring not being held tightly against subflooring or subflooring not sitting tightly against joists. Squeaking floors are discussed extensively in **The Home Reference Book**.
Exposed Tongues	Exposed tongues on tongue-and-groove flooring are usually the result of sanding the floor to create a new wood finish.

Implications

The implications of rot, warped, buckled or stained flooring include—

• cosmetic problems
• trip hazards
• deterioration of the structure below

Squeaks	The implications of squeaks are simply a nuisance. Squeaks do not indicate tructural problems.
Exposed Tongues	Exposed tongues may result in slivers or splinters for people walking in bare or stocking feet. They may also result in exposed nailheads and possible injury. There are obvious cosmetic implications and pieces of flooring may become loose and/or lift as a result.

Strategy

Rot, Warped, Buckled, Stained Floors	Look for stained, warped, buckled or rotted flooring. Pay particular attention to areas below sliding glass doors and windows. Leakage and condensation can combine to cause considerable damage in these areas.

Toilets   Look closely in kitchens and bathrooms for evidence of problems, but focus around toilets. Where possible, go to the floor level below and look up to see if there is evidence of damage to flooring or to ceilings or structural members below the flooring around the toilet. Depending on the extent of the damage and whether or not the problem is active, you may recommend leakage or condensation control measures and flooring replacement. You may also recommend structural repairs.

Squeaks   The first step is to reassure clients that squeaks are common and are not a performance issue. Squeaks can be corrected several ways as discussed in **The Home Reference Book**. The solutions include pulling the finished flooring down against the subflooring and/or pulling the subflooring down against the joists. This can be done from above or below. In many cases, a cost/benefit analysis convinces people to live with the squeaks.

3.1.8 CARPET

- **Rot**
- **Stains**
- **Odors**
- **Buckled**

Causes   Rot, stains and odors may be the result of water problems. Stains and odors can also be the result of spills and/or pets.

Buckled   Buckled carpeting may be an installation or moisture issue.

Implications   The implications of rot, stains and odors may be cosmetic. They may also indicate damage to subflooring and framing below. There may be health implications to stains and odors.

Buckled   Buckled carpeting is a trip hazard.

Strategy

Look For
Moisture   When rot, stains and/or odors are noted, the first step is to determine whether moisture is still present. Again, if dampness is found you can be conclusive. If the carpet is dry, the problem may be intermittent. You may be able to distinguish between pet odors and general dampness. You may want to recommend further investigation. Carpets that are stained or have odors may have to be replaced. The odors may be in the subflooring as well. There are chemicals that can be used to eliminate these odors. In severe cases, some of the subflooring may have to be replaced. Where rot is noted, structural members below may be damaged.

Buckled   Look for carpeting that has lifted up at the middle or edges and may be a trip hazard. Recommend that this be stretched and resecured.

3.1.9 RESILIENT

- **Split**
- **Lifted seams**
- **Open seams**

Causes   These problems are typically mechanical damage or poor installation issues.

Implications Water damage to the subflooring below and trip hazards are the functional implications.

Strategy   Look for splits or tears in resilient flooring. Open or lifted seams are more common on tile floors than sheet goods, simply because there are more seams. Sheet goods typically come in rolls 12 feet wide and there may be no seams in kitchen or bathroom floors.

3.1.10 CERAMIC, STONE AND MARBLE

- **Loose**
- **Grout missing**
- **Cracked or broken**
- **Worn**
- **Stains**

Causes   Tiles that are cracked, broken or loose or have missing grout may be the result of:

- Excessive deflection of the substrate
- Improper installation, including surface preparation, mortar amount and quality, grout amount and quality

Worn   Worn tile may be the result of:

- Normal wear and tear
- Mechanical damage (heavy appliances being dragged across the floor, for example)
- The use of wall tiles on floors (wall tiles are not as durable)

Stains   Stains are typically the result of spills.

Implications Cracked, broken or loose tiles and tiles with grout missing can lead to water damage to the subfloor and may present trip hazards. Worn tiles may only be cosmetic issues, but can be trip hazards if corners are broken or pieces are loose.

Stains   Stains are typically cosmetic issues only.

Strategy   Look for stains and wear and tear. You may be able to see missing grout and cracked or broken tiles. Tapping on tiles also helps to identify loose pieces.

*Conventional
Flooring*

Houses are designed with floor systems that have a considerable amount of deflection. Without special consideration, ceramic tiles over conventional wood floors in houses will often crack. In most cases, this is only a cosmetic problem unless the tiles become loose. Proper installation techniques are described in **The Home Reference Book** and include typically one of these four options:

• A 1¼ inch mortar bed on top of the wood subfloor below the tile
• ³/₄ inch plywood flooring plus ¼ inch underlay below the tile
• A double layer of ⅝ inch plywood or waferboard subflooring below the tile
• Conventional (⅝ inch thick) subflooring with two by two blocking between joists. The blocking should be eight inches on center if the joists are 16 inches on center.

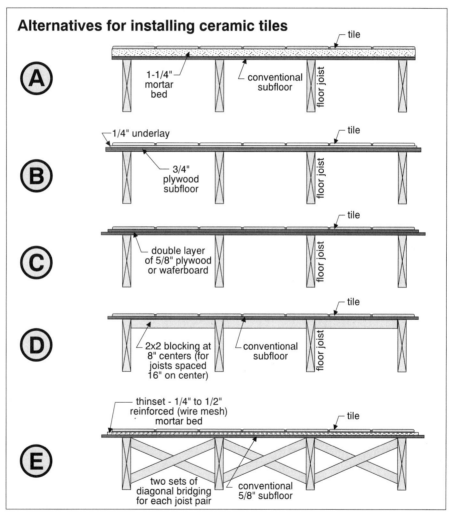

Alternatives for installing ceramic tiles

(A) 1-1/4" mortar bed — conventional subfloor — tile — floor joist

(B) 1/4" underlay — 3/4" plywood subfloor — tile — floor joist

(C) double layer of 5/8" plywood or waferboard — tile — floor joist

(D) 2x2 blocking at 8" centers (for joists spaced 16" on center) — conventional subfloor — tile — floor joist

(E) thinset - 1/4" to 1/2" reinforced (wire mesh) mortar bed — tile — two sets of diagonal bridging for each joist pair — conventional 5/8" subfloor

*The Thinset
Alternative*

We find many cases where ceramic tile is installed on a ¼ to ½ inch thick mortar bed which contains a wire mesh reinforcing lath, all over ⅝ plywood. Two sets of diagonal bridging for each joist pair are all the stiffening that is used. Depending on a number of factors, this approach sometimes works. It is more likely to be successful where joist spans are short or there is 12 inch rather than 16 inch joist spacing.

Use Test Of Time   Where you see this configuration on a 20 year old home and there is no cracking on the grout or tile, you can be comfortable that the system has been successful. In a one or two year old house, you are going to want to be more cautious and say that the installation technique is typical, but not ideal. Cracking of tiles or grout may occur. Point out that this is a cosmetic issue.

Checking The Floor   The easiest way to determine how the ceramic tile has been laid is to remove a floor register on a forced air system. Where this can't be done, you may be able to get a look at the floor around edges or at penetrations for plumbing pipes, for example.

Performance Based Inspection   Where you can't determine exactly how the tile floor has been laid, you can safely default to looking at the condition of the floor. Are there cracked, broken or loose tiles? As long as you document the condition of the floor and your limitations in looking at it, you have done your job. The Standards say we don't have to lift carpets. A throw rug in a kitchen or bathroom may cover cracked tiles. Many inspectors will move these carpets to look at ceramic tiles below.

Insulation
& Interiors
M O D U L E

QUICK QUIZ 2

☑ INSTRUCTIONS

• You should finish Study Session 2 before doing this Quiz.

• Write your answers in the spaces provided.

• Check your answers against ours at the end of this Section.

• If you have trouble with the Quiz, re-read the Study Session and try the Quiz again.

• If you did well, it's time for Study Session 3.

1. List five common flooring materials.

_____

_____

_____

_____

_____

2. List five general problems with floor systems and their implications.

_____

_____

_____

_____

_____

3. List five concrete flooring problems.

_____

_____

_____

_____

_____

_____

_____

4. List six wood flooring problems and their implications.

_____

_____

_____

_____

_____

_____

_____

5. List four carpet problems and their implications.

_____

_____

_____

_____

6. List three resilient flooring problems and their implications.

_____

_____

_____

_____

_____

7. List five ceramic floor problems and their implications.

_____

_____

_____

_____

_____

8. What is a control joint on a concrete slab?

_____

9. Where is rot most likely to occur on a wood frame flooring system?

_____

_____

If you had no trouble with this Quiz, you are ready for Study Session 3.

Key Words:
- *Concrete*
- *Wood*
- *Carpet*
- *Resilient*
- *Ceramic*
- *Trip hazard*
- *Absorbent material*
- *Squeaks*

Insulation & Interiors
MODULE

STUDY SESSION 3

1. You should have completed Study Session 2 and Quick Quiz 2 before starting this Session.

2. This Session covers walls and ceilings.

3. With respect to walls, you should be able to:
- List four common wall finish materials
- Describe a party wall in one sentence
- List four general wall problems and their implications
- List five plaster and drywall problems and their implications
- List four wood wall problems and their implications
- List two party wall problems and their implications
- List two garage wall problems and their implications

With respect to ceilings, you should be able to:
- Describe (in one sentence each) two inspection strategies that help with ceiling inspections
- List four common ceiling finish materials
- List three general ceiling problems and their implications
- List six plaster and drywall ceiling problems and their implications
- List four wood ceiling problems and their implications
- List one metal ceiling problem and its implication

4. Before you start, read Sections 2.0, 3.0 and 9.0 of the Interior Chapter of **The Home Reference Book**.

5. This Session may take you roughly one and a half hours.

6. Quick Quiz 3 is at the end of this Session.

Key Words:

• *Plaster*

• *Drywall*

• *Gypsum*

• *Wood*

• *Party wall*

• *Garage wall*

• *Shadow effect*

• *Nail pop*

• *Sag*

•*Textured ceilings*

S3

► 4.0 WALLS

Function

Wall finishes are decorative. They are part of the look of the interior of the home. Most wall finishes also add rigidity to the structure. Drywall prevents wood frame walls from racking, for example. Finishes also conceal and support insulation and air/vapor barriers. Electrical and mechanical systems are also concealed behind wall finishes.

Materials

Common wall finish materials include—

• plaster or drywall, which can be smooth or textured finish
• wood plank or paneling, which can be solid wood, plywood or hardboard
• masonry or concrete
• fiber cement paneling

There are other wall finish materials, but these ones are the most common.

Plaster and Drywall

Plaster and drywall (modern prefabricated plaster) are very popular as wall finishes with good reason. Plaster and drywall are:

• durable
• chemically inert
• inexpensive
• easy to repair
• easy to paint or wallpaper over
• fire resistant
• rodent and insect resistant
• good at blocking sound

How can you not like a material with all these qualities?! Because these materials are so popular, let's add a little to the discussion in **The Home Reference Book**.

Plaster In The Pre 1900s

Old plaster was made with lime which is calcium oxide, or from crushed limestone which is calcium carbonate. This material was heated to above 1700° or **calcined** to create **quicklime**. Water added to quicklime generates a chemical reaction that results in **slaked** or **hydrated lime** which is used for plaster. The plaster was mixed with water on site to generate the finished product.

Post 1900 Plaster

In the last 100 years, **gypsum** (often in the raw form of alabaster or spar) has been used to make plaster. Gypsum is calcium sulphate. The gypsum is crushed and calcined. It becomes **plaster of Paris** at this point. Additives are put into the plaster of Paris to create plaster. Again, on site we add water to recreate the crystalline structure we recognize as plaster.

Gypsum Lath   As discussed in **The Home Reference Book**, gypsum lath was an interim step between plaster and drywall. These prefabricated plaster boards were covered with paper on both sides. The 16 inch by 32 inch or 48 inch boards were nailed to studs. This lath replaced wood and wire lath. Two or three coat plaster was then added to the gypsum lath.

Drywall   Although drywall was invented much earlier, it became popular in the 1950's. It is also commonly called **Sheetrock**® which is actually a brand name of the U.S. Gypsum Company (now U.S.G. Corporation). Drywall is also called **wallboard, plasterboard, gypboard**, and **gypsumboard**. None of these are brand names.

Composition And Size   Drywall is plaster manufactured in a factory and covered on both sides with treated paper. It is available in panels from $\frac{1}{4}$ inch to $\frac{5}{8}$ inch thick and sizes ranging from four feet by eight feet to four feet by twelve feet. Four-and-a-half-foot-widths are also available.

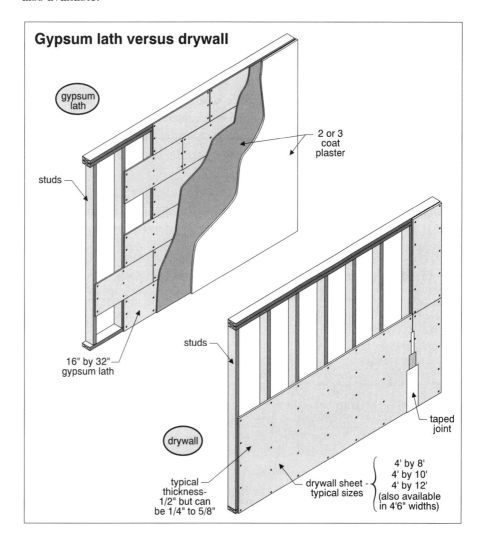

Gypsum lath versus drywall

gypsum lath

2 or 3 coat plaster

studs

16" by 32" gypsum lath

studs

taped joint

drywall

typical thickness- 1/2" but can be 1/4" to 5/8"

drywall sheet - typical sizes

4' by 8'
4' by 10'
4' by 12'
(also available in 4'6" widths)

47

Advantages And Disadvantages

Drywall is more stable than plaster and is smoother than most field applied plaster systems. However, it is thinner and weaker than conventional plaster. The joints are taped and in poor work, the joints are visible. Typically the sides of the drywall sheets are tapered to allow the joint compound and reinforcing paper or mesh to fill the recess and provide a flush surface. The paper drywall tape used in finishing joints performs better than the mesh, according to most.

Limitations

Drywall needs a more uniform, perfectly level substrate than plaster. Drywall is not as flexible as plaster. It's harder to do curved walls successfully with drywall. Again, the joints can be a problem with drywall.

Veneer Plaster

There's another hybrid between plaster and drywall known as **veneer plaster**. It's typically a ⅛ inch plaster finish coat applied over a special wall board. Some people refer to this as **skim coat plaster**. This approach is not widely used.

Wood Paneling

Wood paneling has largely been replaced with plywood and hardboard paneling which is considerably less expensive and easier to apply. Plywood and hardboard paneling can be very thin. The paneling should be at least ¼ inch thick to be applied directly to wall studs on 16 inch centers. Where the paneling is thinner, ⅜ drywall, for example, is typically used behind the paneling. Drywall would similarly be used behind even ¼ inch paneling where the studs are spaced more than 16 inches apart.

General Strategy

Flashlight Parallel To Wall

Home inspectors often shine a flashlight beam parallel to wall surfaces when inspecting walls. The light creates a shadow pattern that highlights flaws, irregularities and patches in the wall.

Push And Tap

Another inspection strategy for walls is to push and tap on the walls. Here we are looking for wall finishes that are loose or pulling away from their substrate. Bulging plaster, for example, is very common in old homes.

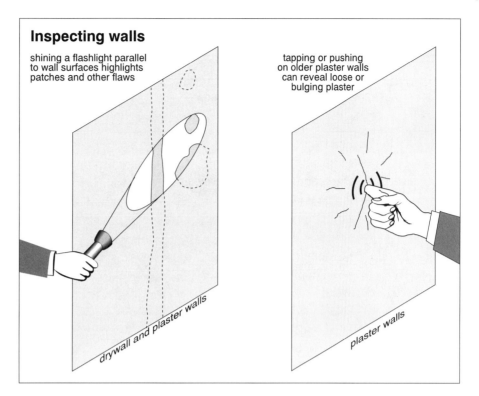

Inspecting walls

shining a flashlight parallel to wall surfaces highlights patches and other flaws

tapping or pushing on older plaster walls can reveal loose or bulging plaster

drywall and plaster walls

plaster walls

Decorating Hides The History

In many cases, fresh paint or wallpaper finishes will conceal considerable movement in walls. This is a limitation that we can't do much about, other than to appreciate that terrific looking interiors do not necessarily mean an absolutely rock solid structure. The newer the finish, the less history we have to rely on.

Drywall Over Plaster

It's very common to replace old plaster with multiple layers of paint and/or wallpaper with new drywall right over the old plaster. This sometimes means adjusting baseboard, window and door casings and other trim and moving electrical boxes, but is often simpler than trying to repair or replace old plaster walls.

Removing Wallpaper May Remove Plaster

When clients are looking at older homes and are thinking about removing wallpaper, they may ask you if that's a problem. Generally speaking, removing wallpaper is not a big problem. However, removing wallpaper on older plaster walls can often pull a good deal of the plaster off the wall. In some cases, it seems as though it's the wallpaper holding the plaster in place. Disturbing the wallpaper creates considerable plaster damage. Incidentally, wallpapering on ceilings creates the same risk. Trying to remove wallpaper from old plaster ceilings can be dangerous. Plaster may fall on you.

Now let's look at some of the specific problems.

4.1 CONDITIONS

Common wall problems include—

1. Water damage
2. Cracks
3. Mechanical damage
4. Inappropriate finishes in wet areas
5. Plaster and drywall
   • Bulging, loose or missing
   • Shadow effect
   • Crumbling or powdery
   • Nail pops
6. Wood
   • Rot
   • Cracked
   • Split or broken
   • Buckled
   • Loose
7. Party walls
   • Not continuous
   • Ice dams
8. Garage walls
   • Not fireproof or gastight

4.1.1 WATER DAMAGE

Cause

Water damage to walls can be caused by:

• roof or flashing leaks
• plumbing leaks
• heating leaks
• air conditioning leaks
• wall, window or skylight leaks
• door leaks (especially sliding glass doors)
• spills from humidifiers, dehumidifiers, watering plants, aquariums, bathtubs, showers, sinks, basins, etc.
• ice damming
• condensation

Implications

Water damage may be simply cosmetic and a decorating issue. If the wall finish material (plaster, drywall, paneling, etc.) is damaged, there may also be damage to the structure behind.

Strategy   We're going to use the same strategy we did when looking at water damage to floors. We'll try to determine –

- the extent
- the source
- whether it's active
- whether there's concealed damage
- what corrective action has to be taken

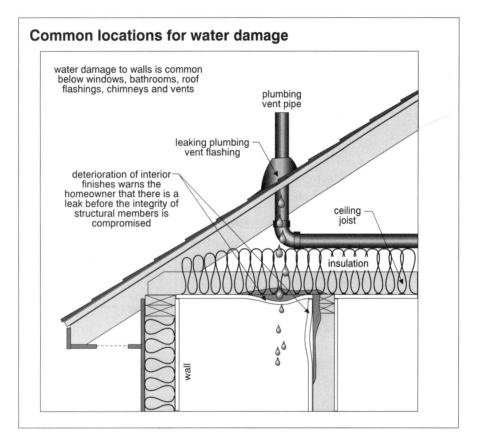

Common locations for water damage

water damage to walls is common below windows, bathrooms, roof flashings, chimneys and vents

plumbing vent pipe

leaking plumbing vent flashing

deterioration of interior finishes warns the homeowner that there is a leak before the integrity of structural members is compromised

ceiling joist

insulation

wall

Common   Water damage to walls is common below windows, bathrooms, roof flashings and
Locations   roof penetrations such as chimneys and vents. In many cases, the plaster or drywall interior finish is much less forgiving than the concealed structural components. This is a good thing in that deteriorated interior finishes will alert people to problems and cause them to take corrective action. Although their intent isn't to protect the structure, the happy result is that concealed wood framing members, for example, are protected.

The Exception   Slow or intermittent chronic moisture from small leaks or condensation is much more likely to cause damage to the structure because the interior finishes may not deteriorate. Moisture may be absorbed and held in framing and insulation materials. Very little moisture may get to plaster or drywall finishes. This has become a bigger problem since polyethylene air/vapor barriers have been used. This plastic can protect the interior finish from moisture damage as a result of leakage or condensation within the walls. This traps the moisture where it won't be noticed. There isn't much you can do about this, but you should understand the possibility.

4.1.2 CRACKS

Causes   Most cracks in interior finishes appear around doors and windows. The majority are cosmetic and are related to incidental movement of the structure. This includes shrinkage and expansion of building materials behind the finishes. Where the movement is significant, cracks in interior walls will typically be only one of several clues.

Implications   Most cracks are simply decorating issues. Cracks associated with movement to the building are, of course, structural concerns.

Strategy   Cracks may appear almost independently of the building age. The presence of cracks is more dependent on when the house was last decorated than the age of the building. Most cosmetic cracking is concealed every time walls are painted or wallpapered.

Check Inside Closets, Under Staircases, Etc.   Many inspectors look closely in areas that may not be decorated regularly, such as closets and areas below stairs. There may be more evidence of the history of the home in these spaces. You have to be careful with the information you get here. Considerable cracking, bulging and loose pieces of plaster may be visible. However, this may indicate neglect rather than any serious problems.

Structural Clues   When trying to determine whether cracks are structural, there are several considerations. We shouldn't drift into a structural inspection discussion, but will list a few things to watch for. Please refer to the Structure Module for more information.

• Do cracks on interior wall finishes also show up on the exterior of the building? This is much more likely on a masonry building than a wood frame building. Wood framing and exterior sidings such as clapboard, aluminum and vinyl siding will conceal much more movement than plaster, drywall or masonry.
• Are windows and doors out of plumb or square?
• Are floors sloped?
• Are there gaps between walls and floors or between walls and ceilings?

- Are the walls themselves visibly out of plumb?
- Is there evidence of movement (up or down, in or out or side to side) of more than ¼ inch in the wall finishes?
- Is the movement visible on interior partitions? Are these load-bearing partitions? How are the partition walls supported? Non-load-bearing partition walls often crack and sag if supported by a single floor joist parallel to the wall, for example. Movement in load-bearing partitions is usually more important.

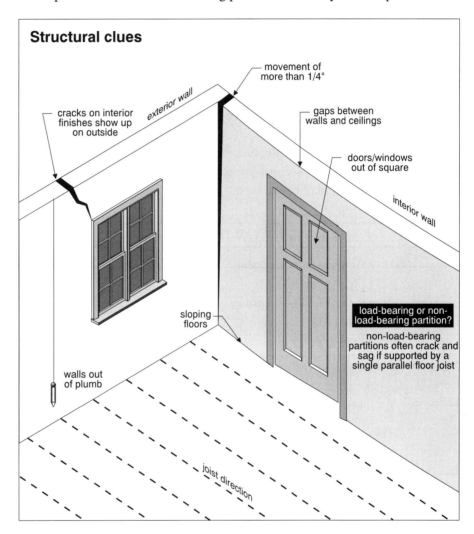

Structural clues

movement of more than 1/4"

cracks on interior finishes show up on outside

exterior wall

gaps between walls and ceilings

doors/windows out of square

interior wall

sloping floors

load-bearing or non-load-bearing partition?

non-load-bearing partitions often crack and sag if supported by a single parallel floor joist

walls out of plumb

joist direction

4.1.3 MECHANICAL DAMAGE

Causes   Mechanical damage is typically the result of impact. This is often furniture being pushed against walls, although it can be several other things.

Implications   Mechanical damage is usually localized and calls for minor repair.

Strategy   Look at wall surfaces and note holes and other evidence of physical damage. Check behind doors, especially where there are no doorstops. Many walls are damaged by door knobs when there is no effective doorstop.

4.1.4 INAPPROPRIATE FINISHES IN WET AREAS

Areas around bathtubs and showers should have smooth, hard, non-absorbent finishes to a height of at least six feet above floor level. This can be ceramic tile, glass, marble, plastic laminate, fiberglass, acrylic and other materials.

Wood Not Appropriate Wood, drywall and plaster around bathtubs or showers are not considered appropriate finish materials.

Causes This is an installation issue.

Implications Absorbent finishes will draw moisture in and are likely to stain or deteriorate. Wood is a good example of this. Plaster or drywall in wet areas will crumble and disintegrate. Textured finishes hold water and foster growth of mold and mildew.

Strategy Look carefully at wood and other absorbent or textured finishes in wet areas. If you see evidence of deterioration, you should recommend replacement. Where there is no evidence, you should let clients know of the possible implications. New occupants of a home may have a very different lifestyle. A wood enclosure around a bathtub may not be a problem until people start using the shower, for example.

4.1.5 PLASTER OR DRYWALL

- **Bulging, loose, missing**
- **Shadow effect**
- **Crumbling or powdery**
- **Nail pops**
- **Poor joints**

Causes Bulging, loose or missing plaster is usually a result of breaking the keys holding the plaster to the lath. This happens with time and vibration. People driving nails into walls with hammers typically damage some keys. People pounding on walls as they run up and down stairs and general vibration in the house can also break plaster keys, allowing the plaster to pull away from the lath.

Shadow Effect The shadow effect is an installation issue. We talk about it in **The Home Reference Book**. It is a cosmetic condition caused on original installation by applying subsequent coats of plaster before previous coats were dry.

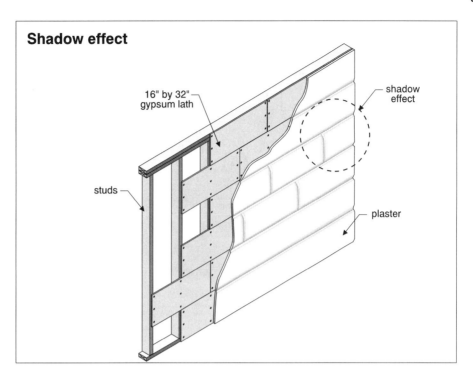

Shadow effect

16" by 32" gypsum lath

shadow effect

studs

plaster

Crumbling Or Powdery	Plaster or drywall that gets very wet or is wet for a period of time will disintegrate. It will become crumbly and powdery.
Nail Pops	Nail pops in new construction are a common result of shrinkage of wood framing members.
Nail Pop Mechanism	Nail pops in drywall occur in newly finished work where wall studs have high moisture content. Drywall is nailed tightly to wall studs. As the wood studs shrink, they pull away from the drywall. The nail will not be pulled with the stud, but the nailhead will remain flush with the drywall surface. When someone leans against the wall, the drywall will be pushed back against the stud. The nail will not move into the stud. The head of the nail will appear to **pop** out, although, in fact, it's not moving, it's the wallboard being pushed back against the stud that creates the nail pop. The solution is simply to drive the nailhead flush and refinish the surface.
Shorter Nails Better	Shorter nails are less prone to popping than long nails. Nails $1\frac{1}{4}$ inch long are suitable for $\frac{1}{2}$ inch drywall.
Screws Are Better	Drywall screws are much better than nails. They are less disruptive on original installation, they hold about three times better and will not pop. Drywall screws are self tapping and have a bugle shaped head. Proper installation avoids tearing the paper at the drywall surface. Screws $1\frac{1}{8}$ inch long are appropriate for $\frac{1}{2}$ inch drywall.

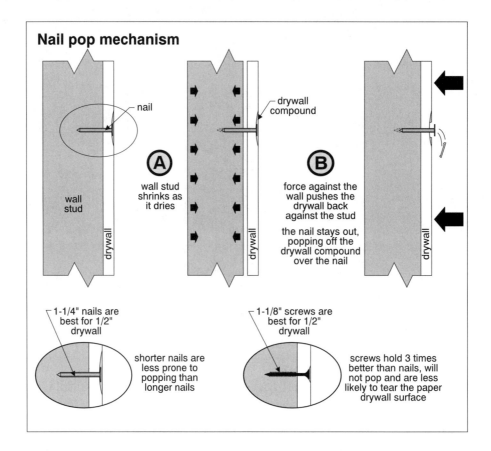

Nail pop mechanism

Drywall Joint Problems   Poor joints in a drywall job are an original installation issue. Causes include –

- inadequate heat in the home during winter construction
- excess moisture in the drywall, drywall compound or the home
- damaged board edges
- tape not fully embedded in compound
- lack of drying time between coats and before painting
- framing members shrinking, warping or twisting

There are several techniques that can minimize or eliminate visible joints. Our interest is simply to recognize the flaws as being joint conditions.

Implications   Bulging, loose or missing plaster is certainly a cosmetic issue. It can also be a safety issue. Falling plaster is heavy. Plaster falling from a wall along a stairwell can hurt someone below.

Shadow Effect   The implications of shadow effect are purely cosmetic.
Crumbling Or Powdery Plaster Or Drywall   The implications of deterioration depend on the amount of water present and whether the problem is active. Concealed damage to the structure is the big risk. If the problem is ongoing, the source has to be controlled first. The structure behind the damaged wall should be investigated, and only then should the plaster or drywall be repaired. If the problem is not active, removing and replacing the damaged section of plaster and drywall may be all that is required.

Nail Pops These are cosmetic issues and easily corrected.
And Poor
Drywall Joints

Strategy We talked about pushing and tapping on walls as a general strategy. We also talked about using a flashlight beam parallel to the wall surface to identify defects. Both of these techniques are valuable in identifying bulging or loose plaster and the shadow effect. Tapping on a wall may also reveal crumbling and powdery plaster or drywall. Poor joints and nail pops may be visible from a distance. The parallel flashlight technique may also be helpful.

4.1.6 WOOD

- Rot
- Cracked, split or broken
- Buckled
- Loose

Causes Wood wall finishes may have deteriorated as a result of moisture. Splitting or cracking may be the result of excessive drying or poor nailing practices. Loose paneling may be the result of improper installation or failed fasteners.

Implications The implications of most interior finish problems are cosmetic. If there is evidence of moisture damage, we're going to have to find out whether there is any concealed damage and whether the problem is ongoing before undertaking repairs.

Strategy A visual inspection of wood paneling is appropriate. The parallel flashlight technique can be helpful, as can pushing and tapping on the wall to determine its tightness.

Don't Use Some home inspectors use moisture meters to look for problems in wood paneling.
Moisture Meter It's not appropriate to use moisture meters with pins, since they will mark the finish.
With Pins Pad type moisture meters can be used, although any moisture meter use goes beyond the Standards.

4.1.7 PARTY WALLS

- **Not continuous**
- **Ice dams**

Causes Party walls are designed to prevent the spread of fire between attached homes. Party walls that are not continuous between adjacent dwelling units are an installation issue in most cases. Occasionally, the cause is a wall penetration that was made after original construction.

Ice Dams Ice dams are a result of original construction as well. Heat loss through the party wall was not adequately considered on original design or construction.

Implications Fire spread is the risk where the party wall is not continuous. Ice dams cause moisture damage to the roof structure and walls.

Strategy It's difficult to see party walls in most cases. You may not know what kind of fire resistance is required for party walls. Does it have to be a **fire separation** or a **fire wall**? In most cases, a fire separation providing a one hour fire resistance rating is adequate. This can be achieved with two by four wall studs covered with ⅝ inch thick Type-X drywall on both sides. The wall studs are typically filled with batt type insulation.

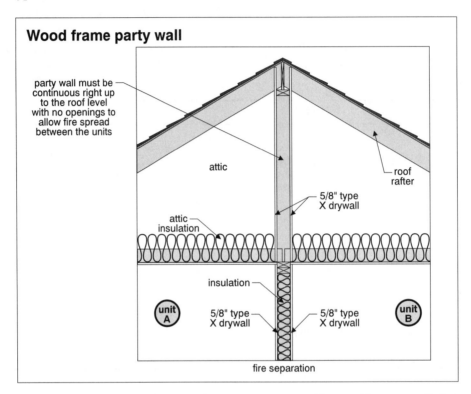

Wood frame party wall

party wall must be continuous right up to the roof level with no openings to allow fire spread between the units

attic

roof rafter

5/8" type X drywall

attic insulation

insulation

unit A

5/8" type X drywall

5/8" type X drywall

unit B

fire separation

Check Attics The most common place for home inspectors to see problems with party walls is in
For Continuity the attic. Party walls should be continuous to the underside of the roof sheathing and should be tightly fit there. There should be no openings in the party wall at the attic level.

Ice Dams Ice dams related to party walls are only an issue in cold climates. Depending on the insulation treatment (which won't usually be visible), there may be considerable heat loss up through party walls. This can melt the snow near the party wall causing water to run into other parts of the roof (including the eaves below the party wall). This may create a significant ice dam problem. We've talked about ice dam problems in the Roofing Module and in the Insulation section of this Module. The water backs up under roof shingles, damaging sheathing and framing below. Damage may also occur to siding and interior finishes.

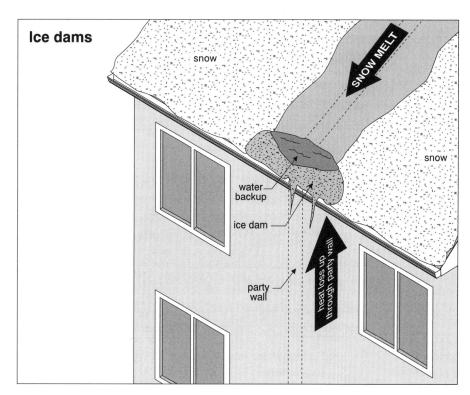

Ice dams

snow

SNOW MELT

snow

water backup

ice dam

heat loss up through party wall

party wall

Hard To Find   This is a problem that's very hard to identify except when the weather conditions are appropriate, and the ice dams have developed. You may, however, see evidence of ice damming. This is typically water damage to roof sheathing on the lower part of the roof only, near the party walls.

4.1.8 GARAGE WALLS NOT FIREPROOF OR GASTIGHT

Walls separating dwellings from garages have to be gastight in Canada and fireproof in the United States. Again, a wood stud wall with appropriate drywall can achieve the necessary fire separation. Drywall joints should be taped and there should be no penetrations through the wall.

Causes   Inappropriate fire or gastightness may be the result of original installation, subsequent mechanical damage or intentional penetrations.

Implications   Exhaust fumes or fire and smoke in the garage can enter the house if adequate fireproofing or gas tightness is not maintained.

Strategy   It's easiest to check this wall from the garage side. Make sure that there is a suitable finish on the garage wall. Look for gaps or other openings in the wall surface where the wall joins the house.

► 5.0 CEILINGS

Function

The function of ceilings is similar to walls. These are primarily decorative. The ceiling finishes typically add some rigidity to the roof structure and help prevent racking. The ceiling finishes also support the air/vapor barrier and insulation in most cases.

Materials

Ceiling materials are similar to wall materials. They can be plaster or drywall, wood, hardboard, plywood, fiber cement or concrete. Acoustic tile ceilings are also common. These tiles can be fiberboard or plastic faced fiberglass. These can be supported on wood strapping or suspended T-bars.

Metal ceilings are rare but may be found, usually in kitchens of old homes. An embossed pattern in the metal is often used to create the look of decorative plaster.

As we discussed with walls, it's not unusual to find that drywall has been provided over old plaster. Acoustic tile ceilings are also often applied below old plaster.

Smooth And Textured

Textured ceilings are very common in some areas because they are quick and inexpensive to install and are much more forgiving of imperfections in drywall. Textured finishes should not be used in kitchens or bathrooms because the large surface areas tend to collect dirt and grease, and are hard to clean.

5.1 CONDITIONS

Let's look at some of the common problems with ceiling finishes.

1. Water damage
2. Cracked, loose or missing sections
3. Mechanical damage
4. Plaster or drywall
    • Bulging, loose or missing
    • Shadow effect
    • Crumbling or powdery
    • Nail pops
    • Poor drywall joints
    • Sag
    • Textured ceilings in wet areas
5. Wood
    • Rot
    • Cracked, split or broken
    • Buckled
    • Loose
6. Metal
    • Rust
7. Lighting poor

5.1.1 WATER DAMAGE

Strategy   Our discussions here are similar to what we talked about on wall systems. The causes, implications and inspection strategies are similar. Many inspectors lift at least some of the tiles on suspended T-bar ceilings. There is often a good deal of information that can be determined looking above these ceilings.

Check Below   We talked earlier about the importance of running plumbing fixtures and then
Fixtures   checking the ceilings below. You should establish an inspection procedure that ensures that you revisit the ceiling below bathtubs and showers after operating these fixtures.

5.1.2 CRACKED, LOOSE OR MISSING SECTIONS

Truss Uplift   Again, this discussion is the same as for walls. However, there is one additional issue – **truss uplift**.

Implications   Truss uplift is a cold weather problem that is not terribly serious from a structural standpoint, but is very disturbing to home owners and can cause considerable cosmetic disruption.

Winter Only   Truss uplift occurs in the winter only. It occurs because the bottom chord of the truss arches up. The ends of the bottom chord rest on the outside wall, but the center portion rises. This usually creates a gap between the top of interior partitions near the center of the house and the ceiling. This gap can be well over an inch. In some cases, the partition wall is rigidly attached to the bottom chord of the trusses. In this case, the partition wall may be pulled up and a gap may appear between the bottom of the wall and the floor.

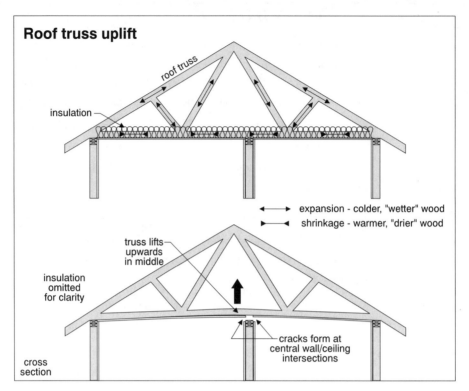

Roof truss uplift

roof truss

insulation

← → expansion - colder, "wetter" wood
► ◄ shrinkage - warmer, "drier" wood

truss lifts
upwards
in middle

insulation
omitted
for clarity

cracks form at
central wall/ceiling
intersections

cross
section

*Raised
Ceiling Or Wall*

These gaps always look like the wall is dropping (if the gap is at the wall ceiling intersection), or like the floor is dropping (if the gap appears at the wall/floor intersection). This is because we are prisoners of gravity, and think that all movement in a vertical plane is down. In this case, however, the movement is up. Either the ceiling is lifting off the wall or the wall is lifting off the floor. Close examination and perhaps the help of a mason's level will allow you to verify this.

Causes

The cause of truss uplift is not perfectly understood. The explanation that makes the most sense to us goes like this.

The cause of the bottom chord arching up is thought to be related to the differential moisture content of the wood framing members in the truss. Let's look at the different environments the truss components see. In cold climates, the bottom chord of the truss is surrounded by insulation. As a result, it is kept relatively warm. The majority of the webs and the top chords are above the insulation and are in a much colder environment. But temperature differences alone do not explain truss uplift.

*Relative
Humidity Is
Key*

Relative humidity is an indication of the amount of moisture in the air relative to the amount the air could possibly hold when it is saturated. Warm air can contain a lot more moisture than cold air. If you take a bundle of air at 70°F and 40% relative humidity, and heat it up, the relative humidity will drop. The amount of moisture in the air stays the same. However, the air is capable of holding much more moisture so at 90°F, the relative humidity might be only 20%.

If you take the same bundle of 70° air at 40% relative humidity and cool it, the relative humidity increases. For example, if you drop the air temperature to about 40°, the relative humidity approaches 100% and you may get condensation. We haven't added moisture to the air. Simply cooling the air has changed the relative humidity.

This is important in understanding truss uplift because wood sees relative humidity. If the relative humidity is low, no matter what the temperature, wood will not tend to draw moisture out of the air. As a matter of fact, moisture will leave the wood and the wood will become dryer. If the wood is surrounded by air with a high relative humidity, moisture will travel from the air into the wood. Now let's look at our attic environment.

Bottom Chord Warm
The bottom chord is in a warm environment. This means that the relative humidity will be low. The air surrounding the bottom chord might be close to 70° and the relative humidity might be 30 to 40%.

Rest Is Cold
The webs and top chords are in a much colder environment, with a much higher relative humidity – perhaps 80%, for example. To put the whole picture together, we need to look at one other thing.

Wood Expands When It's Wet
Most of us are comfortable with the concept that wood swells. From the discussion, it's easy to see that the top chords and webs are going to see more moisture than the bottom chord. As a result, the top chords will want to get longer. They are tightly pinned at the bottom corners. The top chords get longer and the peak of the roof rises slightly. This pulls the webs up and in turn pulls up on the bottom chord. The bottom chord can't rise at the wall corners because it's fixed at the outer edges. As a result, it lifts in the middle. As we discussed, it may lift off the partition walls in the middle of the building, or may pick the partition walls off the floor.

It's A Cyclical Problem
Truss uplift occurs in the winter, but as the attic environment changes and the warmer months come, truss uplift will reverse and the truss will settle back down. The problem repeats itself every winter and corrects itself every summer. The amount of truss uplift may vary depending on weather severity.

Which Trusses Are More Likely To Suffer Uplift?
Trusses with longer spans and trusses with lower slopes are more likely to suffer. Trusses in well insulated attics are more likely to suffer uplift.

Corrective Action
It's not possible to prevent truss uplift. The more tightly we secure partitions to the bottom chord, the more the house is picked up.

Moldings
The solutions typically lie in concealing the movement. One approach is to use a molding attached to the ceiling but not to the partitions. As the ceiling is lifted up during truss uplift, the molding simply slides up the wall. This is sometimes successful in masking cosmetic deficiencies, although the moldings will sometimes scratch the wall surfaces, or the raised moldings will reveal paint of a different color or bare spots on wallpapered sections of the wall.

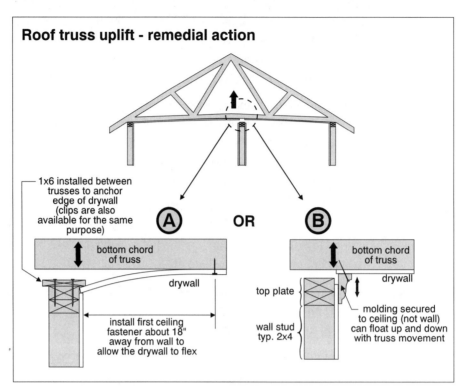

Roof truss uplift - remedial action

1x6 installed between trusses to anchor edge of drywall (clips are also available for the same purpose)

(A) OR (B)

bottom chord of truss

bottom chord of truss

drywall

drywall

install first ceiling fastener about 18" away from wall to allow the drywall to flex

top plate

wall stud typ. 2x4

molding secured to ceiling (not wall) can float up and down with truss movement

---

Float The Drywall

A more recent solution is to allow the ceiling drywall near the center partitions to float. The drywall is not attached to the trusses within roughly 18 inches of the partition walls. The edge of the ceiling drywall is secured to the wall with a 1 by 6 attached to the top plate, or a drywall clip fastened to the top of the wall. When the truss lifts, the edge of the ceiling drywall stays in place since it's not attached to the bottom chord near the partitions. The drywall bends up over 18 inches to the point where its attached to the trusses.

Strategy

When looking at truss roofs, understand that their advantage is their long uninterrupted spans and no need for interior bearing walls, but watch for the uplift problem particularly at interior partitions near the center of the truss span. Truss uplift is much easier to see in the winter, when the trusses have moved, than in the summer when they have relaxed.

Gaps

Look for gaps or evidence of movement at the wall/ceiling intersection and the wall/floor intersection.

Long Spans And Low Slopes

Trusses most likely to lift are those with long spans, low slopes and lots of insulation in the attic. Again, once you recognize the problem, you can set your client's mind at ease because it is not a serious structural issue. However, if you fail to identify it, you can expect a frantic phone call during the winter.

5.1.3 MECHANICAL DAMAGE

Again, the comments are the same as we offered on walls.

5.1.4 PLASTER OR DRYWALL

- **Bulging, loose and missing**
- **Shadow effect**
- **Crumbling or powdery**
- **Nail pops**
- **Poor joints**
- **Sag**
- **Textured ceilings in wet areas**

Most of these comments are the same as wall areas. Let's discuss sags and textured ceilings.

Causes   Sagging ceilings may be caused by wear and tear and vibration, resulting in plaster keys being broken.

Sags   In newer construction, sagging drywall may be an original construction issue. One half inch drywall supported by wood trusses on 24 inch centers will typically sag. The weight of the drywall alone is enough to make it sag as it spans between adjacent trusses. Many builders use one by three strapping (for screws) or two by two strapping (for nails) on 16 inch centers, applied to the underside of the trusses running perpendicular to the joists or truss chords. Others use $^5/_8$ inch drywall to try to avoid the sagging. Some do both.

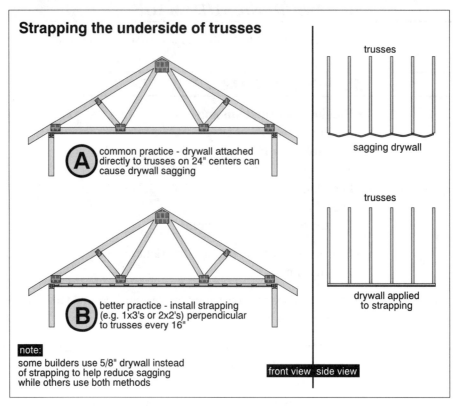

Strapping the underside of trusses

(A) common practice - drywall attached directly to trusses on 24" centers can cause drywall sagging

(B) better practice - install strapping (e.g. 1x3's or 2x2's) perpendicular to trusses every 16"

note:
some builders use 5/8" drywall instead of strapping to help reduce sagging while others use both methods

trusses

sagging drywall

trusses

drywall applied to strapping

front view   side view

Winter Construction

In cold climates, these precautions may not be successful if the construction sequence is poor. Consider a house that is framed, but without interior finishes. Outdoor temperatures are low and it's time to start to finish the inside. We can't put insulation in the attic until we have a ceiling to hold it up. We can't put the ceiling up until we put the polyethylene air/vapor barrier on the underside of the joists or truss chords.

Vapor Barrier But No Insulation

Typically, the air/vapor barrier goes on and the drywall is then installed. The house is usually kept warm enough to allow taping of drywall and to keep workers comfortable. Temporary propane heaters are often used for this. The interior of the home might be at 50°F. The heaters and construction activities generate considerable moisture in the air. The attic area is not heated, of course, but there is no attic insulation yet.

Condensation

The air/vapor barrier is cold and so is the ceiling drywall. The warm, moist air inside the house contacts the cold drywall ceiling. Condensation may develop on the drywall and air/vapor barrier. As the drywall absorbs the moisture, it loses some of its strength. The drywall may sag under its own weight as it absorbs moisture, which increases the weight and slightly reduces the strength.

Adding Insulation Quickly

The trick during winter construction in cold climates is to coordinate the trades so that insulation is added as soon as the ceiling is installed.

What We See You may see sagging or wavy ceilings in relatively new homes, even though the drywall is supported every 16 inches. The cause is often the sequence of construction described above. There are no functional implications. This is purely a cosmetic issue.

Textured Ceilings Where you see textured ceilings in kitchens and bathrooms, there's no serious functional problem. However, these ceilings may already look quite dirty and may be difficult to keep clean. Resurfacing them with a smooth ceiling finish is a possibility, although rarely a priority for a homeowner.

5.1.5 WOOD

- Rot
- Cracked
- Split or broken
- Buckled
- Loose

These issues are the same as we talked about for walls.

5.1.6 METAL – RUSTED

Causes Rust may develop on metal ceilings or on the metal components of suspended T-bar. Rust may be the result of leakage from above or from condensation. Metal ceilings in older houses were typically in kitchens.

Implications In most cases, the implications are cosmetic, although, in severe cases, sections of the ceiling may fall.

Strategy Look for evidence of rust on exposed metal ceilings. If you can lift suspended tiles, check the wiring and T-bar supports for rust.

5.1.7 LIGHTING POOR

All living spaces in houses should have adequate electric lighting. This includes hallways. Best practice in hallways includes three way switches so that the lights can be operated from either end. Note: three way switches are not operated from three different locations, just two.

Causes Inadequate lighting is an installation issue.

Implications Rooms or hallways without adequate lighting are a safety hazard.

Strategy Look for lighting in each room and in hallways and stairs. We'll talk about stair lighting shortly, but essentially a stairway is the same as a hall. Three way switches should be provided to operate the lights from the top or bottom of the stairs. Some jurisdictions allow an exception for stairs leading to unfinished basements, for example.

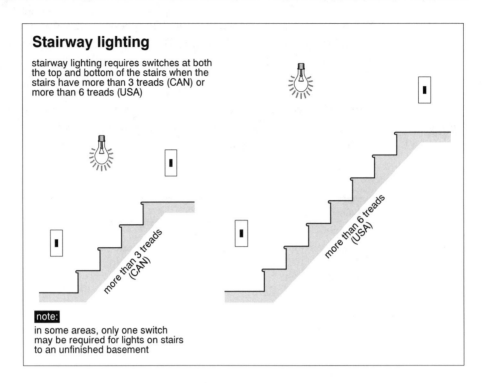

Stairway lighting

stairway lighting requires switches at both
the top and bottom of the stairs when the
stairs have more than 3 treads (CAN) or
more than 6 treads (USA)

more than 3 treads
(CAN)

more than 6 treads
(USA)

note:
in some areas, only one switch
may be required for lights on stairs
to an unfinished basement

Not
Necessarily
At Ceilings

Lighting is often provided at ceiling level, although this is not the case in most
living rooms, many family rooms and some bedrooms. Wall switches often control
one or both halves of duplex receptacles. This allows floor or table lamps to be
turned on when someone enters the room.

Insulation
& Interiors
M O D U L E

QUICK QUIZ 3

☑ INSTRUCTIONS

•You should finish Study Session 3 before doing this Quiz.

•Write your answers in the spaces provided.

• Check your answers against ours at the end of this Section.

• If you have trouble with the Quiz, re-read the Study Session and try the Quiz again.

• If you did well, it's time for Study Session 4.

1. List four common wall finish materials.

    _____

    _____

    _____

    _____

2. Briefly describe a party wall. Include its location, function and possible materials.

    _____

    _____

    _____

3. List four general wall problems.

   _____

   _____

   _____

   _____

4. List five plaster and drywall problems on walls.

   _____

   _____

   _____

   _____

   _____

5. List four problems with wood wall and ceiling finishes.

   _____

   _____

   _____

   _____

6. List two party wall problems.

   _____

   _____

   _____

   _____

7. List two related garage wall problems.

   _____

   _____

   _____

   _____

8. List as many of the implications of as many of the problems in Questions 3 to 7 as you can.

_____

_____

_____

_____

_____

_____

_____

_____

_____

_____

_____

_____

_____

_____

_____

_____

_____

_____

_____

_____

_____

_____

_____

_____

_____

_____

_____

_____

_____

9. Name two special inspection strategies you can use on ceilings.

    _____

    _____

    _____

    _____

10. List four common ceiling finish materials.

    _____

    _____

    _____

    _____

11. List three general ceiling problems.

    _____

    _____

    _____

12. List six plaster and drywall ceiling problems.

    _____

    _____

    _____

    _____

    _____

    _____

13. List one metal ceiling problem.

    _____

14. List as many of the implications of the problems in Questions 10 to 13 as you can.

_____

_____

_____

_____

_____

_____

_____

_____

_____

_____

_____

_____

_____

_____

_____

_____

_____

_____

_____

_____

_____

_____

_____

_____

_____

_____

_____

_____

_____

_____

_____

If you had no trouble with this Quiz, you are ready for Study Session 4.

Key Words:

• *Plaster*

• *Drywall*

• *Gypsum*

• *Wood*

• *Party wall*

• *Garage wall*

• *Shadow effect*

• *Nail pop*

• *Sag*

• *Textured ceilings*

Insulation & Interiors

MODULE

STUDY SESSION 4

1. You should have completed Study Session 3 and Quick Quiz 3 before starting this Session.

2. This Session covers trim, counters, cabinets and stairs.

3. At the end of this Session, you should be able to:
- list seven trim components on the interiors of homes
- list seven countertop materials
- list five trim problems and their implications
- list nine countertop problems and their implications
- list twelve cabinet problems and their implications
- define tread width, rise, run, stringer, winder, guardrail, handrail and baluster as they apply to interior stairs
- list 35 common problems with stairs and their implications

4. Before you start, read Sections 4.0 and 5.0 of the Interior Chapter of **The Home Reference Book**.

5. This Session may take you roughly one and a half hours to complete.

6. Quick Quiz 4 is at the end of this section.

Key Words:

- *Baseboard*
- *Quarter round*
- *Casing*
- *Sill*
- *Chair rail*
- *Plate rail*
- *Cornice molding*
- *Rosettes*
- *Medallions*
- *Tread width*
- *Rise*
- *Run*
- *Stringer*
- *Head room*
- *Landing*
- *Winders*
- *Handrail*
- *Guardrail*
- *Balusters*

► 6.0 TRIM, COUNTERS AND CABINETS

The Function of Trim

Trim is used to cover joints at changes in material and direction. Trim includes baseboard, quarter round, door and window casings and floor thresholds. Trim is also used to protect walls. Chair rails are an example of this. Trim can be used to add architectural appeal and display ornamentation. Plate rails are an example of this kind of trim. Trim can be used to make wall/ceiling transitions and add architectural detail to walls and ceilings. Cornice moldings and rosettes or medallions are examples of these.

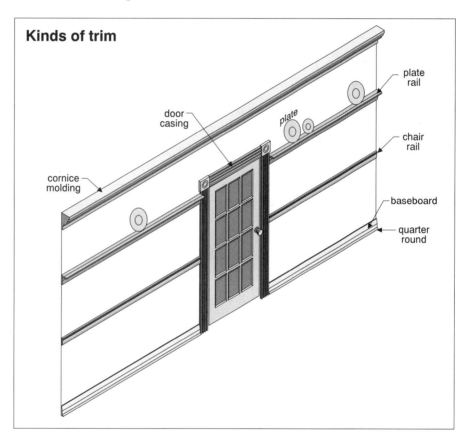

Kinds of trim

Trim Materials

Trim may be made of wood or wood-based products, including fiberboard. Trim may also be marble, ceramics, plaster or polystyrene (foamed plastic). The more fragile materials such as plaster and polystyrene are commonly used for ceiling trim which isn't likely to be mechanically abused.

The Function of Counters and Cabinets

Counters provide working surfaces in kitchens, bathrooms, pantries and bars. Cabinets provide storage facilities in kitchens, bathrooms and other rooms.

Counter Materials

Most counters are installed over particleboard, although there are some exceptions. Counter materials include—

• Plastic laminate
• Wood
• Marble
• Granite
• Synthetic marble
• Stainless steel
• Ceramic tile

Many counters are provided with backsplashes which are often the same material as the counter.

Cabinet Materials

Cabinets may be solid wood, particleboard or metal. Particleboard is often covered with a plastic laminate.

6.1 CONDITIONS

Let's look at some common problems with these systems.

1. Trim problems
   • Missing
   • Water damage
   • Rot
   • Loose
   • Mechanical damage
2. Counter problems
   • Entire top loose
   • Loose or missing pieces
   • Burned
   • Cut
   • Worn
   • Mechanical damage
   • Stained
   • Metal rusted
   • Ceramic grout missing or loose, ceramic tiles missing or loose
   • Substrate rotted

3. Cabinet problems
   - Water damage
   - Rot
   - Stained
   - Mechanical damage
   - Worn
   - Broken glass
   - Defective hardware
   - Stiff or inoperative
   - Not well secured to wall
   - Door or drawers missing or loose
   - Other pieces missing or loose
   - Shelves not well supported
   - Rust

6.1.1 TRIM PROBLEMS

Causes   Missing or loose trim may be an installation problem.

Rot and other water damage are the result of leaks, spills or condensation.

Mechanically damaged or loose trim is typically a result of abuse.

Implications Missing or damaged trim may be simply a cosmetic problem. Loose trim can allow air leakage into building components which may result in condensation damage. Rotted or mechanically damaged trim may conceal rot or other damage to building systems behind.

Strategy   Look for trim to be continuous, intact and well secured. Note missing or damaged trim. Where possible, look behind the trim for other damage.

6.1.2 COUNTER PROBLEMS

Causes   Countertops may be loose because of poor installation. Loose or missing pieces may be the result of water damage or physical abuse. Burned, cut and worn surfaces are normal wear and tear. Mechanical damage is usually the result of an impact which may be a one time event or repetitive. Stained counters are usually the result of liquid penetrating through cuts and plastic laminate, for example. Metal which is rusted is usually the result of a defective metal or strong acid. Ceramic tiles that are loose or missing or have grout that is loose or missing are usually the result of poor installation, mechanical abuse, or excessive deflection in the substrate.

Rot in the substrate is usually the result of leakage around sinks and basins. This often occurs at faucet connections and may be the result of splashing or leakage of the faucet.

Implications   The implications of a loose countertop may be personal injury if the countertop falls. Poor hygiene is the implication of loose or missing pieces, burned, cut or worn surfaces, mechanical damage or stained counters and rust. Loose or missing ceramic tiles or grout may also be hygiene issues as well as cosmetic defects.

Rotted substrate may result in the countertop collapsing or sinks, faucets or basins coming loose.

Strategy   When looking at countertops, grab the edge and try to lift with moderate force. Don't damage the countertop by applying excessive force. Look for:

- Loose or missing pieces, burns, cuts, or worn areas
- Mechanical damage resulting from impact
- Stains on marble, wood and plastic laminates
- Rust on metal countertops
- Loose or missing tiles or grout on ceramics

When you're looking at cabinetry and plumbing fixtures, check the underside of the countertops, especially around sinks and faucets, for evidence of rot.

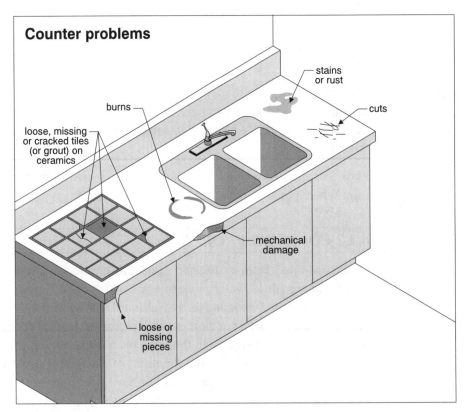

Counter problems

stains or rust

burns

cuts

loose, missing or cracked tiles (or grout) on ceramics

mechanical damage

loose or missing pieces

Move Cutting   While home inspectors are not required to move household goods, sliding a cutting
Boards   board or other articles out of the way so you can see the entire countertop can be revealing.

Ask Client To Help?   Many home inspectors get their clients involved when looking at counters and cabinets in kitchens and bathrooms. The evaluation of these systems is somewhat subjective; allowing clients to open and close cupboard drawers and pull out drawers may be helpful to both you and the client.

6.1.3 CABINET PROBLEMS

Causes   Cabinet problems may be caused by:

- water-damaged, rotted or stained cabinets may be caused by leaks from the roof, plumbing or heating systems, walls or windows.
- damage may also be the result of splashing at sinks and counters.
- mechanical damage, worn cabinetry and broken glass may be the result of normal wear and tear or physical abuse.
- defective hardware or stiff or inoperative drawers and doors may be an original installation issue, a lack of maintenance, or abuse.
- cabinets not well secured to walls are typically an installation issue.
- doors, drawers or other pieces that are missing or loose may be the result of poor original construction or abuse.
- shelves that are not well supported are usually missing support guides or pins.
- rust is more common in metal medicine cabinets than anywhere else in the home.

Implications   The implications of severe water damage may be failure of the cabinets. The implications of damage or wear may be collapse of the cabinet or inoperable doors or drawers, if severe. In many cases, the defects are cosmetic.

Glass   Cracked or broken glass can be a safety issue if glass falls onto people.

Hardware   Defective hardware and stiff or inoperative doors or drawers are functional problems that diminish the usability of the cabinets.

Secured To Wall   Cabinets that are not well secured to a wall are a serious safety issue. Falling cabinets can seriously injure people.

Loose Or Missing Doors   Doors, drawers or other pieces that are missing or loose affect the usability of cabinets.

Weak Shelves   Shelves that are not well supported are usually minor problems, unless they are filled with china before giving way. This can cause damage and injury.

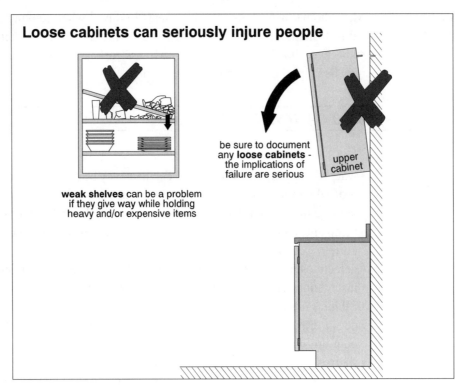

Loose cabinets can seriously injure people

weak shelves can be a problem if they give way while holding heavy and/or expensive items

be sure to document any **loose cabinets** - the implications of failure are serious

upper cabinet

Rust   Rusted medicine cabinets are not hygienic and the cabinet may eventually rust through and become nonfunctional.

Strategy   Where water damage is visible, probe for rot and make sure the cabinet structure is intact.

Let the client know about wear or mechanical damage that affects the usability of the cabinetry.

Recommend that broken or cracked glass be replaced. Operate all doors and drawers, looking for hardware or operational deficiencies. In many cases, adjustment or lubrication is all that is required.

Apply moderate upward force on wall-hung cabinets to ensure they aren't loose. Test that shelves are secure by applying moderate downward force to the front edge.

► 7.0 STAIRS

Function

Stairs allow people to move between different levels of the home.

Materials

Stairs are typically wood, metal or concrete. Let's review a couple of definitions. **Rise** is the vertical distance between two treads. **Run** is the horizontal distance from one riser to the next, measured along the tread. **Tread width** is the depth of the tread from front to back including the nosing.

Rules and dimensions

There are lots of rules and various dimensional constraints for stairs. These vary depending on where you are in North America. **The Home Reference Book** covers some of the rules. We'll recap them here quickly although you'll need to check your local jurisdiction.

Straight treads

- Thickness – minimum one inch, if risers support front of treads.
- Thickness – minimum 1½ inches, if risers do not support treads.
- Rise – maximum 8 to 8¼ inches (depending on jurisdiction).
- Uniformity – three-eighths inch maximum variation in rise in some jurisdictions (other jurisdictions do not specify any allowable deviation).
- Nosing or back-slope – maximum 1½ inches.
- Tread width – minimum 9 to 9¼ inches, depending on jurisdiction.
- Run – minimum 8¼ to 9 inches

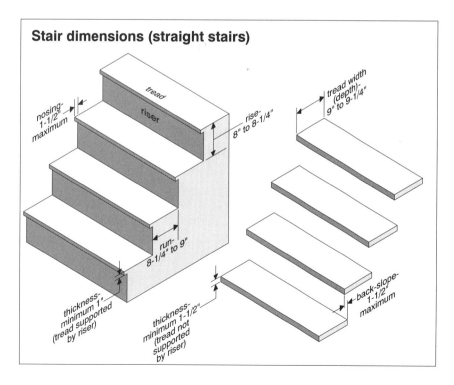

Stair dimensions (straight stairs)

nosing- 1-1/2" maximum

tread

riser

rise- 8" to 8-1/4"

tread width (depth)- 9" to 9-1/4"

run- 8-1/4" to 9"

thickness- minimum 1" (tread supported by riser)

thickness- minimum 1-1/2" (tread not supported by riser)

back-slope- 1-1/2" maximum

Curved treads

• Average tread width – minimum 9 inches
• Tread width at 12 inches from narrow side of stairs – minimum 9 inches
  (another jurisdiction says the tread width has to be at least 9 inches at the
  middle of the step)
• Tread width – minimum 6 inches at narrowest point

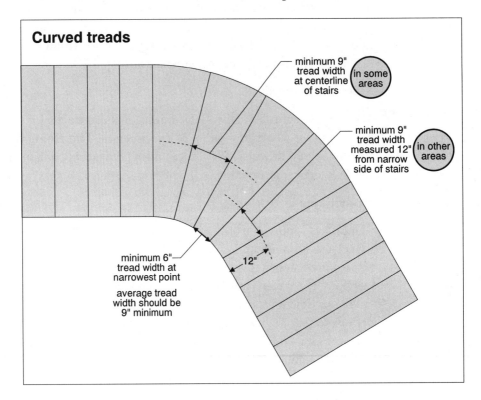

Curved treads

minimum 9"
tread width
at centerline
of stairs

in some
areas

minimum 9"
tread width
measured 12"
from narrow
side of stairs

in other
areas

12"

minimum 6"
tread width at
narrowest point

average tread
width should be
9" minimum

Winders

• Only one set of winders allowed per staircase
• Maximum turn for each winder – 30 degrees
• Maximum turn for any set of winders – 90 degrees (usually three winders)

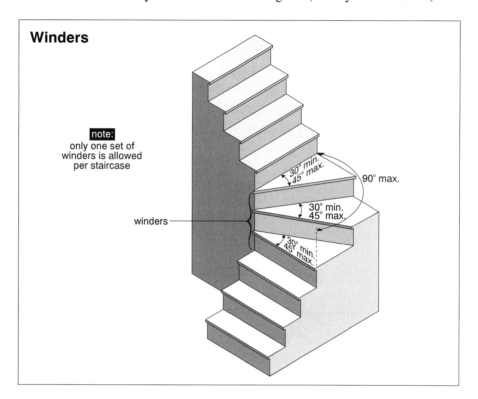

Stringers

- Depth – minimum 2 inches by 10 inches
- Depth left after notching for treads – minimum 3½ inches
- Thickness if supported by wall – minimum 1 inch
- Thickness if not supported by wall – minimum 1½ inches
- Distance between stringers if risers do not support front of treads – maximum 35 inches
- Distance between stringers if risers do support front of treads – maximum 47 inches

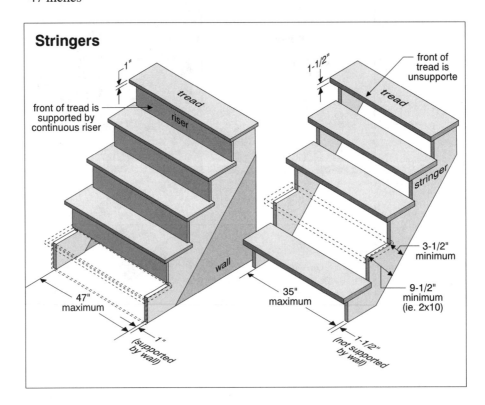

Stairwell width

• Minimum 34 to 36 inches (depending on jurisdiction)

Headroom

• Minimum 6 foot, 6 inches to 6 foot, 8 inches (depending on jurisdiction) measured vertically above a diagonal line drawn through tread nosings.

Landings

• Minimum 36 inches long and same width as stairwell. Note: not needed if door at top of stairs opens away from stairs.

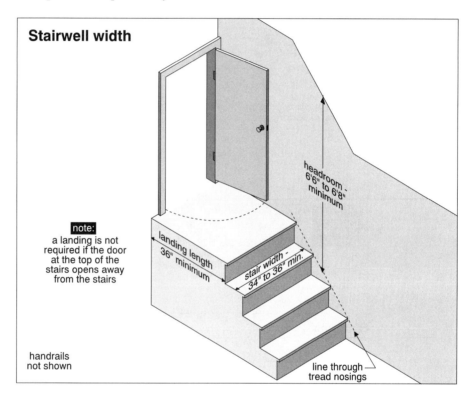

Stairwell width

note:
a landing is not required if the door at the top of the stairs opens away from the stairs

landing length 36" minimum

stair width - 34" to 36" min.

headroom - 6'6" to 6'8" minimum

handrails not shown

line through tread nosings

Handrails and guardrails (guards)

Handrails are something you can hold onto going up or down stairs. **Guardrails** or **guards** keep you from falling off a hallway or over the edge of a landing or balcony. Most people think of handrails as installed on the diagonal, and guardrails as horizontal around the top of openings. A handrail on the open side of a staircase is also a guardrail, keeping you from falling over the side of a staircase.

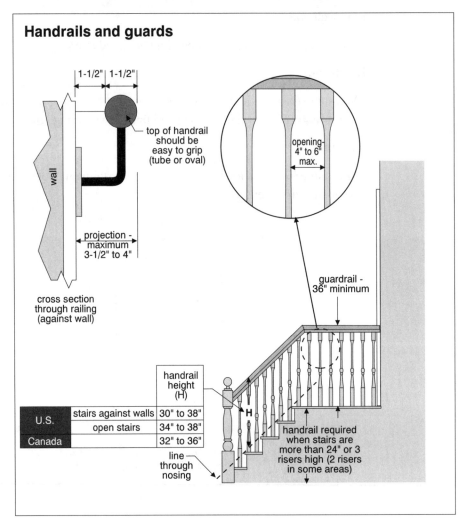

Handrails and guards

1-1/2" 1-1/2"

wall

top of handrail
should be
easy to grip
(tube or oval)

opening -
4" to 6"
max.

projection -
maximum
3-1/2" to 4"

cross section
through railing
(against wall)

guardrail -
36" minimum

handrail
height
(H)

H

U.S.	stairs against walls	30" to 38"
	open stairs	34" to 38"
Canada		32" to 36"

line
through
nosing

handrail required
when stairs are
more than 24" or 3
risers high (2 risers
in some areas)

Handrails

- When required – all stairs more than 24 inches high or three risers high (some jurisdictions); any stairs more than two risers high (other jurisdictions).
- Height range – 34 to 38 inches on open stairs and 30 to 38 inches on stairs against walls (U.S.); 32 to 36 inches (Canada).
- Number needed – on at least one side of staircase. Should be on open side, if there is one.
- Shape of handrail – should be easy to grab, top should be 1½ inches maximum and tapered or round for easy grip (2 x 6 on edge is not a good handrail).
- Distance from wall – should be at least 1½ inches.
- Total projection into stairwell – maximum 3½ to 4 inches (depending on jurisdiction).

Guardrails around top of openings

- Height – minimum 36 inches

Balusters (spindles)

- Should be vertical only; horizontal and diagonal balusters may create an easy climbing grid for children.
- Openings – maximum 4 to 6 inches (don't want children to crawl through)

Summary

Now that we've looked at all the rules, let's look as some of the common problems we find. As you can imagine, most of these are simply disregarding the rules. Again, we should emphasize the need to check your own jurisdiction and make up your own stair checklist.

7.1 CONDITIONS

1. Rot or water damage
2. Mechanical damage
3. Treads
   - Too thin
   - Excessive rise
   - Not uniform
   - Excessive nosing or back slope
   - Inadequate tread width
   - Too many winders
   - Winder angle too big
   - Worn or damaged
   - Sloped
   - Loose or poorly supported
4. Stringer problems
   - Too small
   - Excessive notching for treads
   - Too thin
   - Excessive span between stringers
   - Rot
   - Pulling away from wall or treads
   - Inadequately secured to header
5. Stairwell width inadequate
6. Headroom inadequate
7. Landings
   - Missing
   - Too small
8. Handrails
   - Missing
   - Too high or too low
   - Hard to grasp
   - Loose or damaged

9. Guardrails
   - Missing
   - Too low
   - Loose or damaged
10. Balusters (spindles) for handrails or guardrails
   - Too far apart
   - Easy to climb
   - Loose or damaged
   - Missing
11. Fire stops inadequate
12. Lighting inadequate

7.1.1 ROT

Implications   The implications of these problems and the others we'll address in this section are —

- personal injury. Stairs that are poorly designed are hazards. People may fall down or off the stairs.
- stairs themselves may collapse.

We won't repeat these implications with each discussion.

Strategy   We've talked about rot a number of times. We know the causes and implications. We also know how to look for it. Pay particular attention to the bottom of basement staircase stringers. Watch also for rotted stringers against exterior walls, particularly below grade.

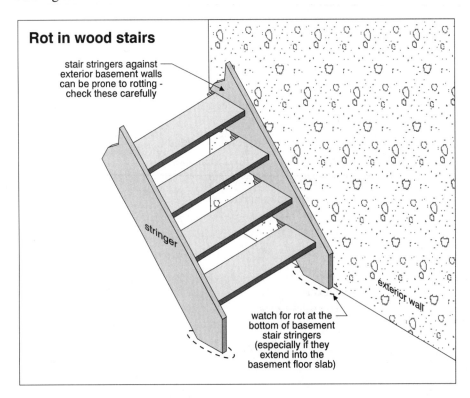

Rot in wood stairs

stair stringers against exterior basement walls can be prone to rotting - check these carefully

stringer

watch for rot at the bottom of basement stair stringers (especially if they extend into the basement floor slab)

exterior wall

7.1.2 MECHANICAL DAMAGE

Strategy   Again, we've talked about mechanical damage. Check for this at the stringers, treads and risers. The header that supports the top of the stringer is not usually visible; if it is, check it for damage as well.

7.1.3 TREAD PROBLEMS

Causes   Problems with tread thickness, rise, uniformity, nosings or width are design issues. Treads that are worn, damaged, loose, sloped or poorly supported are usually wear-and-tear or mechanical damage issues.

Strategy   Most inspectors do not measure all of the critical dimensions, at least after the first few inspections. You'll come to very quickly develop an eye and a feel for stairs as you do inspections. The muscle memory of your body will usually tell you if stairs are not uniform. You'll intuitively sense stairs that are too steep (excess rise) or have inadequate tread width. Have you measured the stairs in your own home? This is a good place to start.

As you step on each tread, you'll notice excessive wear or slope, or if treads are loose.

Look At
Tread/Stringer
Connection
With Flashlight   There are a number of ways that stringers can support treads. Support can be through notched stringers, routed stringers, cleats or any combination.

You should get in the habit of checking the tread/stringer connection as you go up and down stairs. Any movement of the stringer may result in the tread being poorly secured.

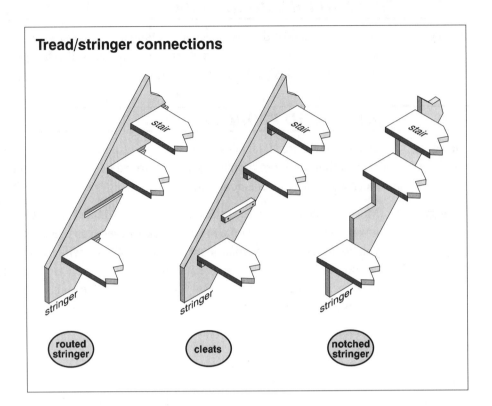

Tread/stringer connections

stair

stair

stair

stringer

stringer

stringer

(routed stringer)   (cleats)   (notched stringer)

7.1.4 STRINGER PROBLEMS

Causes

Stringers that are too small, too thin, excessively notched or have an excessive span are design issues. Rot, pulling away from treads or walls and poor connection to the header are installation or maintenance issues or the result of movement of the structure.

Strategy

Stringers are not always visible. Where they are, check for movement. Standing at the top or bottom of a set of stairs, you can look along the staircase and see movement of either stringer in a lateral plane. This may cause stair treads to become loose.

Watch for stringers that run parallel to walls. In some cases, trim is added to the top of the wall as it pulls away from the wall. The trim simply covers what would otherwise be an unsightly gap. The gap may be visible from a closet below, but it is also visible looking along the stringer from the top or bottom of the stairs.

Header Connection

Again in most cases you won't be able to see how the stringer is supported by the header. In most cases, movement can't occur without the bottom of the stringer moving relative to the floor. Watch for movement here as well.

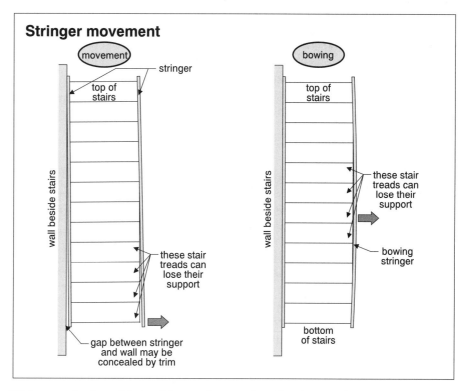

Stringer movement

Rot

We've touched on this but it's worth repeating: Rot is most likely to occur at the bottom of stringers, especially in below-grade stairs.

7.1.5 STAIRWELL WIDTH

You'll get a sense of stairwell width just walking up and down stairs. Basement stairs or stairs to unfinished areas are often narrower. This is acceptable in some jurisdictions and will be common. Similarly, inadequate headroom is often an issue on basement stairs. It's very expensive to rearrange stairs and, while you should advise your client that the stairs are not a standard arrangement, it is not usually practical to rearrange them.

7.1.6 HEADROOM INADEQUATE

Strategy

If you have to duck going up or down the stairs, the headroom is inadequate. In many older homes, the headroom is less than ideal. The headroom for basement staircases is very often less than ideal, but not practical to change.

7.1.7 LANDINGS

Strategy   You shouldn't be standing on a stair to open a door. Most staircases should have landings. Although landings can be omitted at the top of the stairs if the door opens away from the staircase or if there is no door, the best practice includes at least a twelve-inch landing at the top of the stairs. Again, rearranging stairs is very expensive and would only be done as a last resort. You should, however, point out the lack of an adequate landing.

7.1.8 HANDRAILS

Strategy   Handrail problems are obvious and easy to spot with a little practice. Your inspection should include using the handrail as you climb stairs.

Free-standing   Handrails that are not tied into walls, at least at the top and bottom, are most
Railings   susceptible to problems. Many railings are supported at their ends by **newel posts**. If this post is not well anchored to the structure, the handrail can be loose.

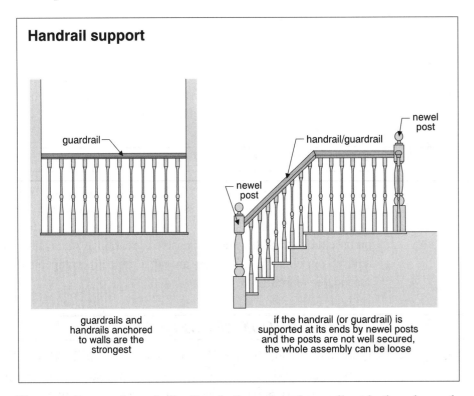

Handrail support

guardrail

newel post

handrail/guardrail

newel post

guardrails and handrails anchored to walls are the strongest

if the handrail (or guardrail) is supported at its ends by newel posts and the posts are not well secured, the whole assembly can be loose

The same is true of guardrails. Guardrails anchored to walls at both ends are the strongest, but it is common to find one or both ends of the guardrail secured to a post.

7.1.10 BALUSTERS (SPINDLES)

Strategy   Balusters or spindles should be vertical, and there should be less than four inches (some areas say six inches) between adjacent balusters. We don't want a child to be able to crawl through. The best inspection includes grabbing each baluster and testing that it is tight. Most home inspectors do not do this; testing a representative sample for tightness is typical.

7.1.11 FIRE STOPS INADEQUATE

In the United States, the underside of staircases has to be covered with half-inch drywall to help protect against fire spreading quickly between floors. In most jurisdictions, fire stops in the floor system are required at the top and bottom of staircases. These will not usually be visible.

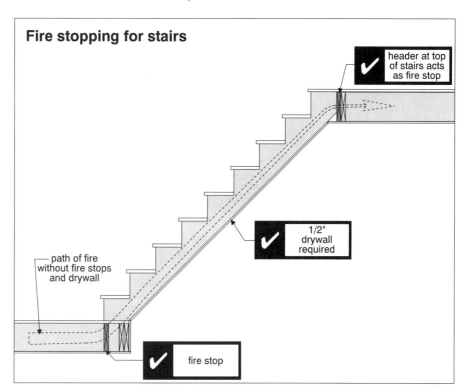

Strategy   Check that the underside of the staircase has a tight and continuous drywall surface.

7.1.12 LIGHTING INADEQUATE

Strategy   All stairwells with more than four risers should have lighting controlled by switches at the top and bottom. These are called **three-way switches**, although they only operate the light from two locations. Three-way switches are not usually required at stairs to unfinished areas, such as basements.

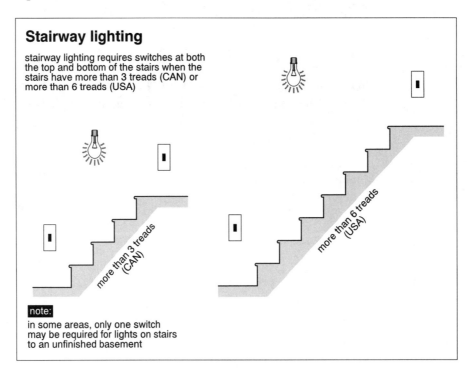

Stairway lighting

stairway lighting requires switches at both the top and bottom of the stairs when the stairs have more than 3 treads (CAN) or more than 6 treads (USA)

more than 3 treads (CAN)

more than 6 treads (USA)

note:
in some areas, only one switch may be required for lights on stairs to an unfinished basement

Operate the light switches at both the top and bottom of stairs to ensure the light responds to both switches properly. Having the light on as you look at the stairs obviously makes a lot of sense.

Insulation
& Interiors
M O D U L E

QUICK QUIZ 4

☑ INSTRUCTIONS

• You should finish Study Session 4 before doing this Quiz.

• Write your answers in the spaces provided.

• Check your answers against ours at the end of this Section.

• If you have trouble with the Quiz, reread the Study Session and try the Quiz again.

• If you did well, it's time for Study Session 5.

1. List two functions of trim on the interior of homes.

_____

_____

2. List seven trim components commonly found in homes.

_____

_____

_____

_____

_____

_____

_____

3. List seven countertop materials you may find in kitchens or bathrooms.

_____

_____

_____

_____

_____

_____

_____

4. List five common problems with interior trim and their implications.

_____

_____

_____

_____

_____

5. List nine common countertop problems.

_____

_____

_____

_____

_____

_____

_____

_____

_____

6. List twelve common cabinet problems.

_____

_____

_____

_____

_____

_____

_____

_____

_____

_____

_____

_____

_____

7. Define each of the following terms in one sentence with respect to stairs:

tread width or depth

_____

rise

_____

run

_____

stringer

_____

winder

_____

guardrail

_____

baluster

_____

8. List as many stair problems as you can.

_____

_____

_____

_____

_____

_____

_____

_____

_____

_____

_____

_____

_____

_____

_____

_____

_____

_____

_____

_____

_____

_____

_____

_____

_____

_____

_____

_____

_____

_____

_____

_____

If you had no trouble with this Quiz, you are ready for the Interim Test.

Key Words:

• ***Baseboard***

• ***Quarter round***

• ***Casing***

• ***Sill***

• ***Chair rail***

• ***Plate rail***

• ***Cornice molding***

• ***Rosettes***

• ***Medallions***

• ***Tread width***

• ***Rise***

• ***Run***

• ***Stringer***

• ***Head room***

• ***Landing***

• ***Winders***

• ***Handrail***

• ***Guardrail***

• ***Balusters***

Insulation
& Interiors
MODULE

STUDY SESSION 5

1. You should have completed Study Session 4, Quick Quiz 4 and the Interim Test before starting this Session.

2. This Session covers windows and doors.

3. By the end of this Session, you should be able to do the following with respect to windows:
   • List four window functions.
   • List four common materials used in frames and sashes.
   • List eight common window types.
   • Describe in three sentences **low-E glass.**
   • Describe in two sentences **gas-filled** windows.
   • List two general problems with windows and their implications.
   • List six frame problems and their implications.
   • List two drip cap problems and their implications.
   • List ten exterior trim problems and their implications.
   • List eight sash problems and their implications.
   • List six interior trim problems and their implications.
   • List six glass problems and their implications.
   • List five hardware problems and their implications.
   • Explain in one sentence how a window location can be a problem.
   • List four storm and screen problems and their implications.
   • Explain in one sentence why window size can be a problem.
   • Explain in two sentences how ice dams occur at skylights.

You should also be able to do the following with respect to doors:
- List five functions of exterior doors.
- List four functions of interior doors.
- List four common materials used in doors.
- List two general problems with doors and their implications.
- List thirteen door and frame problems and their implications.
- List two drip cap problems and their implication.
- List twelve exterior trim problems and their implications.
- List nine interior trim problems and their implications.
- List eight hardware problems and their implications.
- List five garage door problems and their implications.

4. Before you start this Session, read sections 6.0 and 7.0 in the Interior Chapter of **The Home Reference Book**.

5. This Session may take you roughly one and a half hours.

6. Quick Quiz 5 awaits you at the end of this Session.

Key Words:

- *Light*
- *Ventilation*
- *Means of egress*
- *Glazing*
- *Low-E glass*
- *Gas-filled windows*
- *Single and double hung*
- *Casement*
- *Fixed*
- *Awning*
- *Hopper*
- *Jalousie*
- *Slider*
- *Muntins*
- *Mullion*
- *Frame*
- *Sash*
- *Sill*
- *Trim*
- *Drip cap*
- *Lintel*
- *Hinged*
- *Bifold*
- *Pocket*
- *Self-closer*

► 8.0 WINDOWS, SKYLIGHTS AND SOLARIUMS

Function

Windows allow light and ventilation. Windows can add to the architectural appeal of the home, and can provide emergency exits (means of egress).

Materials

Window frames and sashes may be made of wood, vinyl (often polyvinyl chloride), metal (steel or aluminum) or fiberglass. Wood windows may also be vinyl-clad or metal-clad.

Glazing Materials

Conventional glass is the most common, although laminated, tempered and wired-glass may be found. Acrylic is common in skylights. Polycarbonates are used in windows where great strength and security are important.

Glazing Types

Windows may be single-, double- or triple-glazed. Single-glazed windows may have storm windows and screens. Double- and triple-glazed windows and skylights may have additional energy efficiency features, such as low-E glass and gas-filled spaces. We'll talk about those later.

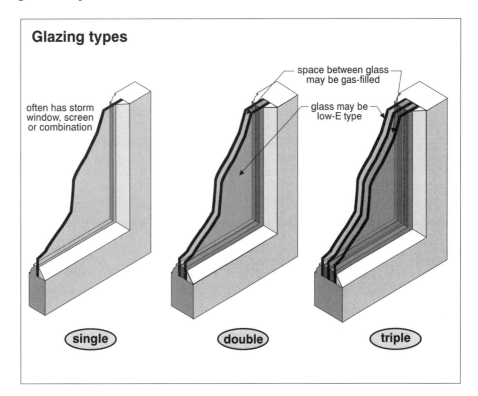

Glazing types

often has storm window, screen or combination

space between glass may be gas-filled

glass may be low-E type

single   double   triple

Window Types

Common window types include—

• Single-hung (only the bottom sash is operable)
• Double-hung (both the top and bottom sash is operable)
• Casement (the windows may swing in, out or pivot)
• Horizontal sliders
• Awning (hinged at top and open outward)
• Hopper (hinged at bottom, may open in or out)
• Fixed (includes conventional glazing and glass block; glass block may be structural)
• Jalousie (typically used only in warm climates, since they do not seal tightly but are good for ventilation)

Window types

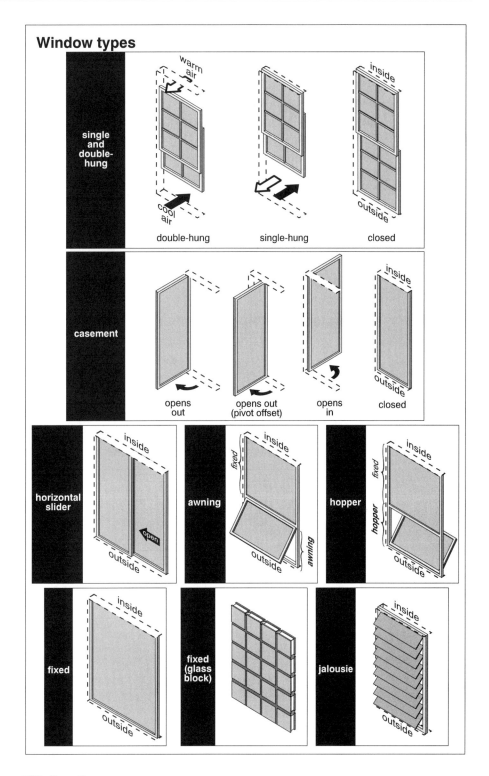

Window Components

There are many window components. It can be a little overwhelming trying to remember all the names. The illustration is probably the best way to learn these. Let's deal with some of the main components.

Frames   The **frame** is the perimeter structure of the window.

Sashes   **Sashes** support the glass and are the movable part of the window.

Muntins   **Muntins** divide the glass into individual **panes** within the sash. In the early days of glass making, it was hard to make large panes of glass. Muntins were used to create large windows. Muntins may be wood, lead, brass, vinyl or decorative tape. Many muntins are decorative and applied adjacent to the glass; they don't actually divide the glass into separate panes. Many of these are clip-in systems that are removable to make it easy to wash the windows. True muntins separate the glass into individual panes. Many modern muntin systems are simply to create an architectural effect. Now that glass can be made in large sheets, it's much less expensive to have fewer large panes of glass than several small ones.

Mullions   **Mullions** are the posts separating two adjacent windows in a group. Mullions are typically wood, vinyl or metal.

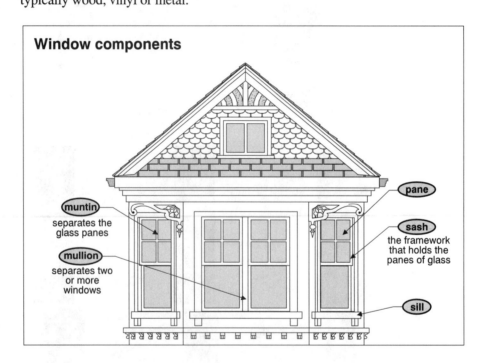

Window components

muntin — separates the glass panes

mullion — separates two or more windows

pane

sash — the framework that holds the panes of glass

sill

Fall Protection

Windows on stair landings should have a minimum sill height of 36 inches. If someone stumbles on the stairs, we don't want them falling out through the window. Where sills are lower, interior grilles or guardrails can be added.

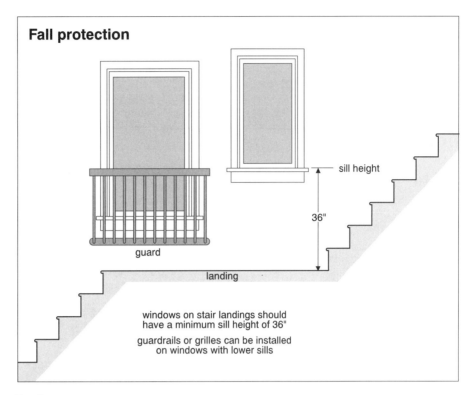

Fall protection

sill height

36"

guard

landing

windows on stair landings should
have a minimum sill height of 36"

guardrails or grilles can be installed
on windows with lower sills

Leakage

Most windows leak both air and water. Air leakage is often unnoticed. The water leakage is usually intermittent and may only occur during wind-driven rains from a given direction. Windows may leak at the top, sides or bottom. Leaks at the bottom are often the most difficult, because they go unnoticed for some time. Water getting into the wall system below can cause considerable concealed damage.

Drain Holes   Windows with tracks at the bottom (horizontal sliders, for example) should have drain holes to allow water in the tracks to escape. Leakage through these is common. Leakage through the bottom corners of wood, metal and vinyl windows is common. Corner joints often fail. Many experts feel that the welded vinyl windows are least likely to suffer this kind of failure.

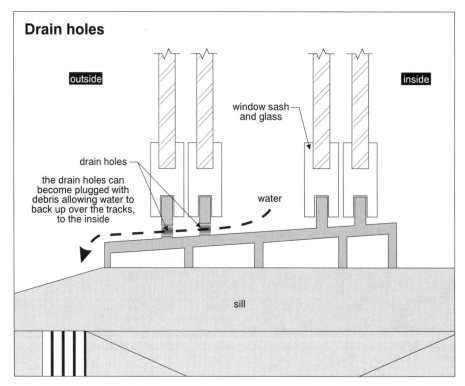

Flashing Pans Some installers use prefabricated pans that form a flashing at the bottom of the window, in anticipation of window leakage. These flashing pans direct water to the outside, rather than allowing it to get into the wall system below the window.

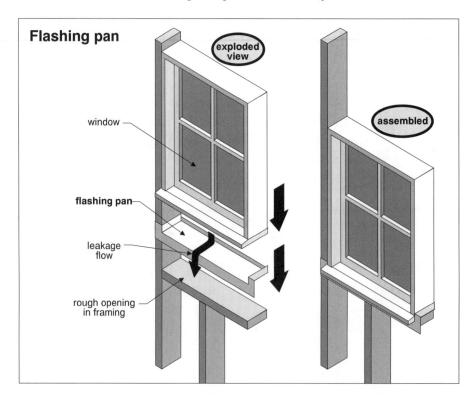

Replacement windows

Replacing windows is a common home improvement project. We are often asked if new windows will save enough money on heating and cooling costs to justify the expense. The short answer is, "No." Windows are wasteful in terms of energy, but even the best modern windows don't dramatically reduce home heating or cooling costs. People who are determined to replace windows will have to settle for the improved comfort resulting from less drafts, easier operation, or improved appearance.

Subjective Sidebar

In our opinion, many traditional homes lose, rather than gain, architectural appeal with replacement windows. However, this is a subjective comment, and we should stay away from these.

Energy efficiency

Considerable attention has been paid to the energy efficiency of windows in recent years. Let's put things in perspective. Current standards for wall insulation often call for R-values of approximately 20. A single-glazed window has an R-value of roughly 1. A double-glazed window has an R-value of roughly 2, and a triple-glazed window has an R-value of roughly 3. Even with the highest energy improvement tricks, R-values of windows do not approach 5. Let's look at some of the ways that window efficiency is improved.

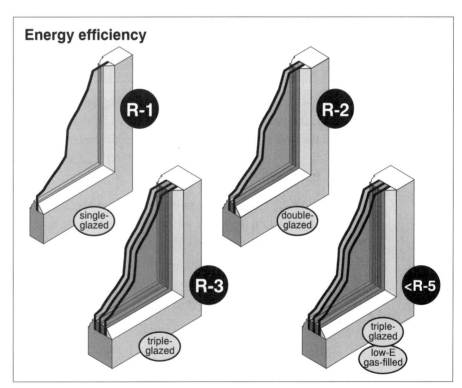

Energy efficiency

R-1 single-glazed

R-2 double-glazed

R-3 triple-glazed

<R-5 triple-glazed low-E gas-filled

Low-E Glass   **Low-E glass** uses coatings that reduce the **emissivity** of windows. The emissivity of a material describes its ability to radiate heat. A heat exchanger and conventional glass have high emissivity. Aluminum foil has low emissivity. We talked about radiant barriers when we talked about insulation. Radiant barriers have low emissivity.

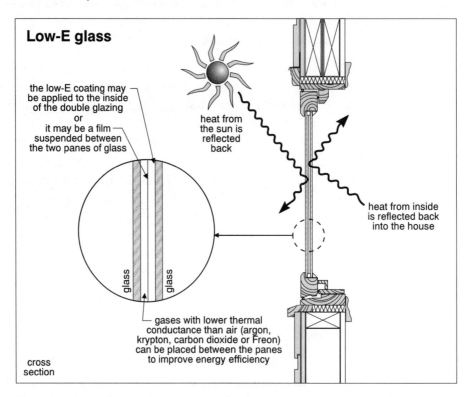

Low-E glass

the low-E coating may be applied to the inside of the double glazing
or
it may be a film suspended between the two panes of glass

heat from the sun is reflected back

heat from inside is reflected back into the house

glass   glass

gases with lower thermal conductance than air (argon, krypton, carbon dioxide or Freon) can be placed between the panes to improve energy efficiency

cross section

Reduced Heat   These low-E coatings reduce heat transfer through the window and reflect the heat.
Flow   In winter, the heat is moving out of the house; the low-E glass reflects some of the heat back into the home. In the summer, the heat typically moves through the window from the outdoors in; the low-E reflects a good deal of this solar heat back outside, helping with the cooling.

Metallic   The coatings are typically metallic and can be applied a couple of ways. The
Coating   coating is usually on the inside of double glazing, so it isn't exposed to the air or elements. Low-E glass can also use a film suspended between the two panes of glass, effectively creating a triple-glazed system.

Blocks   Low-E glass has slightly less light transmission than traditional glass, although
Ultraviolet   it's not usually noticeable. Low-E glass also helps screen out ultraviolet light, resulting in less fading of draperies and furniture, for example.

Gas-filled   Conventional windows have air in the space between the glazing layers. Higher
Windows   efficiency windows use heavy gases, such as argon, krypton, carbon dioxide or Freon®, for example, to improve performance. These gases have lower thermal conductance than air.

These windows are tricky to manufacture because it's hard to keep air out of the space between the glazing. Very good seals are needed around the perimeter to keep the gases in and the air out.

Inspection Implications

You won't usually know by looking whether the glass is low-E and whether the windows are gas-filled. It's not really a big issue for home inspectors, since conventional glass is not a deficiency. This is an upgrade that helps reduce heating costs, but is not a huge issue.

Let's look at some other factors that affect the energy efficiency of windows.

Casements, Awning And Hopper Windows

Casement, awning and hopper windows can be more energy efficient than double-hung and horizontal sliding windows. That's because the sash is pulled tight against a seal, rather than sliding along a seal. There's typically more air leakage in a sliding window.

Move Glass Inboard

Glass is better closer to the inner surface of a wall system than the outer surface. The glass stays warmer and condensation is reduced. Many windows have the glass substantially outboard.

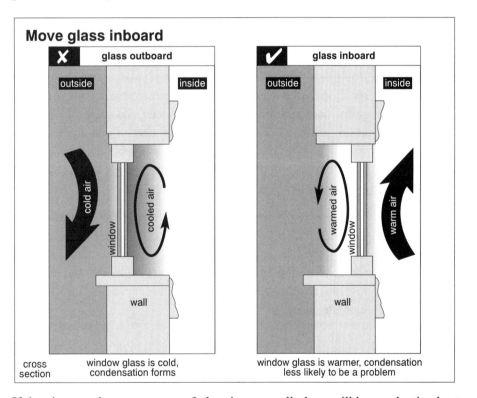

Spacing Between Layers Of Glass

If the air space between panes of glass is too small, there will be conductive heat loss through the window, and the advantage of double-glazing will be substantially reduced. If the air space is too large, convective loops will be set up and, again, the advantage of double-glazing will be lost. The optimum air space appears to be about five-eighths of an inch. Common gaps are roughly one-half inch, which is close to the optimum spacing.

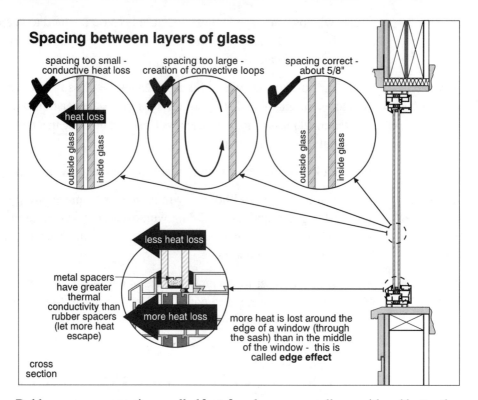

Spacing between layers of glass

spacing too small - conductive heat loss

heat loss

outside glass　inside glass

spacing too large - creation of convective loops

spacing correct - about 5/8"

outside glass　inside glass

less heat loss

metal spacers have greater thermal conductivity than rubber spacers (let more heat escape)

more heat loss

more heat is lost around the edge of a window (through the sash) than in the middle of the window - this is called **edge effect**

cross section

Rubber Spacers And Edge Effect

Rubber spacers, sometimes called **butyl** seals, are generally considered better than metal spacers because of their reduced thermal conductivity. Windows typically suffer more heat loss around the perimeter, and are most vulnerable to condensation at the perimeter. That's because the thermal conductivity of window sashes is greater than the glass itself. This heat loss around the perimeter is typically called the **edge effect**. Some of the early rubber spacers failed to maintain seals as a result deterioration caused by ultraviolet light.

Thermal Breaks In Frames

Metal and vinyl windows typically have a thermal break in the frame and the sashes. A thermal break is an insulating material that isolates the inner half of the frame from the outer half. This reduces thermal bridging and heat loss through the frame. Wood windows do not require thermal breaks, because wood is a much better insulator than metal or vinyl. Wood, of course, has the disadvantage of requiring more maintenance.

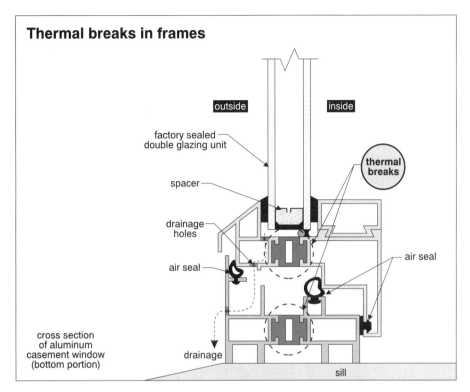

Thermal breaks in frames

outside | inside

factory sealed double glazing unit

thermal breaks

spacer

drainage holes

air seal

air seal

cross section of aluminum casement window (bottom portion)

drainage

sill

Clad Wood Windows Wood's better thermal performance but higher maintenance has led to the development of vinyl-clad and metal-clad wood windows, in an effort to get the best of both materials. Detractors of these types of windows point out that if water finds its way past the cladding, it may be trapped inside and rot the wood. This is something a home inspector would not typically be able to identify.

Let's move on to look at some things that are more meaningful to home inspectors. We'll assume you've read the Windows section of the Interior Chapter of **The Home Reference Book**.

General strategy

When we were inspecting the exterior of the home, we asked you to look at windows. We said your inspection would not be complete until you had looked from the inside. The interior window inspection is really a follow-up to your exterior window inspection.

Upper Floor Windows In two- and three-story homes, your inspection of the window exteriors is usually limited by your distance from the windows. You may be able to open upper floor windows when inside the home and get a look at the exterior. You'll want to determine the condition of sills and seals around the perimeter of the window on the exterior. This is where moisture penetration protection is most important.

Don't Fall Out Sometimes it's not practical to remove screens and glass to get a look outside. You also need to use common sense. Don't risk falling out of the window while straining to get a look.

Caulking Is
Expensive

Many home inspectors are very casual about recommending caulking improvements for homes. There is a substantial expense involved in caulking the entire exterior of a home. If you're going to make this recommendation, tell your client it is costly if done by a contractor.

Beware Of
Steel Casement
Windows

Many early casement windows were single-glazed steel units. These were common in the 1950s and '60s. These windows did not perform well and many have been replaced. They typically have significant air and water leakage and condensation problems. They are often difficult to operate.

Partial
Replacement

In some cases, the sashes have been replaced, but the steel frames and sills remain in place. Since these windows did not have thermal breaks, considerable heat loss and condensation problems may persist. Ideally, the entire window is replaced.

Beware The
Sashless Slider

There is another low quality window that was used in the 1960s, '70s and '80s. This horizontal sliding window has a simple pane of glass as the operable component. There is no sash and no weather stripping. An operating knob may be fastened through a hole in the glass or clipped to the edge of the glass. The seal between adjacent panes is simply glass-to-glass. This is not particularly weather-tight.

These windows suffered considerable condensation and leakage problems. Again, many have been replaced, or provided with an external storm window system, even though the original windows were double glass.

Pull Back
Curtains

A common home inspector mistake is failure to pull back curtains around windows. Window leakage is common and damage often shows up below the windows at either corner. These are the corners that are typically covered by curtains.

Look behind window treatments

water stain

peeling paint

moisture damage often shows up below the lower corners of windows

Operate All Windows

Although the Standards call for operating one window in each room, we recommend you operate as many as possible. We have found that homeowners are often very disappointed when windows do not operate smoothly. It takes a few more seconds in each room, but we recommend you try each operable window.

Sloped glazing

Skylights and solariums have glazing units that are not vertical. These are vulnerable to mechanical damage from hail and falling branches, for example. In many areas, glass has to be laminated, tempered, wired or otherwise strengthened in skylights and solariums. Another common approach is to use acrylic instead of glass.

Strengthened Glass

As a general rule, if the glazing is more than 15 degrees off vertical, some attention has to be paid to strengthening the system to protect against mechanical damage. It's beyond the scope of a home inspection to determine whether glazing has been strengthened since you can't typically determine this by looking. While you may find some clues, there are many cases where you won't be able to be conclusive.

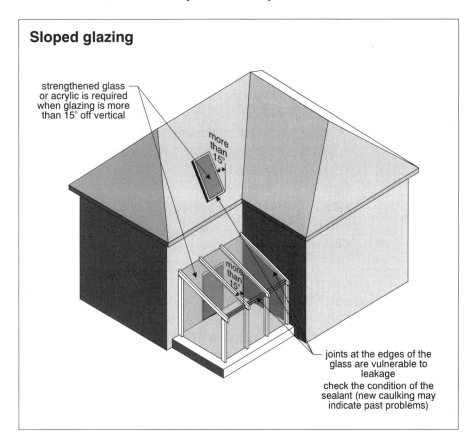

Sloped glazing

strengthened glass or acrylic is required when glazing is more than 15° off vertical

more than 15°

more than 15°

joints at the edges of the glass are vulnerable to leakage

check the condition of the sealant (new caulking may indicate past problems)

Leakage Is Common

Skylights and solariums are frequently troublesome with respect to water penetration. Problems arise at edges of glass panels that are not vertical. Lips are created that catch and hold water. We encourage you to spend a little extra time looking for evidence of leaks around and below these.

Now let's look at some common window problems.

8.1 CONDITIONS

1. Leaks
2. Lintels sagging or missing
3. Frames
    • Rot
    • Rust
    • Racked
    • Deformed
    • Installed backwards
    • Drain holes blocked or missing
4. Exterior drip cap missing or ineffective
5. Exterior trim
    • Missing
    • Rot
    • Rust
    • Damaged, cracked or loose
    • Sills with reverse slope
    • Sill projection inadequate
    • Drip edge missing
    • Glazing compound cracked, missing, loose or deteriorated
    • Caulking or flashing missing, loose, deteriorated, rusted or incomplete
    • Paint or stain needed
6. Sash
    • Rot
    • Rust
    • Inoperable
    • Stiff
    • Won't stay open
    • Sash coming apart
    • Loose fit
    • Weather stripping missing or ineffective
7. Interior trim
    • Rot
    • Stained
    • Missing
    • Cracked
    • Loose
    • Poor fit

8. Glass
    - Cracked
    - Broken
    - Loose
    - Missing
    - Lost seal on double- or triple-glazed
    - Excess condensation
9. Hardware
    - Rust
    - Broken
    - Missing
    - Loose
    - Inoperable
10. Location
    - Sills too low
11. Screens
    - Torn
    - Rust
    - Loose
    - Missing
12. Storm windows missing
13. Too small for egress
14. Ice dams at skylights

8.1.1 LEAKS

Windows may leak air or water. Water leaks are immediate problems.

Causes   Leaks are usually installation problems or manufacturing defects.

Implications   Water leakage is an immediate problem with respect to damage to windows and wall assemblies below, as well as interior finishes.

Air leakage affects heating and cooling costs, and may result in concealed rot damage.

Strategy   We've already talked about being very careful to look around and below windows, especially at the corners. Look also for evidence of leakage at the top of the windows. You will sometimes see rust or water stains across the top of a window.

During warm, dry weather it is sometimes difficult to determine whether water damage around the bottom of a window is the result of condensation or leakage. If you can't be certain, allow for both possibilities.

Skylights
And Solariums

The sloped glazing at skylights and solariums is susceptible to leakage because water gets hung up on the lips and edges supporting the glass. Flashings at curbs and solarium/wall intersections are also common leak spots. Look carefully for stains or water marks on the interior below and around the glazing and flashing. Wood sashes, muntins, mullions and framing members are **very susceptible to rot**. Check carefully, and probe if possible (don't damage finishes).

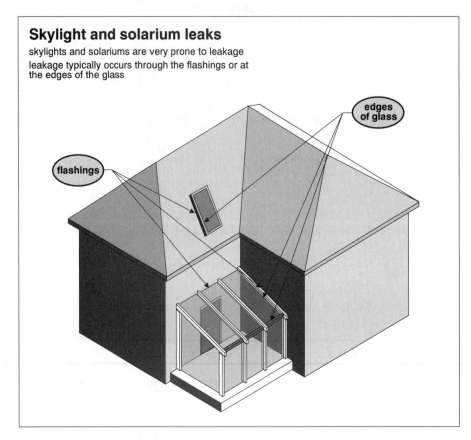

Skylight and solarium leaks
skylights and solariums are very prone to leakage
leakage typically occurs through the flashings or at
the edges of the glass

edges
of glass

flashings

8.1.2 LINTELS SAGGING OR MISSING

This may be more visible from the outside of the building. It's common on large windows or groups of windows. If the lintel or arch is missing or ineffective, you will often see a sag across the top of the window or window group.

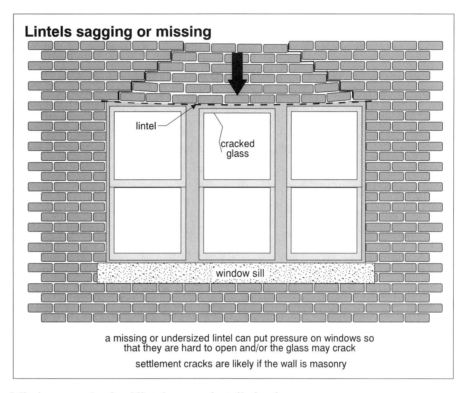

Lintels sagging or missing

lintel

cracked glass

window sill

a missing or undersized lintel can put pressure on windows so that they are hard to open and/or the glass may crack

settlement cracks are likely if the wall is masonry

Cause   Missing or undersized lintels are an installation issue.

Implications   Windows may be difficult to operate and glass may crack. In severe cases, the wall section above may fail.

8.1.3 FRAME

- Rot
- Rust
- Racked
- Deformed
- Installed backwards
- Drain holes blocked or missing

Causes
- Rotted and rusted windows are a maintenance issue.
- Racking is due to structural movement.
- Deformation of window frames is often the result of foaming the windows to insulate and air-seal the gap between the window frame and the house wall. Low-expansion foams, rather than high-expansion foams, are preferred. Foaming should be done in several small steps, rather than all at once, to avoid this problem.

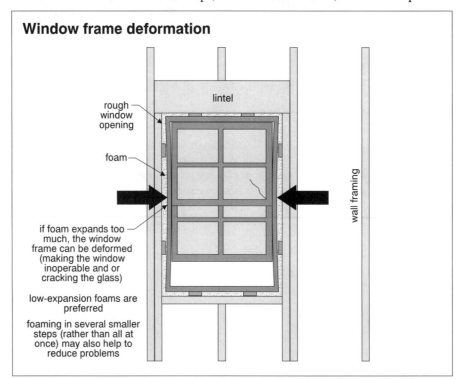

Window frame deformation

lintel

rough window opening

foam

wall framing

if foam expands too much, the window frame can be deformed (making the window inoperable and or cracking the glass)

low-expansion foams are preferred

foaming in several smaller steps (rather than all at once) may also help to reduce problems

- Installing windows backwards is obviously an installation issue.
- Blocked drain holes are a maintenance issue.
- Missing drain holes are manufacturing defects.

Implications
The implications of rot, rust, racking and deformed frames are poor weathertightness and operability. The window may not open or close. It also may crack and, if not corrected, windows may have to be replaced.

The implications of windows installed backwards are water leakage into the building. The sill will probably have a reverse slope and there will be no drain holes in any tracks that may catch water. Drain holes may be present inside the window but, of course, that will just direct water into the house.

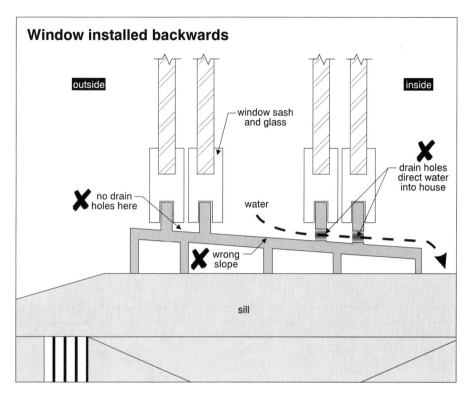

Window installed backwards

outside   inside

window sash
and glass

✘ drain holes
direct water
into house

✘ no drain
holes here

water

✘ wrong
slope

sill

The implication of blocked or missing drain holes is, again, water leakage into the home.

Strategy   If you can get to the exterior sections of the frame, check for rot and rust. Probing wood sections, especially where there are horizontal edges that can collect water, or end grains exposed at corners is a good way to check for this. Don't be too enthusiastic and damage the wood by probing.

Sight Along   Check for racked or deformed frames by sighting along the top, bottom and sides
Frames   of windows.

Sill And Drain   Check the sill slope and look for drain holes to ensure the window has not been installed backwards. Hardware may also provide a clue that the window has been installed backwards.

8.1.4 EXTERIOR DRIP CAP MISSING OR INEFFECTIVE

Causes   Missing or ineffective drip caps are an installation issue.

Implications   The implication is water leakage into the window and wall system.

Strategy   Look for metal caps that protect the tops of every window in wood-frame walls. You won't see this in masonry or brick veneer construction because the windows are recessed into the wall.

Projection   Look for a projection of the cap beyond the window. We are looking for water to
And Slope   fall past, not onto, the window. The cap will be nearly horizontal but should slope to drain water away from the wall.

Width   The drip cap should be slightly wider than the window and should not allow water to drain off the side and get behind siding.

Can Sometimes   If the windows are protected by the roof overhang, no drip cap may be needed.
Be Omitted   As a general rule, if the distance from the top of the window to the soffit is less than one quarter of the soffit width, we don't need a drip cap.

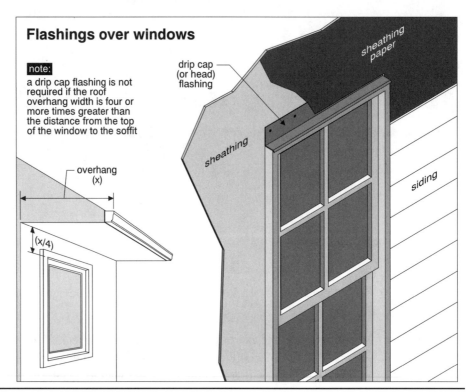

Flashings over windows

note:
a drip cap flashing is not required if the roof overhang width is four or more times greater than the distance from the top of the window to the soffit

drip cap (or head) flashing

sheathing paper

sheathing

siding

overhang (x)

(x/4)

8.1.5 EXTERIOR TRIM PROBLEMS

Exterior trim includes brick molds, casings, sills, muntins and mullions. Problems include —

• missing
• rot
• rust
• damaged, cracked or loose
• sills with reversed slope
• sill projection inadequate
• drip edge missing
• putty (glazing compound) cracked, missing, loose or deteriorated
• caulking or flashing missing, deteriorated, loose, rusted or incomplete
• paint or stain needed

Causes   These are either installation or maintenance issues.

Implications   These problems may impact window performance. Leakage of water and air and operation problems are likely results.

Strategy You may have identified several of these conditions from your exterior inspection. If you look out through windows, you may pick up these from the interior.

8.1.6 SASHES

- Rot
- Rust
- Inoperable
- Stiff
- Sashes won't stay open
- Sash coming apart
- Loose fit
- Weather-stripping missing or ineffective

Implications Window leakage (air and water) and difficulties in operation are the implications.

Strategy Just as we looked at the frames, we should look at sashes for rot and rust. We recommend testing all operable windows if possible. You may find windows are painted closed or do not move easily.

Casement Problems Crank-operated casement windows sometimes do not close tightly. To get the window to close, you may have to grab the top part of the sash and pull it in as you operate the crank. In some cases, you may have to go outside and push the window closed. Obviously, you'll be recommending adjustment or trimming if you encounter these problems. You should make sure you get the window closed tightly, if that's how you found it.

Single- And Double-Hung Windows In **The Home Reference Book** we discussed several ways that double-hung windows were held open. This includes sash cords and counterweights, springs and coiled tape. You should be careful operating single- and double-hung windows. The mechanism that holds the window up when opened may be defective. If the window falls, it can cause serious injury if your hands are below. The window and its glass can also be damaged.

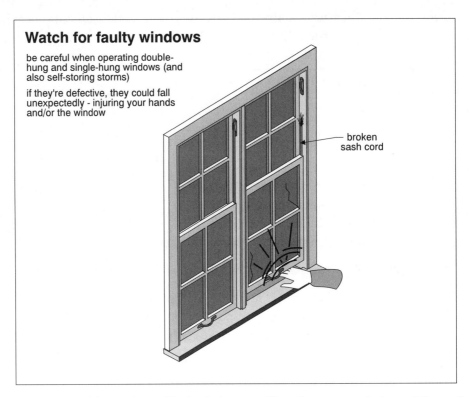

Watch for faulty windows

be careful when operating double-hung and single-hung windows (and also self-storing storms)

if they're defective, they could fall unexpectedly - injuring your hands and/or the window

broken sash cord

Self-storing Storm Windows   The same problem exists with single-hung, self-storing storm windows. Many of these systems use spring-loaded pins that sit in slots or holes to keep the window up. If the pins are not properly seated, the window can fall.

Don't Push On Upper Sash   There is a strong temptation to push on the upper horizontal part (**meeting rail**) of the sash to open a single- or double-hung window. Resist this temptation. Most windows have hardware on the lower rail to open the window. Pushing on the upper rail will eventually lead to separation of the upper rail from the stiles on either side of the sash. The window literally falls apart. Watch for this problem as you operate single- and double-hung windows and don't contribute to the problem. Use the hardware to pull the window open.

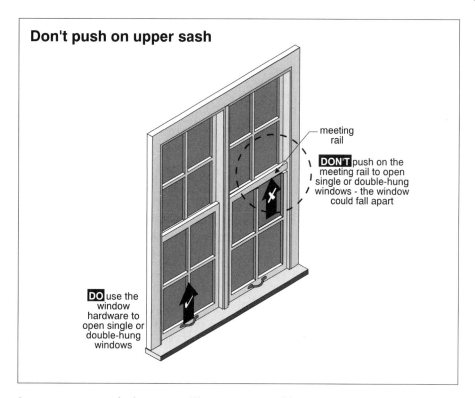

Don't push on upper sash

meeting rail

DON'T push on the meeting rail to open single or double-hung windows - the window could fall apart

DO use the window hardware to open single or double-hung windows

Check Weather Stripping And Fit   As you operate windows, you'll get a sense of how snugly they fit. When the window is closed, look around the perimeter for obvious gaps or discontinuities in the weather stripping.

8.1.7 INTERIOR TRIM

- Rot
- Stained
- Missing
- Cracked
- Loose
- Poor fit

Causes   These are maintenance issues for the most part. Interior trim may be missing because it was never provided.

Implications   These are primarily cosmetic problems, although rot or staining may indicate concealed damage to the structure behind.

Strategy   Look at casings and the interior stool and apron (fancy names for interior sill and trim) for problems. In some cases, the interior stool is used for storing plants, which are watered regularly. This can cause water damage.

8.1.8 GLASS

- Cracked
- Broken
- Loose
- Missing
- Lost seal
- Excess condensation

Causes These window problems are usually maintenance related. Lost seals may be the result of a manufacturing defect. Excess condensation on windows is usually a lifestyle and air quality issue.

Implications Cracked, broken, loose or missing glass can be both a heat loss and heat gain problem, and can be a risk of injury. If the glass is loose, it often rattles whenever someone walks through a room. People may be cut on broken glass.

Lost Seals Lost seals are not particularly serious from an energy efficiency standpoint. The window will still perform reasonably well. However, visibility is often reduced, and the glass may look cloudy, even if there's no condensation present at the moment. Once the seal is gone, condensation will appear and disappear between the panes. This, however, leaves the interior surfaces of the glazing dirty, and the cloudy appearance develops.

Excess Condensation Excess condensation will usually only occur during cold weather. It is the result of high humidity levels in the house. We talked extensively in the Insulation section about how we can control moisture levels inside the house. Eliminating moisture sources and using exhaust fans are obvious steps to control indoor moisture levels.

8.1.9 HARDWARE

- Rusted
- Broken
- Missing
- Loose
- Inoperable

Causes These are maintenance issues for the most part. Missing hardware may be the result of an incomplete original installation.

Implications Windows may not operate at all if hardware is missing or inoperable. Operation may be difficult. As mentioned with double-hung windows, if people push on the meeting rail, rather than using the appropriate hardware, the sash will eventually come apart.

Loss Of Security Window hardware typically includes locking mechanisms. If these are not effective, there is a loss of security for the home.

Strategy As you operate windows, you will recognize hardware problems. If the hardware is working but is obviously rusted or loose, note that in your report.

8.1.10 LOCATION

• Sills too low

Windows at stair landings, for example, should have their sills at least 36 inches above floor level.

Cause   Windows installed with low sills are an installation issue.

Implications If the sills are too low, someone stumbling on the stairs may fall out through the window.

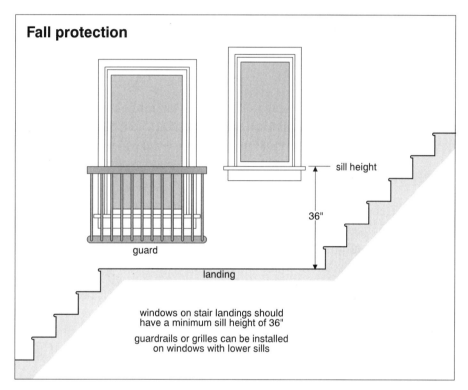

Fall protection

sill height

36"

guard

landing

windows on stair landings should have a minimum sill height of 36"

guardrails or grilles can be installed on windows with lower sills

Strategy   Check that the windows are at least 36 inches above floor level. When they are lower and there is a risk of falling, you can recommend a guard or rail.

8.1.11 SCREENS

• Torn
• Holes
• Rust
• Loose
• Missing

Screens are typically nylon, aluminum or copper.

Causes   Screen problems may be original installation issues, but are most often maintenance and mechanical damage problems. Rusting is a result of the screen's exposure to the environment.

Implications   Screens are designed to keep insects and pests out of the home. If they can't do their job, insects will find their way into the house.

Strategy   As you're looking at windows, it takes no more time to note whether the screens are intact.

8.1.12 STORM WINDOWS MISSING

In many climates, storm windows are usually provided. These may have been replaced with double-glazed windows. If single-glazed windows are noted in a heating climate, you may suspect there are storm windows available.

Causes   Storm windows may never have been provided for the home, or may have been lost or damaged. It's more common, however, that storm windows are stored somewhere, in or near the house.

Implications   Heat loss or heat gain are obvious implications of missing storm windows. Excessive condensation levels on the interior surfaces of primary windows is another implication. Adding storm windows will keep the interior glass warmer and reduce condensation. Storm windows also reduce drafts and may improve comfort as a result.

Strategy   Look for evidence that storm windows have been installed on the outside in the past. There are often clips used to hold them in place. Check basements, garages and other outbuildings as you move through these areas. Very often you will come across storm windows.

Most home inspectors don't go so far as to ensure there is a storm for each window, although you can usually get a sense of this. Tell your client that there are no storm windows in place but there are some stored in the house. Some clients will ask sellers to assure them that there are storm windows for all the windows.

8.1.13 TOO SMALL FOR EGRESS

In many areas, windows have to be available for use as emergency exits. In some jurisdictions, every bedroom has to have a window that can be used as an emergency exit. In other jurisdictions, only one exit window is needed on each floor with bedrooms.

There are several size rules, depending on jurisdiction. Some specify minimum window areas (e.g., 5.7 square feet) with minimum dimensions. We've listed below some of the requirements from various authorities. These are minimum window dimensions:

- 22 inches by 40 inches
- 20 inches by 41 inches
- 24 inches by 34 inches
- 18 inches by 32 inches
- 15 inches by 36 inches

The idea is that someone should be able to crawl out the window if there is a fire that prevents leaving the bedroom through the door.

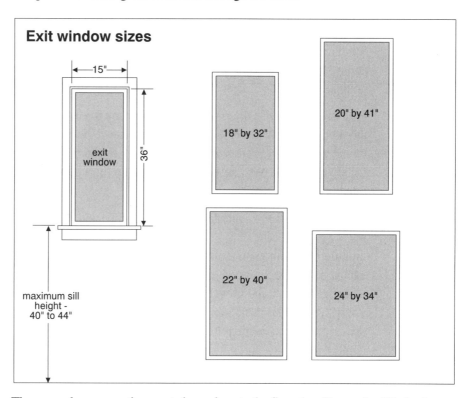

Exit window sizes

├15"┤

exit window

36"

maximum sill height - 40" to 44"

18" by 32"

20" by 41"

22" by 40"

24" by 34"

Sill Height   There are also conventions as to how close to the floor the sill must be. We don't want the window so high on the wall that people could not climb through it. Maximum sill heights for windows used as emergency exits range from 40 to 44 inches.

Keep Bedroom Doors Closed   While we're talking about emergency exits, it's interesting to note that some jurisdictions require every bedroom to have a door. While this seems like a trivial requirement, there's a very good reason for it. If bedroom doors are closed, this greatly reduces the rate at which fires can move through a house. The bedroom door will block smoke and fire spread and reduce the oxygen supply to a fire. The closed bedroom door can give the occupant an extra few precious minutes to awaken and get out of the house.

Code Compliance Inspection?   Home inspectors do not do code compliance inspections. You don't have to refer to the code to let people know the common sense wisdom of providing an emergency escape route. Some inspectors recommend rope ladders and point out bedrooms that do not have a secondary escape route from the house.

Basement Bedrooms   Bedrooms in basements can be dangerous. If a fire starts near the stairwell, there may be no way out of a basement. Many areas require basement bedrooms to have windows that meet the size and height requirements we've talked about. There are also requirements for the size of the window well outside the window, since basement windows are often below grade level. Again, common sense is a much more valuable tool than trying to memorize numbers.

8.1.14 ICE DAMS AT SKYLIGHTS

We talked about skylights in the Roofing Module. We also talked about insulating skylight wells in the Insulation section of this Module.

Causes   Ice dams can occur around skylights because of the heat loss through the glass or acrylic area. Even if the light well and skylight curb are well insulated, there's going to be considerable heat loss through the glazing, whether it's glass or acrylic.

Implications   Heat loss around the skylight will melt snow on the roof, immediately adjacent to the skylight. This melted snow starts to run down the roof and re-freezes. It's easy to understand how, as this water freezes, a dam can be built up around and below the skylight. This can cause water to back up through the roof shingles and leak into the building.

Ice And Water   Some skylights are installed with an apron of Ice and Water Shield, the self-adhering
Shield   modified bituminous membrane often used as eave protection. This apron may extend out six feet around and below the skylight, under the roof shingles. This watertight membrane prevents leakage into the building even if water backs up under the shingles. This membrane is self-healing, so that when roofing nails are driven through it, the membrane won't leak around the nails.

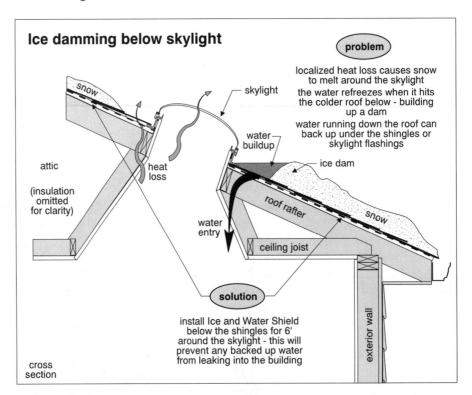

Ice damming below skylight

problem

localized heat loss causes snow to melt around the skylight

the water refreezes when it hits the colder roof below - building up a dam

water running down the roof can back up under the shingles or skylight flashings

snow · skylight · water buildup · ice dam · attic (insulation omitted for clarity) · heat loss · water entry · roof rafter · snow · ceiling joist · exterior wall

solution

install Ice and Water Shield below the shingles for 6' around the skylight - this will prevent any backed up water from leaking into the building

cross section

Strategy   Skylights are vulnerable to leakage. Many inspectors immediately describe leakage as the problem when they see water damage around the skylight. In cold climates, you should allow for the possibility that condensation may be a contributor. The issue that many don't think of, however, is the ice dam. Damage is usually localized below the skylight and slightly to the sides. Damage only occurs when the weather conditions are right.

► 9.0 DOORS

Function

Exterior   Exterior doors are a means of entering and leaving a house. They should provide security and privacy. They should be weather-tight and, if they communicate with garages, should be gas-proof or fireproof, depending on the jurisdiction.

Interior   Interior doors allow passage between rooms. They also afford privacy and some sound protection. As we talked about earlier, they also provide some fire and smoke protection for people in bedrooms.

Materials

Doors   Doors may be made of wood, metal, vinyl or hardboard. Hardboard doors are usually restricted to indoor use.

Cores   Many doors are hollow-cored. The cores on interior doors may have cardboard reinforcement. Exterior metal doors often have polyurethane insulation as cores. These can provide very good insulation levels.

The cores may provide a thermal break for the door. Metal and vinyl doors can transmit a lot of heat from the home to the outdoors if the frame or door is continuous. Thermal breaks reduce heat flow between the inner and outer halves of the door.

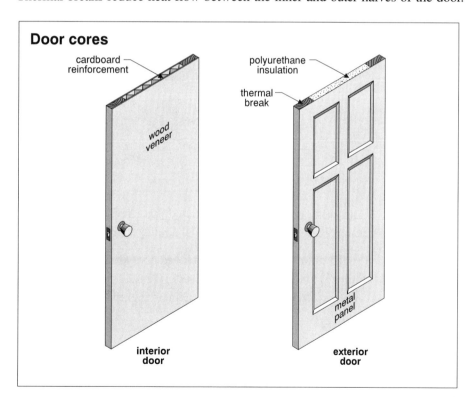

Door cores

cardboard reinforcement
wood veneer
interior door

polyurethane insulation
thermal break
metal panel
exterior door

Sills   Doorsills are typically wood or metal. Essential sill issues include slope, overhang, support and weather stripping.

Door
Operation   Doors may be **hinged** conventionally. They may be **sliding doors** moving in tracks, or suspended from overhead concealed tracks as is typical of **pocket doors**. Doors may also be **bi-fold**, moving in a track.

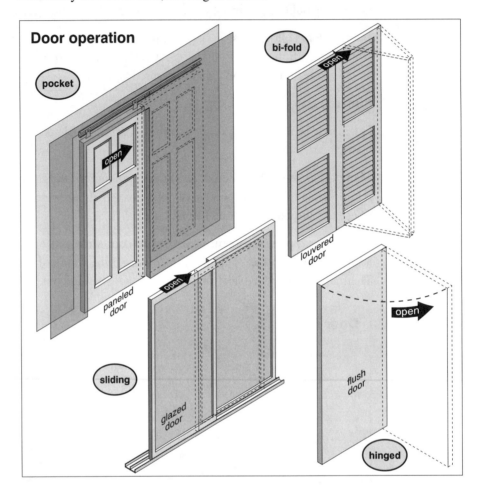

Surfaces   Door surfaces may be flush, paneled, louvered or glazed. Sliding patio doors, for example, are effectively horizontal sliding windows with very little solid material.

French Doors   **French doors** are conventional hinged doors with a difference. These are typically double doors, hinged at either side. Typical French doors are mostly glazing, usually with several muntins breaking the glass up into small panes.

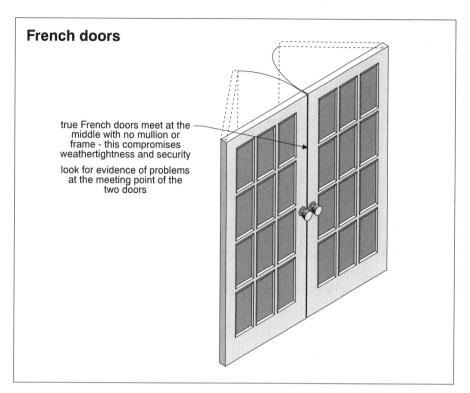

French doors

true French doors meet at the middle with no mullion or frame - this compromises weathertightness and security

look for evidence of problems at the meeting point of the two doors

No Mullions
True French doors meet at the center with no middle mullion or frame. Without a mullion or frame to close against, French doors can be leaky, drafty and less secure than conventional doors. There are several approaches to French door design to maximize the strength, weather-tightness and security. Your inspection should focus on the meeting point between the two doors, looking for evidence of problems. If they are exterior doors, pay particular attention to the possibility of water damage below the doors in the area of the meeting point.

Energy Efficiency

Doors are similar to windows with respect to energy efficiency. Their R-value is usually lower than that of the wall system. Because they open and close, they can also be sources of considerable air leakage.

R-value
The R-value of a solid wood door is roughly 2. The R-value of a double-glazed patio door is also about 2. A wood door with a storm door on the outside may have an R-value of approximately 5. Insulated metal doors can have R-values up to 14. These doors approach the insulation level of good wall systems.

General Strategy

Doors provide useful clues to the condition of the house structure. Because a door is a rectangular component in a rectangular opening, it's a very good reference spot to look for movement in the building. This is true on both interior and exterior doors. As part of your door inspection, you should be looking for doors that are noticeably out of square. It's common to find doors that have been trimmed as their openings shift, so the doors will operate freely. Watch for this trimming, which can indicate the direction and amount of movement in the structure.

Not The
Whole Story
While the movement of a doorframe can be dramatic, it may be relatively trivial from a structure standpoint. An interior partition wall that is not load-bearing may deflect because of heavy loads along the wall, failure to double the parallel joist below, etc. Movement can be dramatic, but may not be a huge structural problem. Don't jump to conclusions, but use this information in the big picture approach.

Operate
Every Door
The Standards indicate that we have to operate one door per room, and operate every exterior door. If possible, we recommend you try every door. Doors that do not open and close are frustrating to homeowners and, while usually not serious, can be the source of nuisance callbacks. Even if people don't complain, they are often irritated they weren't made aware of the problem.

9.1 CONDITIONS

1. Leaks
2. Lintels sagging or missing
3. Door and frame
   • Rot
   • Rust
   • Racked
   • Deformed
   • Damaged
   • Delaminated
   • Loose or poor fit
   • Installed backwards
   • Drain holes blocked or missing
   • Stiff or inoperable
   • Swings open or closed by itself
   • Dark paint on metal exposed to sun
   • Plastic trim on metal door behind storm door
4. Exterior drip cap missing or ineffective
5. Exterior trim
   • Missing
   • Rot
   • Rust
   • Damage
   • Cracked or loose
   • Sills with reverse slope
   • Sill projections inadequate
   • Sill not well supported
   • Sill too low
   • Putty (glazing compound) missing, cracked, loose, deteriorated
   • Caulking or flashing missing, deteriorated, loose, rusted, incomplete
   • Paint or stain needed

6. Interior trim
   - Rot
   - Stained
   - Missing
   - Cracked
   - Loose
   - Poorly fit
   - Floor stained below
   - Doorstops missing or ineffective
   - Guides or stops damaged

7. Glass
   - Cracked
   - Broken
   - Loose
   - Missing
   - Lost seal
   - Excess condensation

8. Hardware
   - Rusted
   - Broken
   - Missing
   - Inoperable
   - Loose
   - Doesn't latch
   - Hinges on exterior of building
   - Self closer ineffective

9. Storm doors and screens
   - Torn or holes
   - Rusted
   - Loose
   - Missing

10. Garage man-door
   - Not fire-rated or exterior type
   - Not weather-stripped
   - Not self-closing
   - Opens into bedroom
   - No six-inch step down to garage

As we discuss these conditions, many are similar to windows. We won't repeat discussions we've already had.

9.1.1 LEAKS

Strategy   Doors can leak both air and water, just like windows. Because doors are often larger and closer to grade level, leakage can be more of a problem. Watch for leakage both at the top and bottom of the doorframe. Check for water damage in the basement, crawlspace or floor below exterior doors.

9.1.2 LINTELS SAGGING OR MISSING

Strategy   We talked about this with windows. This is more commonly a problem with sliding glass doors or French doors, because of the wide opening. Again, it's often easier to see from outside the home.

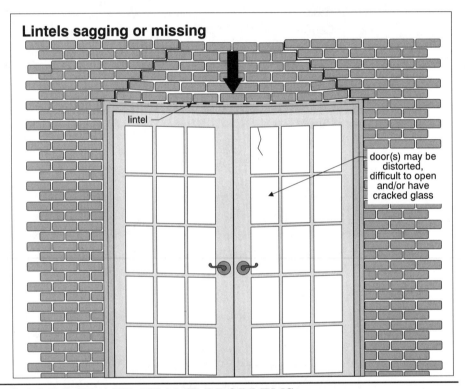

Lintels sagging or missing

lintel

door(s) may be distorted, difficult to open and/or have cracked glass

9.1.3 DOOR AND FRAME PROBLEMS

Strategy   Most of these are similar to the window issues we discussed. Sliding glass doors have tracks at the bottom and should have drain holes. These can be installed backwards. Again, the clues are drain holes on the inside and a reverse slope to the sill.

Stiff Or Inoperable   Doors may be stiff for several reasons. Rollers and wheels on sliders are often damaged. Tracks can be dirty or damaged.

Pocket Doors   Be sure to operate pocket doors. There are frequently track problems and repairs can be difficult and expensive because of access problems. Many pocket doors have recessed hardware and it's very easy for inspectors to walk through an opening and not realize that there is a door into the room.

*Doors Swing
By Themselves*

Doors should stay where they are put. If a door is opened, it should not swing close. If it's half opened, it should not swing in either direction. Doors that are not hung straight will often move on their own. This is a minor issue but is indicative of the quality of workmanship.

*Dark Paint
On Metal
Doors*

Children have been burned touching metal doors that are painted dark colors and exposed to the sun. These doors are often well insulated from behind, and the metal cannot dissipate heat. Dark colors tend to absorb heat, rather than reflect it, and the surface temperature of the metal can be more than 200°F.

*Plastic Trim
On Metal
Doors*

Many manufacturers of insulated metal doors recommend against adding a storm door. The heat buildup between the storm and primary doors can be significant. Raised plastic trim on metal doors will sometimes warp and pull away as a result of the heat buildup. Watch for evidence of this. In rare cases, the door itself can warp to the heat.

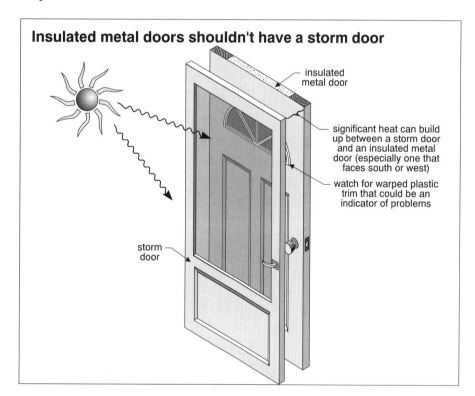

Insulated metal doors shouldn't have a storm door

insulated metal door

significant heat can build up between a storm door and an insulated metal door (especially one that faces south or west)

watch for warped plastic trim that could be an indicator of problems

storm door

9.1.4 EXTERIOR DRIP CAP MISSING OR INEFFECTIVE

This is the same issue as we discussed on windows.

9.1.5 EXTERIOR TRIM

- Missing
- Rot
- Rust
- Damage
- Cracked or loose
- Sills with reverse slope
- Sill projection inadequate
- Sill not well supported
- Sill too low
- Putty (glazing compound) missing, cracked, loose, deteriorated
- Caulking or flashing missing, deteriorated, loose, rusted, incomplete
- Paint or stain needed

Again, we've discussed most of these issues on windows. They are the same for doors. There are some new issues here we should look at.

Sill Not Well Supported
Doorsills are walking surfaces and in this respect are different from windowsills. Sills are typically cantilevered to a certain extent, and may not be well supported by the structure below. As you walk through doors, make sure you stand on the sill. Watch for any movement of the sill under your weight.

Sill Too Low
You should always step up going into a house. The minimum height of the step up ranges from 1½ inches to 6 inches, depending on where you are. Where snow is likely to accumulate, higher steps are better. Find out what's recommended in your area, and check the sill height. This is usually an easy problem to catch, because it feels strange to most people to walk into a house without having to step up.

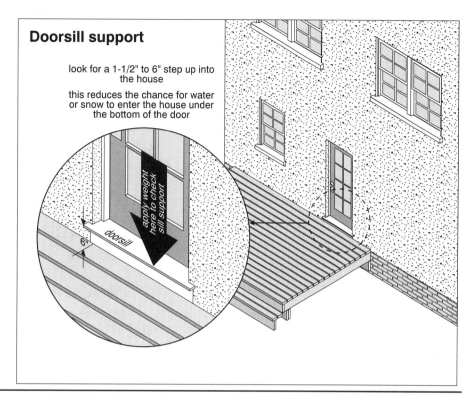

Doorsill support

look for a 1-1/2" to 6" step up into the house

this reduces the chance for water or snow to enter the house under the bottom of the door

apply weight here to check sill support

doorsill

6"

9.1.6 INTERIOR TRIM

- Rot
- Stained
- Missing
- Cracked
- Loose
- Poorly fit
- Floor stained below
- Doorstops missing or ineffective
- Guides and stops damaged

Strategy   Again, most of these we've discussed with respect to windows.

Floors   Leakage around the base of doors is common. Watch for evidence of staining on
Stained Below   the floor below the door. As we discussed earlier, it's ideal if you can get to an
area below the floor and check below the door.

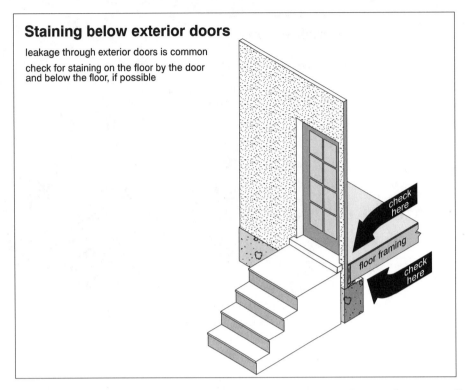

Staining below exterior doors

leakage through exterior doors is common

check for staining on the floor by the door and below the floor, if possible

Doorstops Missing Or Ineffective	Doorstops prevent doors from swinging into walls. Doorknobs can damage wall finishes. There are several types of doorstops, and virtually any door can be provided with a stop.
Opening Distance	Doors typically open either 90 degrees, or 180 degrees. Many designers prefer to have doors open 90 degrees against a wall. Doors that open 180 degrees consume more floor area to allow for the door's travel.
Doors Cover Light Switches	When doors are open, they should not block light switches or other systems that require access. This is not something that most home inspectors would mention to clients, but it does suggest poor attention to detail.
Guides And Stops Damaged	Hinged doors typically close against stops at the sides and top. If these stops are loose, damaged or missing, the door may not close properly or securely. Sliding or rolling doors have end stops. Watch for damage due to excessive force used opening or closing the doors.

Sliding or rolling doors will not operate freely if their tracks or wheels are damaged, dirty, misaligned or obstructed.

9.1.7 GLASS

We've discussed all of these issues with respect to windows. Glass in doors is most often non-operable.

9.1.8 HARDWARE

- Rusted
- Broken
- Missing
- Inoperable
- Doesn't latch
- Hinges on exterior of building
- Self-closer ineffective

Strategy

We talked about most of these when we talked about windows.

Latch Mechanism

Doors should latch positively. Many rely on a tongue and strike plate system. Many doors operate but don't latch because of hardware problems. These can include broken springs, misalignment, missing pieces, etc. You don't need to troubleshoot the problem but you do need to identify it. Exterior door latching problems are more serious than problems on interior doors.

Watch For Hinges On Exterior Of Building

Exterior doors should have hinges on the inside for security reasons. It's very easy to remove a door by popping hinges. Exterior hinges are an indication of low quality workmanship and are a security issue. A majority of hinges are not designed for exposure to the exterior. Look for deterioration.

Self-closer Ineffective

Doors from houses into garages typically require self-closers. Many of these need adjustment. Open the door fully and let it go. Does it close properly?

Many storm doors have self-closers. Again it's common to find problems with these. Open the door and watch to see if it closes and latches properly.

9.1.9 STORMS AND SCREENS

- Torn
- Holes
- Rust
- Loose
- Missing

We talked about these in the Windows section. The same comments apply.

9.1.10 GARAGE MAN-DOOR

- Not fire-rated or exterior type
- Not weather-stripped
- Not self closing
- Opens into bedroom
- No 6-inch step down into garage

Not Fire-rated Or Exterior Type   In some areas, doors between houses and garages must have a fire rating. Twenty minutes is common. In other areas, exterior-type doors are adequate. A one-and-three-eighths-inch-thick solid core door is usually acceptable.

Not Weather-stripped   Garage doors separate the indoors from the outdoors, and should therefore be weather-stripped. Another good reason for weather stripping is to keep automotive exhaust fumes out of the home.

Not Self-closing   Many jurisdictions require that garage doors be self-closing. They don't want people leaving the door open into the house. If an automobile in the garage is running, carbon monoxide fumes can enter the house. This is a life safety issue.

Test The Door   Operate the self-closing system to make sure it works, by opening the door and letting it go.

Opens Into Bedroom   Because garage doors may not be perfectly tight, it is considered risky to have a garage door that opens into a sleeping area. This is not allowed in many areas. Watch for it and describe the risk to you clients if you see it.

No 6-inch Step Down Into Garage   It's a good idea to have to step down into the garage as you leave the house. This is because the garage floor may have water on it that we don't want to run into the house. Low-lying gasoline fumes will have a more difficult time getting into the house if there is a step up from the garage.

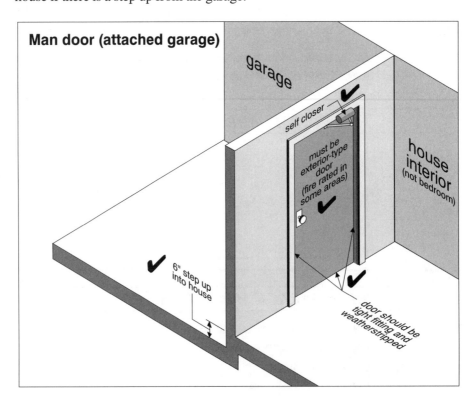

Man door (attached garage)

garage

self closer

must be exterior-type door (fire rated in some areas)

house interior (not bedroom)

6" step up into house

door should be tight fitting and weatherstripped

Insulation & Interiors
M O D U L E

QUICK QUIZ 5

☑ INSTRUCTIONS

- You should finish Study Session 5 before doing this Quiz.
- Write your answers in the spaces provided.
- Check your answers against ours at the end of this Section.
- If you have trouble with the Quiz, re-read the Study Session and try the Quiz again.
- If you did well, it's time for Study Session 6.

1. List four window functions.

   _____

   _____

   _____

   _____

2. List three common materials used in window frames and sashes.

   _____

   _____

   _____

3. List eight common window types.

_____

_____

_____

_____

_____

_____

_____

_____

4. Low-E glass traps heat in the home in winter and keeps heat out in the summer.

True ☐   False ☐

5. Gas filled glass is more energy efficient than conventional double glazing.

True ☐   False ☐

6. List two general window problems.

_____

_____

7. List six frame problems.

_____

_____

_____

_____

_____

_____

8. List two exterior drip cap problems and their implications.

_____

_____

9. List ten exterior trim problems.

_____

_____

_____

_____

_____

_____

_____

_____

_____

_____

10. List eight sash problems.

_____

_____

_____

_____

_____

_____

_____

_____

11. List six interior trim problems.

_____

_____

_____

_____

_____

_____

12. List six glass problems.

_____

_____

_____

_____

_____

_____

13. List five window hardware problems.

_____

_____

_____

_____

_____

14. What is the danger of the window being too low on a stairwell or a landing?

_____

15. List at least four common problems with storms and screens.

_____

_____

_____

_____

16. Under what circumstances might a window be considered too small?

_____

17. Skylights can cause ice damming on roofs.
True ☐   False ☐

18. List five functions of exterior doors.

_____

_____

_____

_____

_____

19. List four common door materials.

_____

_____

_____

_____

20. List five common problems with garage doors.

_____

_____

_____

_____

_____

If you had no trouble with this Quiz, you are ready for Study Session 6.

Key Words:

- *Light*
- *Ventilation*
- *Means of egress*
- *Glazing*
- *Low-E glass*
- *Gas filled windows*
- *Single and double hung*
- *Casement*
- *Fixed*
- *Awning*
- *Hopper*
- *Jalousie*
- *Slider*
- *Muntins*
- *Mullion*
- *Frame*
- *Sash*
- *Sill*
- *Trim*
- *Drip cap*
- *Lintel*
- *Hinged*
- *Bifold*
- *Pocket*
- *Self-closer*

Insulation & Interiors

MODULE

STUDY SESSION 6

1. You should have completed Study Session 5 and Quick Quiz 5 before starting this Session.

2. This Session covers wet basements and crawlspaces.

3. At the end of this Session, you should be able to:
   • List two sources of water in basements and crawlspaces
   • List three contributing factors to wet basements and crawlspaces
   • List six implications of basement and crawlspace problems
   • List twenty signs of moisture in basements and crawlspaces
   • List six corrective actions for wet basements

4. Before you start this Session, read Section 10.0 of the Interior Chapter of **The Home Reference Book**. This one is important. It may be worthwhile to read it twice.

5. This Session may take you roughly one and a half hours.

6. Quick Quiz 6 is at the end of this Study Session.

Key Words:

- *Surface water*
- *Roof water*
- *Ground water*
- *Soil type*
- *Water table*
- *Foundation depth*
- *Cold pours*
- *Honeycombing*
- *Efflorescence*
- *Rot*
- *Mold*
- *Mildew*
- *Rust*
- *Perimeter drainage tile*
- *Drainage membrane*
- *Sumps*
- *Pumps*

► 10.0 WET BASEMENTS AND CRAWLSPACES

Introduction

Dampness or water in basements and crawlspaces is the source of more complaints about home inspectors' performance than any other issue. As we discuss it, you'll get some sense of why this happens. If you haven't read Section 10.0 of the Interior Chapter of **The Home Reference Book**, please do it now. This provides an excellent overview of the issue.

Have you read it? Okay – let's move on. Since this whole section is devoted to a single house problem, we'll get right into a discussion of the causes.

Causes   Water in basements or crawlspaces is typically from one of two sources.

- Surface water from rain or snow that lands on the ground around the house or on the roof.
- Groundwater – water in the earth below grade level.

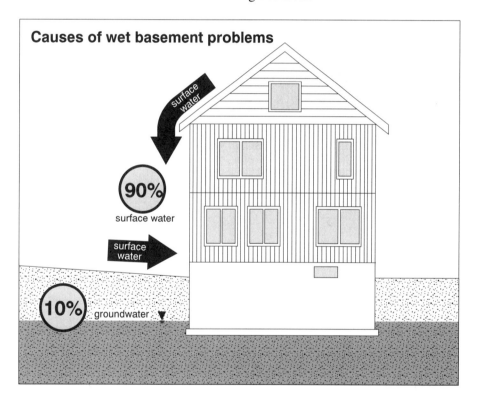

Causes of wet basement problems

90% surface water

surface water

10% groundwater

More than ninety percent of the wet basement and crawlspace problems are a result of surface water. So far, this sounds pretty simple. Let's look at what can go wrong.

Roof Water Control

The roof of the house presents a large, impervious surface. Water that lands on the roof does not soak in – it runs off and accumulates around the outside of the house unless we control it. Gutters and downspouts are designed to collect this roof water and carry it to a safe discharge point. In the Exterior Module we talked about how gutters and downspouts could carry water to a below-grade drainage system, or discharge water above grade several feet away from the house. Either system is acceptable if it is working. As we saw in the Exterior Module, lots of things go wrong with gutters and downspouts.

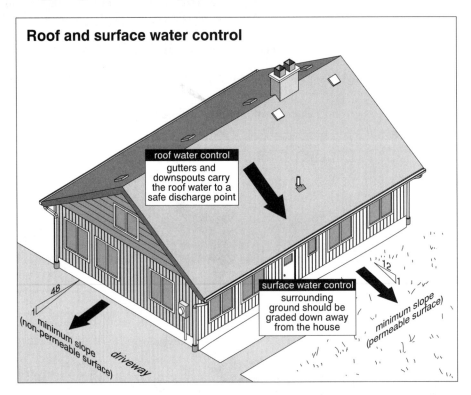

Roof and surface water control

roof water control
gutters and downspouts carry the roof water to a safe discharge point

surface water control
surrounding ground should be graded down away from the house

48
1
minimum slope (non-permeable surface)
driveway

12
1
minimum slope (permeable surface)

Surface Water Control

Home inspectors generally refer to this as **grading**. The strategy is simple. If the ground slopes down away from the building, surface water will run **away from** rather than **toward the home**. If the grade slopes toward the house, water will accumulate against the foundation walls.

Basement Windows And Stairs

Basement window wells and exterior basement stairwells are natural collection points for surface water. Grading should be down away from these, so surface water is not funneled into basement window or stairwells. The wells themselves will receive a certain amount of rain or snow directly (depending on your climate, of course). These wells should be provided with drains to allow the water to escape to a safe location. In some cases, window wells or stairwells are provided with coverings that keep the rain out.

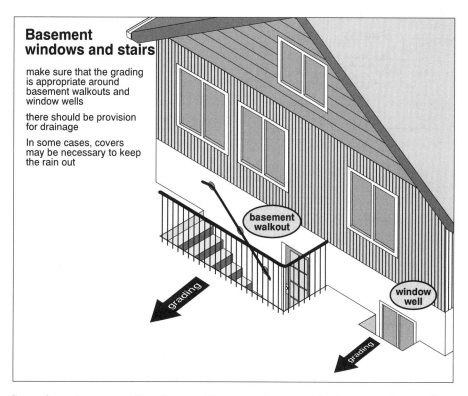

Basement windows and stairs

make sure that the grading is appropriate around basement walkouts and window wells

there should be provision for drainage

In some cases, covers may be necessary to keep the rain out

basement walkout

grading

window well

grading

Basement Window Wells   Sometimes basement flooding problems can be traced back to window wells or stairwells. We often find high water marks on the basement window glass. This usually means that the drain for the window well is obstructed. If the drain is covered (often by a collection of wet papers or leaves in the bottom of the well), water may quickly rise in the well during a heavy rain and pour in through the basement window. It's nice when the solution to a wet basement problem is as simple as removing debris from basement window wells.

Swales   Sometimes it's difficult to create a natural slope away from the house. A house built on the side of a hill, for example, will have water collect against the high side. Gentle troughs in the ground surface can be created to divert the water. These swales are shallow ditches that collect the water several feet away from the building and divert it around one or both sides of the home. Swales can be left as gentle recesses, or filled with gravel to form a level surface.

Catch Basins   In some situations, such as a house at the bottom of a hill, there is no place that water can be diverted. Catch basins may be necessary to collect water at low spots several feet away from the home. These may be drained to a storm sewer, French drain, or other suitable area.

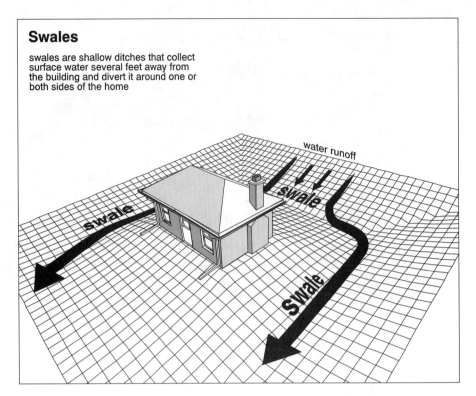

Swales

swales are shallow ditches that collect
surface water several feet away from
the building and divert it around one or
both sides of the home

Foundation Flaws   Virtually all foundation systems have flaws that will allow water to penetrate. Houses are not built like boats. The more flaws there are, the more severe wet basement problems can be.

Surface Water Control Is Key   We should emphasize, though, that if no water collects against the outside of the foundation wall, the basement will be dry no matter how leaky the foundation wall is. That's why the best line of defense is the **surface water control** we talked about with gutters and downspouts and lot grading.

Types Of Flaws   Flaws in poured concrete foundation walls typically fall into one of five categories:

- **Cracks** as a result of concrete shrinkage or settlement.
- **Form tie holes** (sometimes called **tie rod holes**). These are holes left when the forms are removed from both sides of the concrete. These holes should be plugged and sealed, both inside and out. Sometimes they are not sealed; sometimes the seals are not effective.
- **Cold pour**. A cold pour in the foundation is created when the concrete is poured at two different times. The foundation will sometimes be partly poured and then, as a result of construction delays, the balance of the concrete is poured a day or so later. This creates a weak spot at the junction of the two pours. Leakage through cold pours is common.
- **Honeycombing.** Honeycombing is a series of gaps or voids in the concrete. It's usually a result of improper or inadequate tamping and compacting of the concrete when it is poured. The surface of the concrete has a series of voids from air bubbles that look like a honeycomb. The result is a weak spot, vulnerable to leakage.

• **Holes**. Intentional holes through foundations for electric conduits, oil fill and vent pipes, and gas and water pipes, are all potential leakage spots.

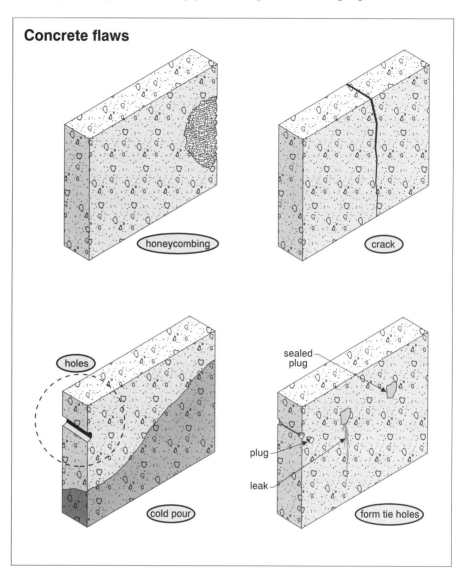

Concrete flaws

honeycombing

crack

holes

cold pour

sealed plug

plug

leak

form tie holes

We'll look at some of the corrective actions shortly.

Where Cracks Appear

Cracks in poured concrete foundations most often appear around windows. Some say that up to 75 percent of cracks are at the corners of windows. This makes sense because a window opening is a weak spot in the foundation wall. Stresses concentrate around openings like this, and cracks frequently develop here. Cracks running down vertically or diagonally from basement windows are very common.

Beam Pockets

Perhaps ten percent of cracks occur around **beam pockets**. Beam pockets are also weak spots in foundation walls because some of the concrete is missing, so the beam can be accommodated.

Joints In Split
Level Homes

Roughly ten percent of all foundation cracks occur at the joints in side-split and back-split houses where the elevation of the foundation changes.

While these are common crack areas, they are not the only ones. Keep these in mind when looking at a wet basement that is finished. It can be difficult to know where the water is coming in. Knowing where cracks are likely to develop can be helpful.

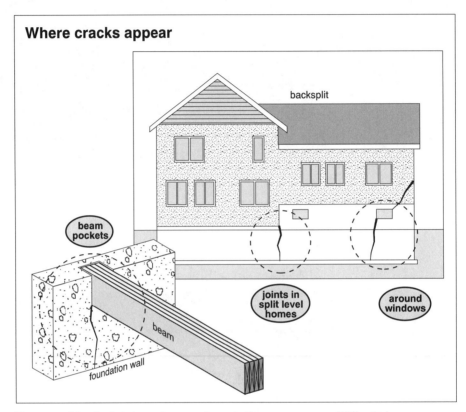

Where cracks appear

backsplit

beam pockets

beam

foundation wall

joints in split level homes

around windows

Masonry
Foundations

Concrete block, brick and stone foundations are more difficult in one sense.

Crack
Locations

Masonry foundation walls behave somewhat differently than poured concrete foundations. We'll focus on concrete block foundations because they are the most common type.

Cracks

Cracks can develop in block foundations. These are typically the result of settlement, lateral soil pressure or heaving, rather than shrinkage.

Mortar Joint
Failures

Mortar may shrink, crack or fall out of block walls creating leakage spaces. Concrete block is somewhat more porous than poured concrete. When the original dampproofing fails, water may find its way more easily through block than poured concrete.

Parging   Concrete block has a rough surface texture compared to poured concrete. Dampproof coatings cannot be successfully applied to concrete block directly because the coating can't completely fill the surface voids. Block walls are typically **parged** with mortar to provide a smooth surface for the dampproofing material. If this parging cracks or pulls away from the block, the dampproofing system is bypassed. Parging may or may not be visible. It typically only extends up the foundation wall to grade level.

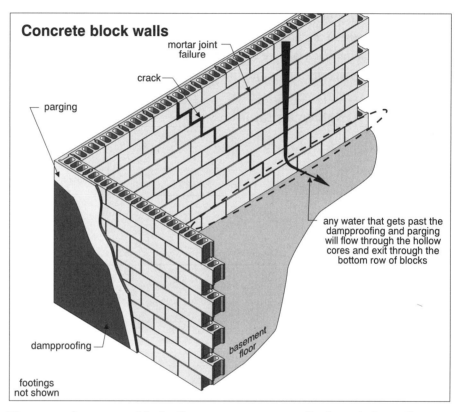

Concrete block walls

mortar joint failure

crack

parging

any water that gets past the dampproofing and parging will flow through the hollow cores and exit through the bottom row of blocks

dampproofing

basement floor

footings not shown

Concrete   The spaces in concrete block allow water to move easily through the wall system
Block Is   once it penetrates the surface. A crack in a concrete block foundation wall will not
Hollow   necessarily leak to the interior straight through the crack. The most typical leakage is along the bottom of the wall over all or most of the wall length. This is because the water gets into the voids and runs along the wall. It's easier for the water to move through the block spaces than to fight its way through the crack. While some water will leak through the crack, it's usually only at the bottom. Water often leaks in through the bottom row of blocks over the entire wall length.

High Water   Some wet basement and crawlspace problems are the result of water in the ground
Table   below grade. The **water table** (the upper limit of the ground water) typically fluctuates seasonally and as areas are developed.

Hydrostatic
Pressure

If the water level is at or above the basement or crawlspace floor level, the water will exert pressure on the floor and any parts of the wall that are below the water table. Water will find its way into the home through any flaws in the basement floor or lower parts of the wall. Because houses are not built like boats, there will always be some flaws allowing water in. These flaws occur around drains and pipe and column penetrations through the slab.

Heaved
Floors

If the basement is reasonably tight and the hydrostatic pressure is significant, the basement floor may heave upward. This allows water to flood the basement and relieve the pressure. This condition is far less common than surface water problems, but is a more difficult one to control. Water collection and pumping systems may work, although the pumps may have to be powerful and very reliable.

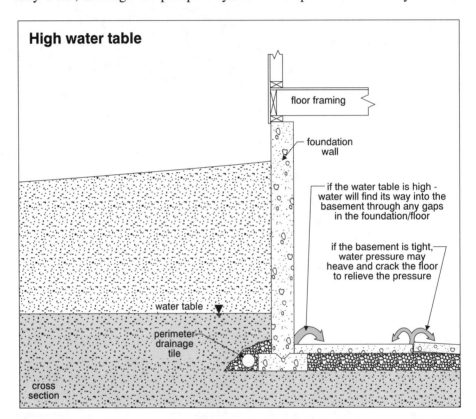

High water table

floor framing

foundation wall

if the water table is high - water will find its way into the basement through any gaps in the foundation/floor

if the basement is tight, water pressure may heave and crack the floor to relieve the pressure

water table

perimeter drainage tile

cross section

Very High
Water Tables

In some areas, the water tables are so high that homes are built either as slab-on-grade or are on piers to keep the foundations above the water. There are other reasons, of course, why houses may be slab-on-grade or built on piers. In warm climates, there's no need for deep foundations because frost will not heave buildings. Where surface soils have very little bearing capacity, piers or piles are often used to support homes on stable soils well below grade.

Contributing Factors

Basements in some houses will be wet while others will not, despite tremendous similarities. Some of the factors that affect basement and crawlspace moisture problems include—

• soil type
• depth of foundation
• height of water table

Soil Type

Sandy soils are free draining and allow water to fall through them quickly. Clay soils tend to hold water up. Clay soils are much more likely to create wet basements because surface water accumulates around the house.

Depth Of Foundation

Deeper foundation walls are exposed to more damp soil. There is also greater risk of encountering a water table, the deeper we go. If the area outside the foundation wall becomes saturated, hydrostatic pressure can force water through flaws in the foundation and, in severe cases, can push the foundation wall inward. This is common with freezing problems in cold climates. Saturated soil adjacent to the foundation expands when it freezes and pushes foundation walls inward.

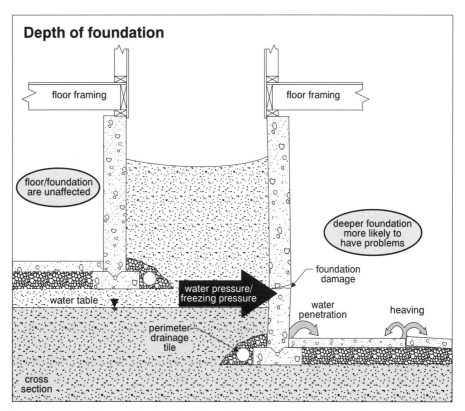

Depth of foundation

floor framing — floor framing

floor/foundation are unaffected

deeper foundation more likely to have problems

foundation damage

water pressure/ freezing pressure

water penetration

heaving

water table

perimeter drainage tile

cross section

Height Of Water Table

We touched earlier on the fact that foundations are vulnerable to leakage if they are at or below the water table. We also indicated that water tables fluctuate from time to time. The higher the water table in a given area, the greater the risk that it will rise above the basement or crawlspace floor level.

Implications   The implications of wet basements or crawlspaces include—

- **Nuisance**. Wet basements can damage storage and lead to uncomfortably high interior humidity levels.
- **Damaged carpeting, wall coverings and furnishings** in finished basement areas.
- **Odors and molds**. If basements are chronically wet, mold and mildew may generate unpleasant odors. Some molds are toxic and there can be a health risk associated with damp, musty basements.
- **Structural deterioration**. Although not common, it is possible for foundation materials to deteriorate as a result of chronic moisture penetrations. We have seen concrete block foundations deteriorate as a result of chronic moisture. Cinder block foundations (made from rock slag rather than concrete) are not usually used below grade because they are susceptible to water damage. We have seen considerable damage to cinder blocks used below grade.
- **Electrical shock or fire hazard**. Wires behind finished walls may corrode or generate short circuits if they become wet. This creates the risk of electrical shock and fire.
- **Damaged insulation.** Where basement or crawlspace walls are insulated on the interior, chronic moisture problems can ruin the insulation.

Strategy   Now we have a sense of the causes and implications of wet basements. What do we tell our clients?

No Crystal Ball Gazing Allowed   We said in **The Home Reference Book** that 98 percent of basements get wet at some point. Some new home warranty statistics suggest that 38 percent of homes have water problems in basements within the first five years of their life. There is no shortage of statistics to suggest that wet basements are a common problem. Our difficulty is that we are visiting the house at one point in time and there is often no evidence of a problem. Decorating and finishes may cover the evidence. It may also be because concrete can dry up very nicely and not show any evidence of problems.

Can't Predict Frequency   While we know that virtually all basements get wet at some point, it's very difficult to know whether a basement is going to get wet:

- Every time it rains
- Every time it rains with the wind coming from a certain direction
- Every time there is more than one inch of rain in an hour, for example (rainfall intensity)
- Every time it rains more than two inches in total over a week (rainfall volume)
- When there is a 100 year flood (a rain volume that only occurs once every 100 years, according to weather office statistics)
- Only when the ground is frozen (frozen ground can't absorb moisture)
- Only when gutters or downspouts are clogged or leak
- Only when window wells are clogged with debris
- Only when the perimeter drainage tile system fails

Severity

It's difficult to know how often a basement will get wet and under what rainfall conditions. It's also impossible to know how wet it will get. Will the basement:

- Be damp?
- Be wet?
- Have water on the floor? Will it be a trickle, stream or torrent?
- Flood? Is there likely to be ½ inch or six inches of water in the basement floor?

Report Your Limitations

You get the idea. Predicting basement leakage is a tough game. On one hand, you don't want to be alarmist and point out problems that don't exist. This makes you very unpopular with sellers, for example. On the other hand, you often have suspicions but can't substantiate them. This makes for interesting reporting. It's one of the reasons we rely on Section 10.0 in **The Home Reference Book** to explain the situation to our clients.

What to Watch for

Let's list some of the clues you can watch for.

- Water or dampness on the walls or floor
- Efflorescence on the walls or floor
- Rot, stains or water marks on doors, walls, windows and basement stair stringers
- Rust at baseboard nails, carpet tack strips, columns or appliances
- Odors, mold and mildew
- Rot
- Loose floor tiles (tap on the tiles)
- Damaged basement storage (sagging, stained cardboard boxes, for example)
- Storage kept off floor on pallets or shelves
- Patches in walls
- Patches in floors, especially around the perimeter (possibly indicating perimeter drainage tile inside the foundation)
- A trough at the wall/floor intersection around the perimeter (peripheral drain)

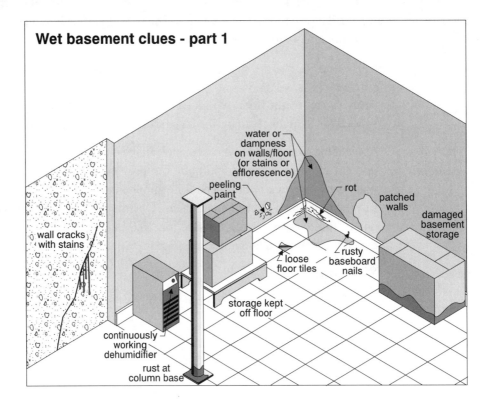

Wet basement clues - part 1

- water or dampness on walls/floor (or stains or efflorescence)
- peeling paint
- rot
- patched walls
- damaged basement storage
- wall cracks with stains
- loose floor tiles
- rusty baseboard nails
- storage kept off floor
- continuously working dehumidifier
- rust at column base

- Sump pumps operating continuously
- A full sump
- Two spare sump pumps on hand
- An auxiliary electrical supply (battery or generator) for sump pumps
- A high water level alarm on the sump
- Crumbling plaster, drywall or masonry
- Peeling paint
- Wall cracks with stains
- Recent exterior excavation on the outside
- Evidence of new dampproofing material on the exterior
- Evidence of a drainage layer or membrane (we'll describe these shortly)
- A dehumidifier in the basement running constantly (Dehumidifiers don't solve leaking basement problems. They may actually increase efflorescence on walls by drawing more moisture through the walls.)

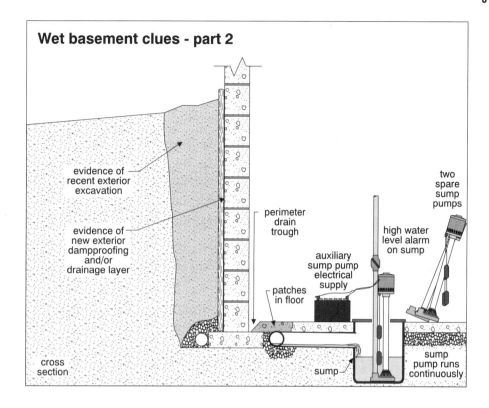

Wet basement clues - part 2

evidence of recent exterior excavation

evidence of new exterior dampproofing and/or drainage layer

perimeter drain trough

patches in floor

auxiliary sump pump electrical supply

high water level alarm on sump

two spare sump pumps

cross section

sump

sump pump runs continuously

Should Be Easy

With all these possible indicators, calling out wet basements should be easy. It isn't. Let's look at some of the common situations that can fool you. There are lots of sources of water in basements that may have nothing to do with leakage from the exterior. These include—

Things That Can Fool You

- Sweating pipes. Cold water pipes can cool the warm, moist air around them and develop condensation that drips constantly. Over time, this can look like a leak.
- Leaking appliances such as dishwashers and water heaters.
- Leaking plumbing fixtures including toilets, bathtubs, shower stalls, sinks, and basins.
- Leaks from hot water heating systems.
- Leaks from central air conditioning or heat recovery ventilator condensate lines.
- Leaks from above grade walls, windows and doors. Water often collects on the tops of foundation walls and runs down the inside, looking like a leak from the exterior. The problems are typically flaws in siding, doors, windows or flashings. (The clue here is that the foundation wall is stained **above** grade level as well as below.)
- Sewer backup. Basements can flood as a result of water coming back up through floor drains. Backwater valves are used to prevent sewer backup where problems are chronic.
- Condensation and mold in cold rooms – common around the top part of the rooms. Doors to cold rooms should be well weatherstripped. Cold rooms should be vented to the outdoors.

Look At
History

There are lots of things that can look like wet basements but are not. In many houses, you will see the history of a number of events. Some of them may include exterior leakage and others will be caused by the things we've just listed. You're going to need some experience to differentiate among the many things that may be going on. The more you see and the more you think while you look, the better you'll get at it.

Analyzing
The Sump
And Pump

Many homes have sumps and pumps in the basement. Not all of these are designed to correct or prevent wet basement problems. In some areas, sumps and pumps are provided because storm sewers may be at a higher elevation than the basement floor. This means that water from a perimeter drainage tile system, and possibly from gutters and downspouts, may have to be collected and pumped up to a storm sewer.

Sumps and pumps may also be used for things like laundry tubs and floor drains, although this is less common and not accepted practice in modern construction.

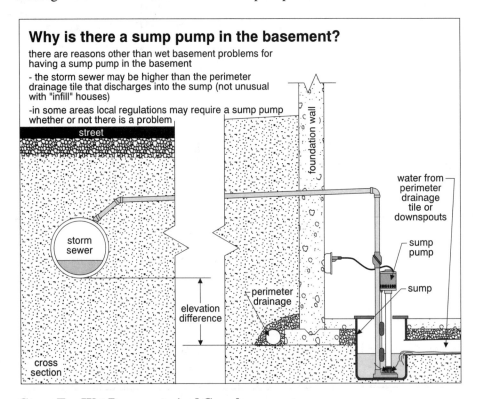

Cures For Wet Basements And Crawlspaces

Home inspectors shouldn't venture too far into the world of repair. However, wet basements are so common that this is a hard one to avoid. As you can see from our **Home Reference Book** discussion, we suggest a sequential approach. Many homeowners are not patient enough to follow this approach and most contractors are unwilling to risk callbacks by trying inexpensive but uncertain repair methods. Many contractors would rather use a more aggressive approach and be sure that the problem is solved. Although this is more expensive, it's hard to blame contractors for not wanting to expose themselves to complaints.

Step By Step   Our step by step approach works like this:

Control
Surface Water

1. Control surface water

a. Control water coming off the roof. Make any necessary repairs and additions to gutters and downspouts.

b. Control surface water around the house. Establish a positive slope down away from the house and from basement window wells and stairwells. Use swales and catch basins if necessary. Note: this is not necessarily inexpensive work.

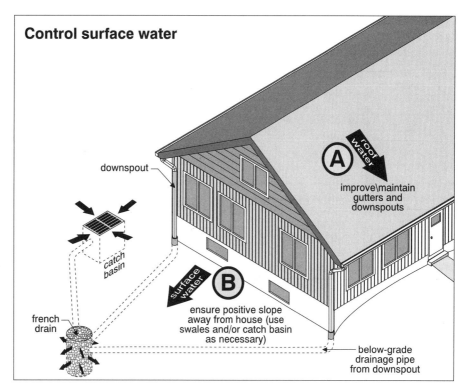

Control surface water

downspout

roof water

(A) improve\maintain gutters and downspouts

catch basin

surface water

(B) ensure positive slope away from house (use swales and/or catch basin as necessary)

french drain

below-grade drainage pipe from downspout

Patch Cracks

2. Patch obvious flaws. Patching poured concrete foundation walls from the inside is a possibility. With concrete block foundation walls, repairs have to be undertaken from the outside. Patching the inside of a crack on a block foundation wall will simply allow the water to run through voids in the blocks to other parts of the wall.

Epoxy And
Polyurethane

Since the 1970s, epoxy resin has been used extensively to patch cracks in foundation walls. More recently, polyurethane injection has become popular. Again, these approaches can only be applied to poured concrete walls.

Exterior patching material can take several forms. Traditional bituminous dampproofing can be used, sometimes with plastic film reinforcement. New high-tech rubberized asphalt products can be used.

Drainage
Layers

More recently, drainage layers or membranes have been used against the foundations without patching the cracks. These can be plastic egg crate type grids, oriented strand semi-rigid fiberglass board, or grooved polystyrene insulation, for example. The approach is simple. If there are voids adjacent to the foundation wall, water will fall quickly through them to the perimeter drainage tile where it can be carried away. The success of this approach depends on the performance of the perimeter drainage tile.

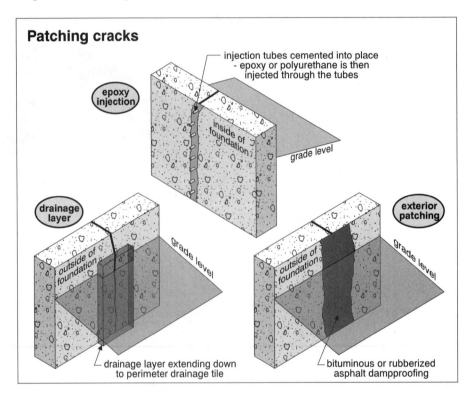

Patching cracks

injection tubes cemented into place
- epoxy or polyurethane is then
injected through the tubes

epoxy injection

inside of foundation

grade level

drainage layer

outside of foundation

grade level

drainage layer extending down
to perimeter drainage tile

exterior patching

outside of foundation

grade level

bituminous or rubberized
asphalt dampproofing

Insulation
Advantages

Some of these drainage layers are also insulations as we've discussed. This can help reduce heat loss from the home and protect the foundations from frost damage. The insulation can also protect the foundation by cushioning the impact of backfill. They are typically sealed at the top.

So Far, So
Good

Early experience with these drainage membranes is good. It remains to be seen whether these membranes and the perimeter drainage tile systems they rely on will perform over the long term.

3. **An interior drainage system** around the perimeter may be appropriate if the exterior perimeter drainage tile is clogged or missing. The interior tile may only go along one wall or around the entire home. The tile is surrounded by gravel, and the concrete floor is re-poured over the tile, leaving a one-half inch space between the slab and foundation wall, so water can get to the drainage tile. A sump and pump are typically used to carry the water to a safe location. Holes are often drilled in the bottom row of concrete block walls to allow water into the drainage tile.

May Affect Footing This approach is less expensive than excavating and addressing the problem from the outside. It can be disruptive and it can also create a structural problem if the footings are relying on a concrete floor slab for lateral support. Cutting away the floor and excavating down beside the footings to put in a drainage tile can create an unstable situation. Specialists should be engaged to advise on this approach.

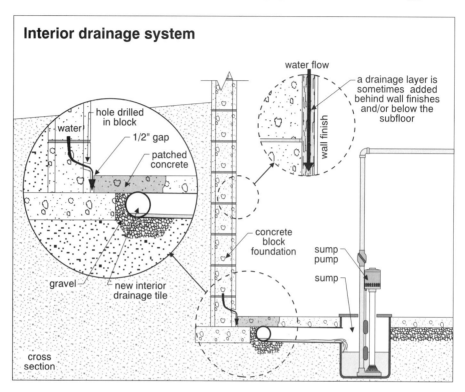

Perimeter Tile May Not Be Necessary In some cases where high water tables are the issues and leakage up through the basement floor is chronic, perimeter drainage tile may not be necessary. If the concrete floor slab has a granular layer below (common in modern construction), there may be a better answer if the granular material below the floor slab will allow water to move freely through it. A sump and pump put into the basement floor and opened to the granular drainage layer, may allow water to flow quickly to the sump. It can be pumped to a safe location. This is obviously less expensive and less disruptive to the footings than a perimeter drainage tile approach inside the foundations.

Interior Trench Basement floors may have trenches cut around the perimeter to collect water. A trench carries water to the floor drain. This doesn't stop the leak, but directs the water to the floor drain.

Drainage Layer On Walls This is sometimes combined with a drainage membrane on the foundation walls. The drainage membrane may be covered by insulation and finishes.

Drainage Layer Under False Floor A variation is to lay a drainage membrane on the concrete floor without cutting trenches. The membrane is covered with subfloor and finished flooring. Water runs across the floor to the drain without damaging finishes.

These systems can be successful, although if the foundation wall is prone to deterioration, this can continue unnoticed. Also, if water stands anywhere in the system, the development of molds and other unhealthy growths may be a problem.

Holes In Block Foundations   Where the outside perimeter drainage tile is thought to be in working order, holes are sometimes drilled through block foundations from the inside, near floor level. The holes on the inner face of the block are then sealed. The idea is that water can escape from the block wall into the drainage tile. The risk here is that if the drainage tile is clogged and flooded, the situation may get worse instead of better.

4. **Exterior excavation and adding or replacing perimeter drainage tile** is a possibility. This last resort approach is both disruptive and expensive. Up until the last 10 or 15 years, this work has typically been accompanied by applying a new dampproofing layer to the exterior of the foundations prior to backfilling.

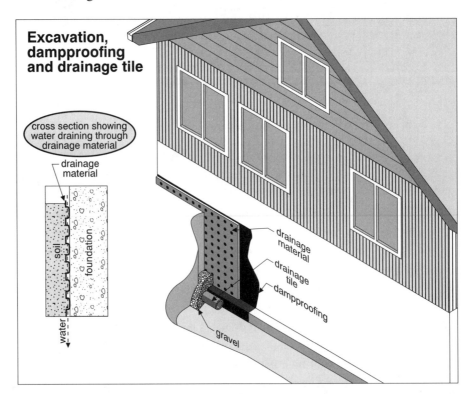

Summary

Bentonite Clay Injections   In some areas, the injection of special clays is promoted as a solution to wet basements and crawlspaces. The philosophy is that these clays are moisture sensitive and as they get wet will expand to fill all the voids in the soil. If the voids in the soil are filled with clay, water can't enter them. In practice, we have not seen good results with this system.

Using Moisture Meters

The use of moisture meters goes beyond the Standards. Many home inspectors use moisture meters in finished basements to detect moisture in carpeting, tack strips, baseboards, and wall finishes, for example.

Client Sensitivity

Adjusting client expectations is part of any home inspection. It is as important with respect to wet basements and crawlspaces as with any component of the inspection. Your clients have to understand that all basements leak at some point. If we control surface water we improve matters dramatically. Where ground water is an issue, there are strategies that can be used, but the risks are higher. There are no long term guarantees and if nature unleashes a hurricane or torrential rains, there may be no way to keep basements dry.

Determine Client Expectations

Many clients have lived in houses with basements and are aware of the realities. To other clients, the possibility that there may be water in the house comes as a shock and is very unsettling. You are wise to discuss this issue with your client for a minute or so to get a sense of their expectations. If they are unrealistic, you had better make an effort to adjust them.

Insulation & Interiors
M O D U L E

QUICK QUIZ 6

☑ INSTRUCTIONS

• You should finish Study Session 6 before doing this Quiz.

• Write your answers in the spaces provided.

• Check your answers against ours at the end of this Section.

• If you have trouble with the Quiz, re-read the Study Session and try the Quiz again.

• If you did well, it's time for Inspection Tools, Inspection Checklist and Inspection Procedure.

1. Most wet basement problems come from what two sources?

    _____

    _____

2. What three contributing factors involving foundation depth and conditions outside the house can affect basement problems?

    _____

    _____

    _____

3. List six implications of wet basements.

_____

_____

_____

_____

_____

_____

4. There are at least 20 common signs that basements have been wet or are vulnerable to wetness problems. See how many you can list.

_____

_____

_____

_____

_____

_____

_____

_____

_____

_____

_____

_____

_____

_____

_____

_____

_____

_____

_____

_____

5. Describe four corrective actions to wet basement problems. Try to arrange them in order of least expensive to most expensive.

_____

_____

_____

_____

_____

6. The most common complaint against home inspectors is what house problem?

_____

7. Dehumidifiers are an effective way to correct wet basement problems.
   True ☐   False ☐

If you had no trouble with this Quiz, you are ready to have a look at the Inspection Tools, Inspection Checklist and Inspection Procedure.

Key Words:

- **Surface water**
- **Roof water**
- **Ground water**
- **Soil type**
- **Water table**
- **Foundation depth**
- **Cold pours**
- **Honeycombing**
- **Efflorescence**
- **Rot**
- **Mold**
- **Mildew**
- **Rust**
- **Perimeter drainage tile**
- **Drainage membrane**
- **Sumps**
- **Pumps**

▶ 11.0 INSPECTION TOOLS

The following tools are recommended as a minimum for an Interior inspection.

Ladder

To access ceilings and skylights.

Flashlight

To look into dark areas and scan walls and ceilings for bulges and cracks.

Screwdriver or Carpenter's Awl

To tap on floors and walls looking for loose tiles and to probe for rot.

Tape Measure

To check staircase headroom, guardrail heights, etc.

Your Senses

As always, your eyes are your main inspection tool. Your nose, your ears and your sense of touch are also important. You'll be sniffing for odors, pushing and tapping on walls and listening for plaster keys and other falling debris, as well as listening to the sound produced when you tap on materials.

We are tempted to add that your most important sense is your common sense.

Optional items (beyond Standards):

Level and Plumb Bob

To measure the amount by which components are out of level or out of plumb.

Moisture Meter

To check for moisture in walls, floors and ceilings, and to check stains to determine if they are wet.

Humidity Meter

To check moisture levels in the air in crawlspaces, basements and living spaces.

► 12.0 INSPECTION CHECKLIST

Location Legend	N = North	S = South		E = East	W = West
CS = Crawlspace	1 = 1st Floor	2 = 2nd Floor	3 = 3rd Floor	B = Basement	

FLOORS				
LOCATION	FLOORS – GENERAL	LOCATION	CARPET ON FLOORS	
	• Water damage		• Rot	
	• Trip hazard		• Stains	
	• Mechanical damage		• Odors	
	• Loose or missing pieces		• Buckled	
	• Absorbent materials in wet areas		• Lifted at seams or edges	
	CONCRETE FLOORS		RESILIENT FLOORING	
	• Cracked		• Split	
	• Settled		• Lifted seams	
	• Heaved		• Open seams	
	• Water on floor			
	• Efflorescence		CERAMIC TILE, STONE, MARBLE, ETC.	
	• Slopes away from drain			
	• Hollow below .		• Grout loose	
			• Grout missing	
	WOOD FLOORS		• Tiles cracked	
			• Tiles broken	
	• Rot		• Tiles missing	
	• Warped		• Tiles worn	
	• Buckled		• Stains on tiles	
	• Stained			
	• Squeaks			
	• Exposed tongues			

WALLS				
LOCATION	WALLS – GENERAL	LOCATION	MASONRY OR CONCRETE WALLS	
	• Water damage		• Spalling	
	• Cracks		• Mortar missing or deteriorated	
	• Mechanical damage		• Efflorescence	
	• Inappropriate finishes in wet areas			
	• Truss uplift		PARTY WALLS	
	PLASTER OR DRYWALL WALLS		• Not continuous – incomplete	
			• Not continuous – penetrated	
	• Bulging		• Ice dams	
	• Loose or missing			
	• Shadow effect		GARAGE WALLS AND CEILINGS	
	• Crumbling or powdery			
	• Nail pops		• Not fireproof	
	• Poor joints		• Not gastight	
	WOOD FLOORS			
	• Rot			
	• Cracked, split or broken			
	• Buckled			
	• Loose			

CEILINGS			
LOCATION	**CEILINGS – GENERAL**		**WOOD CEILINGS**
	• Truss uplift		• Rot
	• Water damage		• Broken, cracked or split
	• Stains		• Buckled
	• Cracked, loose or missing sections		• Loose
	• Mechanical damage		
			METAL
	PLASTER/DRYWALL CEILINGS		
			• Rust or loose
	• Sag		
	• Shadow effect		**POOR LIGHTING**
	• Crumbly or powdery		
	• Poor joints		• Rooms
	• Nail pops		• Hallways
	• Textured ceilings in poor locations		

COUNTERS			
LOCATION	**COUNTER TOPS**	LOCATION	**TRIM**
	• Entire top loose		• Missing
	• Loose or missing pieces		• Water damage
	• Burned		• Rot
	• Cut		• Loose
	• Worn		• Damaged
	• Damaged		• Counter problems
	• Stained		• Entire top loose
	• Rusted metal		• Loose or missing pieces
	• Missing grout on ceramics		• Burned, cut or worn
	• Loose ceramic tiles		• Mechanical damage
	• Rotted substrate		• Stained
			• Rise or run not uniform
	CABINETS		• Excessive nosing
			• Tread width too small
	• Water damage		• Winders – too many
	• Rot		• Winders – too big an angle
	• Stained		• Stringers – too small
	• Damaged		• Stringers – excessive notching
	• Worn		• Stringers – too thin
	• Broken or cracked glass		• Excessive span between stringers
	• Stiff or inoperative		• Rot at bottom
	• Not well secured to wall		• Pulling away from wall or treads
	• Doors or drawers missing or loose		• Inadequately secured to header
	• Other pieces missing or loose		
	• Shelves not well supported		
	• Rust on medicine cabinets		
	• Drawers – missing or defective stops		

STAIRS			
LOCATION	**STAIRS – GENERAL**		**HANDRAILS**
	• Rot		• Missing
	• Damaged		• Too low
			• Too high
	TREADS		• Hard to hold
			• Loose
	• Worn or damaged		• Damaged
	• Sloped		
	• Loose		**STAIRWELL WIDTH**
	• Poorly supported		
	• Thickness inadequate		• Too narrow
	• Rise excessive		• Headroom inadequate
	GUARDRAILS		**LANDINGS**
	• Missing		• Missing
	• Too low		• Too small
	• Loose		
	• Damaged		**FIRE SAFETY**
	SPINDLES OR BALUSTERS		• Drywall missing or incomplete on underside of stairs
	• Too far apart		**LIGHTING**
	• Easy to climb		
	• Loose		
	• Damaged		• Missing
	• Missing		• Not controlled by three way switch

WINDOWS			
LOCATION	**EXTERIOR DRIP CAP**	LOCATION	**SASHES**
	• Missing		• Rot
	• Ineffective		• Rust
			• Inoperable
			• Stiff
	EXTERIOR TRIM		• Won't stay open
			• Sash coming apart
	• Missing		• Loose fit
	• Rot		• Poor weatherstrip
	• Rust		
	• Damaged, cracked or loose		**INTERIOR TRIM**
	• Sills with reverse slope		
	• Inadequate sill projection		• Rot
	• No drip edge		• Stained
	• Poor fit		• Missing
	• Putty (glazing compound) cracked, missing, loose or deteriorated		• Cracked
	• Caulking or flashing missing, deteriorated, loose, rusting or incomplete		• Loose
	• Paint or stain needed on underside of stairs		
			HARDWARE
	WINDOWS, SKYLIGHTS AND SOLARIUMS – GENERAL		
			• Rusted
	• Air leaks		• Broken
	• Water leaks		• Missing
	• Lintels sagging or missing		• Inoperable
			• Loose
	FRAME		• Sills too low on stairs or landing
	• Racked		**SKYLIGHTS AND SOLARIUMS**
	• Rot		
	• Rust		• Evidence of ice dams
	• Deformation		• Special glazing not provided (more than 15% off vertical)
	• Installed backwards		
	• Drainholes blocked or missing		**STORMS AND SCREENS**
	GLASS (GLAZING)		• Missing
			• Torn or holes
	• Cracked		• Rusted
	• Broken		• Loose
	• Missing		
	• Lost seal on double or triple glazing		**MEANS OF EGRESS**
	• Excess condensation		
	• Loose		• Missing
			• Too small

179

DOORS			
LOCATION	**DOORS – GENERAL**		**INTERIOR TRIM**
	• Air leaks		• Rot
	• Water leaks		• Stained
	• Lintels sagging or missing		• Missing
			• Cracked
	DOOR AND FRAME		• Loose
			• Poorly fit
	• Rot		• Floor stained below
	• Rust		• Doorstops missing or ineffective
	• Racked		• Guides and stops missing or damaged
	• Deformation		
	• Damaged		**GLASS (GLAZING)**
	• Delaminated		
	• Loose or poor fit		• Cracked
	• Installed backwards		• Broken
	• Drain holes blocked or missing		• Missing
	• Stiff		• Lost seal on double or triple glazing
	• Inoperable		• Excess condensation
	• Swings open or closed by itself		• Loose
	• Dark paint on metal exposed to sun		
	• Plastic trim on metal door behind storm		**HARDWARE**
	EXTERIOR DRIP CAP		• Rusted
			• Broken
	• Missing		• Missing
	• Ineffective		• Inoperable
			• Loose
	EXTERIOR TRIM		• Hinges on exterior
			• Self-closer missing
	• Missing		• Ineffective
	• Rot		
	• Rust		**STORMS AND SCREENS**
	• Damaged, cracked or loose		
	• Sills with reverse slope		• Missing
	• Inadequate sill projection		• Torn or holes
	• No drip edge		• Rusted
	• Putty (glazing compound) cracked, missing, loose or deteriorating		• Loose
	• Caulking or flashing missing, deteriorated, loose, rusting or incomplete		
	• Paint or stain needed		**GARAGE DOORS**
	• Sill not well supported		
	• Sill too low		• Not fire rated or exterior type
			• Not weatherstripped

BASEMENTS			
LOCATION	**WET BASEMENTS AND CRAWLSPACES**	LOCATION	**WET BASEMENTS AND CRAWLSPACES**
	• Mildew		• Auxiliary power for sump pump
	• Rot		• Crumbling plaster, drywall or concrete on walls
	• Loose floor tiles		• Peeling paint
	• Damaged storage		• Wall cracks with water stains
	• Storage kept off floor		• Recent exterior excavation
	• Floor drain missing		• New dampproofing material
	• Floor drain at high spot		• Drainage membrane
	• Floor patched around perimeter (drainage tile system?)		• Dehumidifier in basement
	• Trough at wall/floor intersection		• Poor grading (control of surface water)
	• Peripheral drain		• Poor gutters and downspouts (control of roof water)
	• Sump full		• Cold pours in concrete walls
	• Sump pump operating continuously		• Honeycombing in concrete
	• Spare pump on hand		• Water on floor
			• Dampness on floor or walls
			• Efflorescence
			• Rot
			• Stains
			• Water marks
			• Rust
			• Odors
			• Mold

► 13.0 INSPECTION PROCEDURE

Primarily Visual The inspection of building interiors is primarily visual. You will be operating doors and windows, cabinet doors and drawers and as you move through the interior. You will also be operating plumbing fixtures, space heaters, electrical switches and testing electrical receptacles. Many inspectors carry a flashlight, but keep at least one hand free. You'll want it for pushing and tapping on suspect surfaces.

A Double Tour Approach Some home inspectors perform the entire interior inspection in a single tour. We've talked about the wisdom of a double tour approach that involves doing most of the components on the first tour and doing a test of the heating system, for example, as a second tour. This second tour gives you a chance to make sure you haven't missed any rooms or areas and to recheck anything that deserves a second look. A ceiling below a shower stall is a good example of something that deserves a second look after you've run water.

Multi Discipline The interior inspection is unusual because you are looking at several different systems. In most parts of a home inspection, you are concentrating in one area. Here you have to look at not only interior finishes, but some exterior and roofing issues, plumbing, heating, air conditioning, electrical and structural issues. The interior inspection can challenge your organizational abilities and your mental flexibility.

Checklist Versus Passive Approach Some inspectors follow a point by point guide to walk them through their inspections. Others take a more passive approach that allows the room to communicate with the inspector. We recommend a combination of these two approaches. It's another advantage of the two pass system.

Function, Not Form One of the difficult things to focus on during an inspection is the function of interior components, rather than the form. Wall, floors and ceilings can look great, but be very delicate. Finishes can look horrible, but be quite sound. While the beauty may be only skin deep, the substance goes deeper.

Clients Join You We've said that if clients are going to follow you anywhere throughout the house, they'll follow you during your interior inspection. This is a good opportunity to communicate your findings and to learn about clients' expectations. Many inspectors involve the clients in the inspection, opening and closing doors and windows and even testing electrical receptacles. Many inspectors ask clients to operate cabinet doors and drawers.

Ask About Plans This is also a good time to ask clients if they have plans to make changes to the house. This can cut short a very awkward discussion about a low quality rear addition. Life becomes much easier when the client tells you that they are going to tear off that ugly rear addition. Many inspectors ask about plans for the house during the opening interview, but this is also a good time.

Use All The Clues	As you go through the interior of the home, there are lots of subtle clues. Changes in wall and ceiling height can tell you about additions and remodeling projects. Changes in wall, floor and ceiling finishes may also suggest changes to the home. Similarly, different window and trim styles can help date different parts of houses or ages of renovation projects. Has the house been substantially rearranged? What are the implications of this? Sometimes if you dig deep enough, you'll find that a house was substantially rebuilt as a result of fire or other damage. This may be only a matter of interest, but it can provide valuable information as well.
Don't Assume The Obvious	There is a strong tendency to assume that everything is typical. You have to discipline yourself to look for the unusual and unexpected. Don't be lulled into expecting everything to be all right and then seeing it as you assumed it. Keep an open mind, especially when the house looks great.
Don't Crystal Ball Gaze	Home inspectors have to be careful to report what they see, not what they assume. When people ask if they can remove the wall to wall carpeting and expose the hardwood flooring, you may feel very clever that you have confirmed the hardwood flooring. However, you can't tell much about the condition of the old hardwood. It may not be suitable for exposing, even with refinishing.
Is The Basement Dry?	When people ask about how dry the basement is, you should be careful with your answers. It may be dry today, but do you know it will always be dry? Of course not.
Attic Free Of Rot?	Unless you can see every square inch of the attic, don't describe it as being rot free. Similarly, don't extrapolate to concealed areas what you see in exposed areas. If you see evidence of R-20 insulation in a wall, don't assume that all walls in the house are similarly insulated. If this turns out to be a small addition or remodeling project, you're going to look very foolish. In short, report what you see, not what you think, unless you classify it as an opinion rather than a fact.
What You Can't See	Clients deserve to know that you couldn't get into the roof space for the flat roof or get into the crawlspace under the rear addition. This needs to be clearly documented in your report. The absence of the limitation will lead people to the reasonable conclusion that everything was inspected and is fine.
Floor Materials	It becomes a subconscious characteristic of home inspectors that as they walk through a room, they sense whether they are on wood or concrete flooring. In some cases, you'll have to bounce on the balls of your feet to verify what your are on. There will be the odd time you get fooled. However, this is a good habit to develop.

How Level Is Level?

We do not use spirit levels or marbles on hardwood floors to determine whether components are plumb and level. It's always safe to say that walls are not plumb and floors and ceilings are not level. At some level, even if it's microscopic, this has to be true. What you are looking for is something that falls within the tolerances of the average person. Your eyes, your sense of touch and your sense of balance are generally enough to quantify this. The older the house, the more out of plumb and out of level things tend to be. Homeowners also tend to be more understanding and forgiving of these characteristics in older houses. You need to know the house and know your clients' expectations by the time you get to the end of the inspection. What's perfectly normal to you may be alarming to them. We recommend that you have this discussion during the inspection, rather than over the phone after they move in and are upset.

Multiple Layers Of Finishes

Older houses may be expected to have multiple layers of plaster and drywall on walls and ceilings. There may also be multiple layers of flooring. Again, we recommend that you let clients know about these things and the fact that you might not find them.

Stripping That Wallpaper

We touched earlier on the fact that removing wallpaper can bring down plaster and drywall with it. Letting your client know before the fact is a very good idea.

What Does This Stain Mean?

Checking stains to see whether they are wet is a worthwhile exercise. You can use your hand or some type of moisture meter. Again, don't get lulled into a false sense of security. If the stain is wet, you can be conclusive about an ongoing problem. If the stain is dry, you can't be conclusive. The problem may be intermittent and it just happens to be dry right now. Don't tell them there is no active problem because the stain is dry.

Stairs

We talked about lots of rules for stairs with lots of numbers. Most home inspectors use a more passive approach on stairs after they have acquired experience. As you walk up and down stairs, your muscle memory will tell you whether they feel right or not. Stairs that are unsafe usually make you feel awkward in some way as you move up or down. Let your senses raise red flags about stairs.

Counters In Kitchens And Bathrooms

We often invite clients to look at these things for themselves. They are usually more diligent than you may be in moving cutting boards, toasters and kettles to look at the conditions of countertops. Ensure things get put back where they belong.

Countertop Security

Make sure you check that the counter is well secured to the base cabinets. Lifting with moderate pressure is recommended. Don't use excessive force.

Cabinets

The big issue on cabinets is how well they are secured to the wall, particularly upper cabinets. Again, use moderate pressure to check these. Watch for shelves that are not well secured and drawers that pull right out of the cabinets. Stops are obviously missing or ineffective.

Doors And Windows	Operate as many doors and windows as you can:

- Are windows easy to open and close?
- Do windows stay put when they are opened?
- Do they lock when they are shut?
- Do doors latch properly?
- Have they been trimmed so they will close?
- Does this indicate considerable building movement?
- Move drapes around windows and sliding doors. Look for water damage below doors and windows.

Double Or Triple Glazing?	Sometimes you can tell how many panes of glass are in windows by looking at the spacer. With practice, you can also use the reflection of a ring, screwdriver or pen, for example, held close to the glass to see how many glazing layers there are. If there are two reflections, you usually have double glazing. Triple glazing will usually yield three reflections of a shiny object. Practice before you rely on this technique.

Wet Basements And Crawlspaces	Our final note on inspection procedure is to be careful with basements and crawlspaces. As we said in the text, there are more complaints against home inspectors about wet basements and crawlspaces than any other single issue. Look carefully, document what you can see and allow for the possibility (certainty?) that leaks will occur when conditions are right.

Insulation
& Interiors
M O D U L E

FIELD EXERCISE 1

☑ INSTRUCTIONS

You should have completed everything up to and including 13.0 INSPECTION PROCEDURE before starting this Exercise.

This Field Exercise will be partly research and partly inspection. We're also going to have a report writing exercise at the end. You should allow yourself about four hours for this Field Exercise.

Let's start with the research side of things.

A. Research

You'll want to talk to at least two building supply houses and two builders. They may be new home builders or renovators. Here's a list of questions you can ask. You may come up with more of your own.

1. What are the most common floor finishes used in homes in this area? What are the pros and cons of each material? What common flooring problems do you see?
2. What are the most common wall and ceiling finishes used? What are the pros and cons of each? What are the most common problems you see with wall and floor finishes?
3. What are the most common countertop materials used? What are the pros and cons of each? What problems do you see most frequently with countertops?
4. Are stairs in this area mostly site built or prefabricated? Are there significant advantages to either approach? What are the most common stair problems that you see?

5. What are the most common styles of windows used in this area? What are the pros and cons of each style? What problems do you see with each? What window materials are commonly used in this area? What are the pros and cons of each material? What problems do you see with them?

6. What are the most common styles of doors in this area? Are there significant pros or cons for any of them? What are the most common door installation problems that you see? What is the most common exterior door material? Do any of the materials have particular problems?

7. Do skylights and solariums create special problems? If so, what are they?

8. Are wet basements or crawlspaces a common problem in this area? Why is that? What's the best way to prevent wet basements and crawlspaces?

B. Inspection

This is an exercise you can do almost every day. Good home inspectors are always aware of their surroundings. You can start this one at home and then find four other houses, preferably of different styles and ages. Let's start by identifying the materials. See if you can fill out this chart for each of the five houses, identifying the materials in each case.

Note: You will probably find that you have more than one material in each category for each home.

	1	2	3	4	5
Floors					
Walls					
Ceilings					
Countertops					
Cabinets					
Window Styles					
Window Materials					
Door Styles					
Door Materials					

Now use the inspection strategies you've learned and the Inspection Checklist to look for problems in each of these systems.

Document problems in the chart attached.

	1	2	3	4	5
Walls					
Ceilings					
Trim					
Counters					
Cabinets					
Stairs					
Windows					
Skylights					
Solariums					
Doors					
Basement & Crawlspace Leakage					

Look at your completed chart. For each component, which problems were the most common? This should give you some valuable insight into what to watch for on inspections.

C. Report Writing

Let's finish by selecting the house that had the most problems in the last section of Part B above. Write a narrative report for a client about the interior of the house. include—

- A description of the material or system
- The problem or problems you noted
- The implications of these problems
- The recommended action for improving the situation (e.g., repair, replace, provide, improve, monitor, investigate further)
- Any limitations to your inspection

Once you've completed the Exercise, you are ready to move onto the Final Test.

► 14.0 BIBLIOGRAPHY

TITLE	AUTHOR	PUBLISHER
Construction Principles Materials and Methods	Harold B. Olin, John L. Schmidt, Walter H. Lewis	*Institute of Financial Education and Interstate Printers and Publishers Inc.*
Canadian Building Digests: (listed below)	Various	*Institute for Research in Construction National Research Council of Canada*
CBD1 – Humidity in Canadian Buildings	N. B. Hutcheon	*ibid.*
CBD4 – Condensation on Inside Window Surfaces	A.G. Wilson	*ibid.*
CBD9 – Vapour Barriers in Home Construction	G.O. Handegord	*ibid.*
CBD16 – Thermal Insulation in Dwellings	W.H. Ball	*ibid.*
CBD23 – Air Leakage in Buildings	A.G. Wilson	*ibid.*
CBD25 – Window Air Leakage	J.R. Sasaki and A.G. Wilson	*ibid.*
CBD36 – Temperature Gradients Through Building Envelopes	J.K. Latta and G.K. Garden	*ibid.*
CBD42 – Humidified Buildings	N.B. Hutcheon	*ibid.*
CBD44 – Thermal Bridges in Buildings	W.P. Brown and A.G. Wilson	*ibid.*
CBD48 – Requirements for Exterior Walls	N.B. Hutcheon	*ibid.*
CBD50 – Principles Applied to an Insulated Masonry Wall	N.B. Hutcheon	*ibid.*
CBD57 – Vapour Diffusion and Condensation	J.K. Latta and R.K. Beach	*ibid.*

CBD72 – Control of Air Leakage is Important	G.K. Gardon	*ibid.*
CBD104 – Stack Effect in Buildings	A.G. Wilson and G.T. Tamura	*ibid.*
CBD110 – Ventilation and Air Quality	A.G. Wilson	*ibid.*
CBD111 – Decay of wood	M.C. Baker	*ibid.*
CBD130 – Wetting and Drying of Porous Materials	P.J. Sereda and R.F. Feldman	*ibid.*
CBD142 – Space Heating and Energy Conservation	D.G. Stephenson	*ibid.*
CBD149 – Thermal Resistance of Building Insulation	C.J. Shirtliffe	*ibid.*
CBD161 – Moisture and Thermal Considerations in Basement Walls	C.R. Cracker	*ibid.*
CBD166 – Plastic Foam	A. Blaga	*ibid.*
CBD167 – Rigid Thermal Plastic Foams	A. Blaga	*ibid.*
CBD168 – Rigid Thermal Setting Plastic Foams	A. Blaga	*ibid.*
CBD 175 – Vapour Barriers: What Are They? Are They Effective?	J.K. Latta	*ibid.*
CBD231 – Moisture Problems in Houses	A.T. Hansen	*ibid.*
CBD245 – Mechanical Ventilation and Air Pressure in Houses	C.Y. Shaw	*ibid.*
Moisture: **Moisture Movement; Historical Perspective on North American Wood Frame Construction; Roof Ventilation and Construction; Moisture Investigation (papers)**	Joseph W. Lstiburek	*Presented at ASHI®/CAHI Annual International Conference, 1990*

• **Moisture Problems** • **Drywall Application** • **Indoor Air Quality** • **Door and Window Installation** • **Ventilation: Health and Safety Issues** • **Guide to Residential Exhaust Systems (booklets – part of the Builders' Series)**	Compilation	*Canada Mortgage and Housing Corporation*
Troubleshooting Guide to Residential Construction	Compilation from the Editors of The Journal of Light Construction	*Builderburg Group Inc.*
Insulation Materials (booklet)	Compilation	*Small Homes Council – Building Research Council, University of Illinois at Urbana Champaign*
CSA F-326 Residential Mechanical Ventilation Systems	Compilation	*Canadian Standards Association*
Provide Fresh Air and Control Humidity in a Tighter House	Compilation	*Ontario Ministry of Energy*
Drywall Construction Handbook	Various	*U.S. Gypsum Co.*
Recommended Specifications for the Application and Finishing of Gypsum Board	Various	*Gypsum Association*
Where and How to Insulate Basements	Compilation	*Ontario Ministry of Energy*
Energy Saving with External Basement Wall Insulation	Compilation	*Canada Mortgage and Housing Corporation*
Canadian Home Builders Association Builders' Manual	Compilation	*Canadian Home Builders Association*
Certified Home Ventilating Products Directory	Compilation	*Home Ventilating Institute*
Residential Energy – Cost Savings and Comfort for Existing Buildings	John Krigger	*Saturn Resource Management*

House Ventilation and Air Quality Reference Guide	Compilation	*Ontario Hydro*
Canadian Wood Frame House Construction	Various	*Canada Mortgage And Housing Corporation*
Insulation Aftermath	John Eakes	*Ontario Marketing Productions Limited*
Complying with Residential Ventilation Requirements	Compilation	*Canada Mortgage and Housing Corporation*
Consumers Guide – Keeping the Heat In	Compilation	*National Resources Canada*
Humidity, Condensation and Ventilation in Houses	Compilation	*National Research Council of Canada*
Condensation in the Home: Where, Why and What to Do About It	John W. Sawers	*Canada Mortgage and Housing Corporation*
Journal of Light Construction (various articles)	Various	*Builderburg Group Inc.*
Avoiding Moisture Problems (a paper)	Jeffrey C. May, M.A.	*Presented at American Society of Home Inspectors 1998 Annual Conference*
1997 ASHRAE Handbook FUNDA-MENTALS	Compilation	*American Society of Heating, Refrigeration and Air-Conditioning Engineers, Inc.*
National Building Code of Canada	Compilation	*National Research Council of Canada*
CABO One and Two Family Dwelling Code	Compilation	*The Council of American Building Officials*
Uniform Building Code	Compilation	*International Conference of Building Officials*

Note: Copies of building codes available from:

Southern Building Codes Congress International (SBCCI), 900 Montclair Road, Birmingham, AL, 35213

Council of American Building Officials (CABO), 5203 Leesburg Pike, Suite 708, Falls Church, VA, 22041

Building Official Codes Administrators International (BOCA), 4051 West Flossmoor Road, Country Club Hills, IL, 60478

International Conference of Building Officials (ICBO), 5360 Workman Mill Road, Whittier, CA, 90601

Review copies of the International Building Code – 2000 (IBC) are available at any code office. This is a new code document that may replace all of the existing ones.

► ANSWERS TO QUICK QUIZZES

Answers to Quick Quiz 1

1. 1. Walls, ceilings and floors
    2. Steps, stairways, and railings
    3. Counter top and a representative number of installed cabinets
    4. A representative number of doors and windows
    5. Garage doors and their operators

2. All exterior doors

3. A representative number, ideally one per room

4. A representative number, usually one per room

5. These don't have to be inspected.

6. 1. Paint, wallpaper and other finish treatments
    2. Carpeting
    3. Window treatments
    4. Central vacuum systems
    5. Household appliances
    6. Recreational facilities or another dwelling unit

7. Carpet, wallpaper, furniture and storage can all conceal interior finishes

8. Two

9. Remodeling or renovations

10. No

11. No

12. Yes, although it is not explicitly stated in the Standards

Answers to Quick Quiz 2

1. 1. Concrete
    2. Wood
    3. Hardwood or softwood
    4. Carpet
    5. Resilient
    6. Ceramic and quarry tile
    7. Stone and marble

S1

2.

Problem:	Implications:
Water damage	1. Cosmetic 2. Rot, staining or other damage to finish 3. Rot or other damage to structural components
Trip hazard	1. Personal injury
Mechanical damage	1. Unevenness or loss of continuity in the system
Loose or missing pieces	1. Trip hazard 2. Moisture entry to subflooring
Absorbent materials in wet areas	1. Premature deterioration 2. Rot damage to subflooring 3. Odors and air quality issues

3. Problem:
1. Cracked
2. Settled, heaved
3. Water and efflorescence
4. Slope away from drain
5. Hollow below

4.

Problem:	Implications:
Rot	1. Cosmetic problems 2. Trip hazards 3. Deterioration of the structure below
Warped	1. Cosmetic problems 2. Trip hazards 3. Deterioration of the structure below
Buckled	1. Cosmetic problems 2. Deterioration of the structure below
Stained	1. Cosmetic problems 2. Trip hazards 3. Deterioration of the structure below
Squeaks	1. A nuisance only
Exposed tongues	1. Slivers, splinter or exposed nailheads and possibly injury 2. Cosmetic

5.

Problem:	Implications:
Rot	1. Cosmetic problems 2. Possible damage to subflooring below 3. Health implications
Stains	1. Cosmetic problems 2. Possible damage to subflooring below 3. Health implications
Odors	1. Cosmetic problems 2. Possible damage to subflooring below 3. Health implications
Buckled	1. Trip hazard 2. Cosmetic problems

6.

Problem:	Implications:
Split	1. Water damage to subflooring 2. Trip hazard
Open seams	1. Water damage to subflooring 2. Trip hazard
Lifted seams	1. Water damage to subflooring 2. Trip hazard

7.

Problem:	Implications:
Loose	1. Water damage to subflooring 2. Trip hazard
Grout missing	1. Water damage to subflooring
Cracked or broken	1. Water damage to subflooring 2. Trip hazard
Worn	1. Cosmetic 2. Trip hazard if pieces are loose
Stains	1. Cosmetic

8. Provide stress concentration points so that if the slab cracks, it will crack here

9. Around plumbing fixtures, especially toilets

Answers to Quick Quiz 3

1. 1. Plaster or drywall
    2. Wood plank or paneling
    3. Masonry or concrete
    4. Fiber cement paneling

2. Party walls are located between attached dwelling units, and act as a sound and fire separation. Common materials are masonry block and wood frame with drywall.

3. 1. Water damage
    2. Cracks
    3. Mechanical damage
    4. Inappropriate finishes in wet areas

4. 1. Bulging, loose or missing
    2. Shadow effect
    3. Crumbling or powdery
    4. Nail pops
    5. Poor joints

5. 1. Rot
    2. Cracked, split or broken
    3. Buckled
    4. Loose

6. 1. Not continuous
    2. Ice dams

7. 1. Not fireproof
    2. Not gas-tight

8. 1. Cosmetic
    2. Damage to structure behind
    3. Decorating issues
    4. Structural movement or settling
    5. Minor repair
    6. Staining and deterioration
    7. Mold or mildew growth
    8. Life safety from falling plaster
    9. Concealed damage
    10. Fire spread where party wall not continuous
    11. Moisture damage
    12. Exhaust fume entry into the house from garage

9.   1.   Shine flashlight along ceiling
         2.   Lift tiles on suspended ceilings

10.   1.   Plaster or drywall
         2.   Wood, hardboard or plywood
         3.   Fiber cement or concrete
         4.   Acoustic or suspended tile

11.   1.   Water damage
         2.   Cracked, loose or missing sections
         3.   Mechanical damage

12.   1.   Shadow effect
         2.   Crumbling or powdery
         3.   Nail pops
         4.   Poor drywall joints
         5.   Sag
         6.   Textured ceilings in wet areas

13.   Rust

14.   1.   Cosmetic
         2.   Mechanical damage
         3.   Damage to structure behind
         4.   Failure of ceiling finishes
         5.   Safety hazards
         6.   Staining and deterioration
         7.   Mold and mildew growth
         8.   Decorating issues

Answers to Quick Quiz 4

1.   1.   Cover joints at changes of material and direction
         2.   Protect walls

2.   1.   Baseboard
         2.   Quarter round
         3.   Door casing
         4.   Window casing
         5.   Chair rails
         6.   Plate rails
         7.   Cornice moldings
         8.   Rosettes or medallions

3.
1. Plastic laminate
2. Wood
3. Marble
4. Granite
5. Synthetic marble
6. Stainless steel
7. Ceramic

4.

Problem:	Implications:
Missing	Cosmetic
Water damage	Rot; damage to building systems behind
Rot	Damage to building systems
Loose	Damage to building systems behind
Mechanical damage	Cosmetic or air leakage

5.
1. Entire top loose
2. Loose or missing
3. Burned
4. Cut
5. Worn
6. Mechanical damage
7. Stained
8. Metal rusted
9. Substrate rotted

6.
1. Water damage
2. Rot
3. Stained
4. Mechanical damage
5. Worn
6. Broken glass
7. Defective hardware
8. Stiff/inoperative
9. Not well secured to wall
10. Door or drawers missing/loose
11. Shelves not well supported
12. Rust

7. Tread width or depth
- the width of the place your foot steps on

Rise
- the height between steps

Run
- the horizontal offset between steps

Stringer
- the supports for the treads

Winder
- a tread that tapers to a point

Guardrail
- a railing to prevent you falling into a stairwell or off an open stairway

Baluster
- the vertical spindles on a guardrail or handrail

8. GENERAL:
1. Rot/water damage
2. Mechanical damage

TREADS:
3. Too thin
4. Excessive rise
5. Not uniform
6. Excessive nosing or back slope
7. Inadequate tread width
8. Too many winders
9. Winder angle too big
10. Worn or damaged
11. Sloped
12. Loose or poorly supported

STRINGERS:
13. Too small
14. Excessive notching for treads
15. Too thin
16. Excessive span between stringers
17. Pulling away from wall or treads
18. Inadequately secured to header
19. Stairwell width inadequate

LANDINGS:
20. Headroom inadequate
21. Landings missing
22. Too small

HANDRAILS:
23. Missing
24. Hard to grasp
25. Loose or damaged
26. Too low or too high

GUARDRAILS:
27. Missing
28. Too low

BALUSTERS:
29. Too far apart
30. Easy to climb
31. Loose or damaged
32. Missing
33. Fire stops inadequate
34. Lighting inadequate

Answers to Quick Quiz 5

1. 1. Light
2. Ventilation
3. Architectural appeal
4. Emergency exit

2. 1. Wood
2. Vinyl
3. Metal

3. 1. Single hung
2. Double hung
3. Casement
4. Horizontal sliders
5. Awning
6. Hopper
7. Fixed
8. Jalousie

4. True

5. True

6. 1. Leaks
2. Lintels sagging/missing

7. 1. Rot
2. Rusted
3. Racked
4. Deformed
5. Installed backwards
6. Drain holes blocked/ missing

8. 1. Missing - the implication is water leakage
2. Ineffective - the implication is water leakage
3. Water leakage

9. 1. Missing

    2. Rot

    3. Rust

    4. Damaged, cracked, loose

    5. Sills with reversed slope

    6. Sill projection inadequate

    7. Drip edge missing

    8. Putty cracked, missing, loose, deteriorated

    9. Caulking/flashing missing, loose, rusted, incomplete

    10. Painted or stain needed

10.
    1. Rot

    2. Rust

    3. Inoperable

    4. Stiff

    5. Sashes won't stay open

    6. Sash coming apart

    7. Loose fit

    8. Weather stripping missing or ineffective

11.
    1. Rot

    2. Stained

    3. Missing

    4. Cracked

    5. Loose

    6. Poor fit

12.
    1. Cracked

    2. Broken

    3. Loose

    4. Missing

    5. Lost seal

    6. Excess condensation

13.
    1. Rusted

    2. Broken

    3. Missing

    4. Loose

    5. Inoperable

14. Someone stumbling on the stairs could fall through the window

15.
    1. Torn

    2. Holes

    3. Rust

    4. Loose

    5. Missing

16. If the window is required for egress

17. True

18.
    1. Means of entry and exit

2. Security
3. Privacy
4. Weather-tight
5. Gasproof or fireproof if they connect to a garage

19. 1. Wood
2. Metal
3. Vinyl
4. Hardboard

20. 1. Not fire-rated or exterior type
2. Not weatherstripped
3. Not self-closing
4. Opens into bedroom
5. No six inch step down to garage

Answers to Quick Quiz 6

1. 1. Surface water from rain or snow
2. Groundwater

2. 1. Soil type
2. Depth of foundation
3. Height of water table

3. 1. Nuisance
2. Damaged interior finishes and furnishing
3. Odors and molds
4. Structural deterioration
5. Electrical shock or fire hazard
6. Damaged insulation

4. 1. Water or dampness on walls or floor
2. Efflorescence on walls or floor
3. Rot, stains or water marks on walls, doors, windows, basement stairs
4. Rust at baseboard nails, carpet tack strips, columns, appliances
5. Odors, mold, mildew
6. Rot
7. Loose floor tiles
8. Damaged basement storage
9. Storage off floor
10. Wall patches
11. Floor patches
12. Trough or trench around floor perimeter
13. Sump pumps operating continuously
14. Full sump
15. Two spare sump pumps
16. Auxiliary electric supply for pump
17. High water alarm on sump
18. Crumbling plaster, drywall or masonry
19. Peeling paint

20. Wall cracks with stains
21. Recent excavation
22. Evidence of drainage membrane
23. New dampproofing
24. Dehumidifier running constantly

5. 1. Control surface water – grading, gutters downspouts, driveways, etc.
2. Patching flaws – crack repairs, drainage layers
3. Interior drainage system – trench around interior floor slab, with a sump pump
4. Exterior excavation and dampproofing – very expensive, replace or add weeping tile

6. Wet basements

7. False